3

Environmental biology
of
agaves and cacti

Environmental biology of agaves and cacti

Park S. Nobel
University of California
Los Angeles

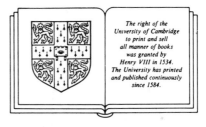

The right of the
University of Cambridge
to print and sell
all manner of books
was granted by
Henry VIII in 1534.
The University has printed
and published continuously
since 1584.

Cambridge University Press

Cambridge
New York   New Rochelle
Melbourne   Sydney

Published by the Press Syndicate of the University of Cambridge
The Pitt Building, Trumpington Street, Cambridge CB2 1RP
32 East 57th Street, New York, NY 10022, USA
10 Stamford Road, Oakleigh, Melbourne 3166, Australia

First published 1988

Printed in the United States of America

Library of Congress Cataloging-in-Publication Data
Nobel, Park S.
Environmental biology of agaves and cacti / Park S. Nobel.
p. cm.
Includes index.
1. Agave – Physiology. 2. Cactus – Physiology. 3. Agave – Ecology.
4. Cactus – Ecology. I. Title.
QK495.A26N63 1988
584'.25 – dc 19   87–17314

ISBN 0 521 34322 4

British Library Cataloguing in Publication data applied for.

Contents

Preface

Agaves and cacti capture the imagination of nearly everyone. Who can resist admiring their unique shapes or empathizing with their tolerance of droughts and high temperatures? How do these interesting and economically important desert succulents respond to specific environmental factors? We shall see that the monocotyledonous agaves are actually quite similar to the dicotyledonous cacti in their reactions to the environment. The aims of this book are quite ambitious: (1) to help interpret the environmental responses of agaves and cacti; (2) to consolidate the present level of our knowledge of these succulents for comparison with other plant groups; (3) to show how modeling can be used to analyze relations between morphology, microclimate, and productivity; (4) to serve as an example for the study of the environmental responses of relatively uninvestigated groups of plants; and (5) to provide a basis for evaluating whether certain arid, semiarid, and other regions can be successfully exploited for the cultivation of these succulents in the future. Although well over 600 references are cited, considerable emphasis is placed on research emanating from my laboratory since 1975.

To set the stage, we will describe the many uses of as well as early environmental research on agaves and cacti together with their taxonomy, morphology, and anatomy, including pictures of over thirty species that have played important roles in environmental research (Chapter 1). The physiological key to their ecological success is the water-saving consequences of Crassulacean acid metabolism (Chapter 2). The responses of agaves and cacti to three physical factors of the environment are discussed next: water relations, including a discussion of such topics as rain roots, capacitance, stomatal responses, C_3/CAM shifts, seedling establishment, and reproduction (Chapter 3); temperature aspects of gas exchange and survival limits, together with a model relating morphology, tissue temperature, and species distribution (Chapter 4); and photosynthetically active radiation, with a presentation of models describing radiation interception and the resulting morphological adaptations of agaves and cacti (Chapter 5). Nutrient responses are reviewed, and the importance of nitrogen is demonstrated (Chapter 6). To help integrate the environmental effects, an environmental productivity index is proposed and then used to discuss productivity and its morphological correlates for agaves and cacti; productivity is also predicted over wide geographical areas to help evaluate the agronomic potential of these succulent plants (Chapter 7).

The level of presentation is between that of an elementary textbook and the research literature. Approximately 100 photographs, over 200 figure panels, and about 30 tables are used to illustrate the state of our research knowledge for agaves and cacti. Because emphasis is on principles of wide biological applicability, certain aspects of universal appeal are developed in considerable detail. The book can therefore serve as a text for a course on the environmental responses of organisms, as well as meeting the needs of plant physiologists, ecologists, and agronomists. Another intended audience is animal ecophysiologists and modelers. The succulent plant enthusiast/hobbyist should also find the material comprehensible and useful. To facilitate understanding by readers of such different backgrounds, every major scientific term is italicized and defined the first time used, and the first entry in the index indicates the definition of that term. We should also emphasize that much of the information on the environmental responses culminating in productivity predictions is pertinent to decision makers in Latin America as well as many other arid or semiarid regions where agaves and cacti can be grown. Thus, the book summarizes the present state of our knowledge and clearly points to the future, with respect to both needed research and authenticated utilization of these plants.

Thanks are due to many. First of all, my research on desert succulents has been generously supported by the Ecological Research Division (ERD) of the United States Department of Energy, the U.S. National Science Foundation, and the University of California at Los Angeles. Experiments specifically designed for this book and capably performed by Terry L. Hartsock were also supported by ERD. Many graduate students, postdoctoral fellows, and visiting scientists have contributed to the overall research effort of my laboratory, as detailed in the literature cited. A tremendous debt of gratitude is expressed to Marjorie Macdonald for her excellent typing. The figures were skillfully drawn by Hildy Heinkel, Amy Roberts prepared the references, and the photographs were taken by the person cited in the figure caption, or by me. The following individuals also made important critical comments that greatly enhanced the final manuscript: Dr. Wade L. Berry, Dr. Arthur C. Gibson, Dr. Barry A. Prigge, Dr. Paul J. Schulte, Augusto C. Franco, Loraine U. Kohorn, Michael E. Loik, Gretchen B. North, Mark T. Patterson, Cheryl C. Swift, and David T. Tissue. As you peruse this book, I hope you discover and share my enthusiasm for the special environmental responses of these remarkable plants.

Los Angeles   Park S. Nobel
30 November 1987

1 Introduction

Environmental biology deals with how organisms respond to their environment. This includes responses to physical factors such as air temperature, relative humidity, precipitation, wind speed, and solar radiation – factors generally familiar from weather reports. It also encompasses these factors in very local regions, or *microhabitats* (Fig. 1.1), where a seed may chance to fall and hence seedling establishment ensue. Seeds of *Agave deserti* or *Ferocactus acanthodes** can germinate in late summer in the Sonoran Desert, when the maximum air temperature reported by the regional weather bureau or registered on the digital electric sign of a local bank may be 35°C. But this is of little consolation to a small seedling in bare ground where the soil surface temperature can exceed 70°C (158°F)!

The environment represents all the external conditions that can affect the growth and survival of an organism. Thus, we will also consider the *edaphic* or soil factors, ranging from nutrients to water. It is not sufficient to know how much rainfall there is; we must also know how a particular precipitation event affects the energy of the soil water in the vicinity of a root. Water moves energetically downhill,

* The full, unambiguous, scientific name of an organism is an italicized Latin binomial, indicating genus and species, followed by the *authority* (usually abbreviated), which indicates the person who first applied the current binomial. The complete names, including authorities for all the species mentioned in this book, are collected together in the section Taxonomy and Morphology (Table 1.1 for agaves and Table 1.2 for cacti).

Figure 1.1. Seedling of *Agave deserti* from the northwestern Sonoran Desert growing in the special microhabitat provided by a nurse plant, the desert bunchgrass *Hilaria rigida*. The agave seedling is 4 cm tall and 4 years old. Germination occurred in August or September 1982, and the photograph was taken by David T. Tissue in July 1986. The site, known as Agave Hill, is in the Philip L. Boyd Deep Canyon Desert Research Center near Palm Desert, California, at 33°38′N, 116°24′W, and an elevation of 850 m (Ting and Jennings, 1976).

so the *water potential* must be lower in the root than in the adjacent soil for water uptake to occur. We will also consider why a succulent agave or cactus filled with water does not lose this precious commodity back to a drying soil and thus become desiccated to the point of death. The topic of *water relations*, including water uptake by roots, requires a detailed consideration of the components that contribute to water potential together with the observed root properties, such as the rehydration of existing roots and the induction of new lateral roots after rain.

Understanding the consequences for water uptake of the observed root distribution in the soil can be aided by sets of equations that are combined into a *model* to describe plant behavior. We will find that models help interpret the shapes of agaves and cacti with respect to the interception of solar radiation; models can indicate the influences of spines on stem temperatures, which affect the ranges over which cacti can occur naturally. Another objective in our consideration of the environmental biology of agaves and cacti will be to integrate all the physical factors so that we can predict *productivity* in different regions – after all, a plant will not be successful in a particular environment if it does not have a net positive uptake of carbon dioxide (CO_2) from the atmosphere and the incorporation of the carbon from such CO_2 into organic compounds. As is already apparent, we are building up a specialized vocabulary in this chapter, some terms of which have rather specific definitions. For example, *desert* is an anthropocentric term referring to dry places that are often generally hot for part of the year, *arid* refers to regions with less than 250 mm of annual rainfall (ignoring certain complications between the timing of rainfall and the ambient air temperature), and *semiarid* generally refers to regions with 250 to 450 mm of annual rainfall (250 mm = 0.25 m $\cong$ 10 inches).

There are many reasons for choosing agaves and cacti for our study, not the least of which is their intriguing shapes. Indeed, many people are attracted to them because of their *morphology*, or external form and structure (Fig. 1.2). The often spectacular and bizarre morphology of these desert succulents can also lead to interesting effects on their distribution. Other people are attracted by the frequently severe nature of their habitats. Survival of agaves and cacti in such regions is often related to their fleshy massiveness, or *succulence*. We shall see that such water storage capability also occurs on a cellular level and is related to the photosynthetic pathway used by most of them, known as *Crassulacean acid metabolism*, or *CAM* for short. Plants exhibiting CAM open their *stomata* (surface

pores) and have a net uptake of CO_2 mainly at night, when the lower tissue temperatures and higher ambient relative humidity lead to less water loss than for daytime stomatal opening. Thus, CAM can be crucial for growth and survival in arid habitats and has evolved many times in diverse *taxa* (groups of related plants). Indeed, another reason for studying agaves and cacti together is that these unrelated taxa have evolved remarkably similar responses to a particular set of environmental conditions, as we shall see when individual physical factors are considered. Over the years many different species of agaves and cacti have been used for food, fodder, fiber, fences, fuel, medicinal purposes, and ornamental horticulture, and many agaves and cacti are currently cultivated over wide areas in arid and semiarid regions of the world. As more marginal lands in arid and semiarid regions are, of necessity, brought into cultivation in the future, these species will assume even greater economic importance.

Uses

Research on agaves and cacti prior to the twentieth century did not employ the equipment currently considered indispensable for environmental biology. Nevertheless, certain agronomic principles were clearly recognized by native Indians in what is now North America, especially Mexico. Such Indians supplemented their diets with the raw fruits from many species of cacti and the roasted inflorescences, stems, and folded leaves of agaves (Sauer, 1965); roasting broke down the *glucans* (polymers of glucose, such as the common glucan *starch*) and other hexose polymers into digestible sugars (e.g., glucose and fructose).

Based on evidence from fossilized human feces, Callen (1965) has shown that both agaves and cacti were consumed by humans at least 9,000 years ago (Fig. 1.3). The characteristic stomata and *druzes* (aggregate crystals of calcium oxalate radiating from a central point) of cacti helped identify the consumed cacti as a *Stenocereus* (organ-pipe cactus) and an *Opuntia* (termed "prickly-pear" cactus based on fruit shape for those opuntias with characteristically flattened pads known as *platyopuntias*). The *Agave* was identified

Figure 1.2. Variation in stem morphology of four species of cacti occurring *sympatrically* (meaning, at the same site): (**A**) *Ferocactus acanthodes* (a barrel cactus), (**B**) *Opuntia acanthocarpa* [deerhorn cholla; note the red diamond rattlesnake (*Crotalus ruber*) near the center of the picture], (**C**) *Mammillaria dioica* (known as the fishhook cactus because of the shape of its spines), and (**D**) *Opuntia basilaris* (beaver-tail cactus). The site is the same as for Figure 1.1.

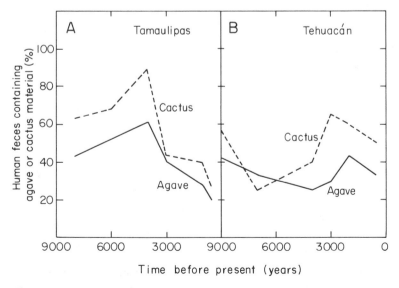

Figure 1.3. Percentage of fossilized human feces that contained material identified as coming from an agave or a cactus. Samples were collected from caves in Tamaulipas, a state in northwestern Mexico, and Tehuacán, a state in southern Mexico. Modified from Callen (1965).

based on the pattern of *epidermal* (outermost) leaf cells around its stomata. On average, about half of the human feces contained cactus stem material, and nearly as much contained agave leaf material, throughout the entire period studied of nearly 9,000 years (Fig. 1.3), indicating the occurrence of desert succulents in the diet of prehistoric Indians from Mexico for an extremely long period. Archeological evidence also indicates that artifacts made from agave fibers and tools from agave leaf tips have been in use from 9,000 years before the present onward (Gentry, 1982). For instance, the hard fiber from *Agave lechuguilla* was used over 8,000 years ago for sandals (Crane and Griffin, 1958). Use of agaves and cacti undoubtedly led to their cultivation, beginning at least 6,000 years before the present (Sánchez-Mejorada R., 1982). For instance, cacti were apparently vegetatively propagated by prehistoric people in the Caribbean by inserting stem segments into specially raised soil mounds, which provided locally well-drained soil suitable for these desert succulents. Agaves were cultivated in the southwestern United States about 1,000 years ago by the Hohokam, who apparently used rocks as a mulch to prevent loss of soil water from around the plants (Fish et al., 1985).

Beginning in the fifteenth century, many of the ancient agronomic practices and uses of these plants were described by the invading Europeans, who disseminated agaves and cacti worldwide. In this section we will summarize some of these uses of agaves and cacti. In the next section we consider certain early research that helps illustrate the environmental "strategies" used by desert succulents.

AGAVES
Beverages
In the late fifteenth and early sixteenth centuries, the culturing of agaves became widespread (Sanchez Marroquin, 1979; Gentry, 1982). As the Spaniards colonized regions in the northcentral part of present-day Mexico, they forced the Nahuatl Indians to move with them and to cultivate *Agave salmiana* (Fig. 1.4A) and a few other species (commonly called "maguey") in warm and arid climates for *pulque* (Fig. 1.4B), a fermented beverage produced from the sap that collects near the top of the stem after the central spike of folded leaves is excised from the larger plants (Fig. 1.4C). Actually, the sweetish sap is often consumed directly as the beverage *aguamiel* ("honey water"). These two beverages presumably were drunk in prehistoric times. Indeed, the Aztecs discovered the use of pulque during their migration to the Valley of Mexico at the end of the twelfth century (Gentry, 1982).

Internationally, agaves are better known for the distilled beverages *mescal* (also spelled *mezcal*), produced from *A. salmiana* and at least eight other species (e.g., *A. angustifolia* in Oaxaca), and *tequila*, produced primarily from *A. tequilana* (Fig. 1.5). These beverages were developed after Europeans introduced distillation into North America in the sixteenth century. A mescal factory tends to be a cottage (even bootleg) industry, where the piñas or cabezas (Fig. 1.5B), generally weighing 25–50 kg (but which can be up to 170 kg), are traditionally crushed by a large millstone turned by a burro, oxen, or, more recently, a tractor. The resulting mash is fermented and then distilled in small stills, sometimes with a second distillation (Bahre and Bradbury, 1980; Tello Balderas, 1983; Tello Balderas and García-Moya, 1985). Tequila, which is generally double distilled, tends to be produced in large, modern factories near its namesake town of Tequila, Jalisco, where its manufacture began in 1621. Its production is more formal than that of mescal; in addition, the better grades of tequila are aged for various periods before distribution (Valenzuela, 1985).

Fiber
Beginning in the nineteenth century, fiber-bearing agaves, mainly *Agave sisalana* (sisal; Fig. 1.6), were exported from Mexico, forming the basis of major industries in Indonesia and the Philippines in the nineteenth century and in East Africa in the twentieth century. To protect its fiber industry, Mexico had prohibited the export of its native *A. sisalana*, but some plants were available from Florida. In 1893, 1,000 *bulbils* (plantlets produced on the inflorescence of certain agaves and which are suitable for vegetative reproduction) of such *A. sisalana* were sent to Germany; about 200 survived this trip and 62 sur-

Figure 1.4. (**A**) *Agave salmiana,* which is used for both pulque and mescal, growing near Salinas de Hidalgo, San Luis Potosí, Mexico. The leaves and inflorescences have been cut on the large plants, which leads to the accumulation of more organic compounds such as sugars in the stem. (**B**) A "pulquero" using a gourd to remove aguamiel from a large agave near Talpalpa, Jalisco, Mexico (photographed by Ana G. Valenzuela). (**C**) The sap collects in a basin atop the stem formed by cutting through many leaves.

vived the ensuing trip to German East Africa, now Tanzania (Lock, 1962). Within five years, these plants had multiplied into 63,000 plants, which laid the foundation for what was to become one of the most widespread cultivations of any CAM plant. Indeed, plantations of *A. sisalana* currently occupy much of Tanzania and part of Angola, Kenya, Mozambique, and Uganda. These African countries supply about half of the world's hard fiber, another 20% or

Figure 1.5. (**A**) Fields of *Agave tequilana* growing near Tequila, Jalisco, Mexico. Plants in the foreground are five years old. (**B**) The author holding the stem and attached leaf bases, which is called a piña because it resembles a pineapple (or a "cabeza," meaning head). This piña was harvested from an eight-year-old plant and will be taken to a local factory for roasting for 24–36 h followed by shredding and then fermenting of the mash for 36–48 h as a prelude to distillation to obtain tequila.

so coming from *A. sisalana* grown in Brazil (Gentry, 1982). By the early twentieth century, *A. sisalana* was also successfully cultivated in India, Southeast Asia, many Pacific islands, and Australia in regions of sufficient rainfall and well-drained soils (Smith, 1929). Besides commercial uses as a hard fiber, primarily as a twine for bales of hay, for rope, and for sacks, a large handicraft industry has arisen around *A. sisalana*, ranging from baskets to upholstery to dart boards (Kenya Sisal Board, 1984; Baker, 1985).

Much hard fiber is also produced from agaves in Mexico (about 20% of the world's supply). This fiber comes mainly from *A. four-croydes* (henequén, Fig. 1.7) grown in plantations in the Yucatán peninsula of eastern Mexico and secondarily from *A. lechuguilla* (Fig. 1.8), which has been collected from the wild in at least seven states in northeastern

Mexico on more than 130,000 km^2 at some time during the last 200 years (Taylor, 1966; Sheldon, 1980; Garcia de Fuentes and de Sicilia, 1984; Ramírez, 1985). The older unfolded leaves of *A. fourcroydes* are harvested one or two times per year; for *A. lechuguilla* the young leaves still folded about the central spike are harvested for fiber, and new central spikes are then produced so that the same plants can be harvested every one to five years. These species occur in quite different habitats: *A. fourcroydes* is grown in the unique, shallow limestone soils of Yucatán, which has no surface streams or rivers, although the mean annual temperature is mild (27°C) and annual precipitation regularly exceeds 1,000 mm (Smith and Cameron, 1977); *A. lechuguilla* is exposed to the rainfall vagaries of the Chihuahuan Desert.

Over twenty specific commercial uses

Figure 1.6. *Agave sisalana,* which has been cultivated worldwide for the hard fiber in its leaves known as "sisal" or "sisal hemp": (**A**) Abandoned sisal plantation near Thika, Kenya; (**B**) Mature plants from which lower leaves have been harvested; and (**C**) Empty, rusty carts for hauling sisal from field near Nairobi, Kenya.

exist for the fibers of agave from Mexico, primarily for woven objects, ropes, or as stuffing (Gentry, 1982). The fiber has been used for paper and, more recently, in construction material (Belmares, Castillo, and Barrera, 1979; Cruz-Ramos, Orellana, and Robert, 1985; Padilla R. and Fuentes R., 1985). *Agave fourcroydes* began to dominate the economy of the Yucatán peninsula in the mid-nineteenth century and became the main export crop of Mex-

Figure 1.7. (**A**) *Agave fourcroydes* growing near Mérida, Yucatán, Mexico. The six-year-old plants are growing at the experimental plots of Centro de Investigaciones Agrícolas de la Peninsula de Yucatán (CIAPY). Measurements are being made by Carloz Oropeza. (**B**) Woodcut of a fiber plantation of *A. fourcroydes* in Yucatán (adapted from Royal Gardens, Kew, 1898). (**C**) Fiber drying in sun near Mérida (photographed by Luis del Castillo).

ico near the turn of the century (Camp, 1980). Also, at least some annual income for up to 500,000 people in northeastern Mexico can come from *A. lechuguilla* (Sheldon, 1980). However, the advent of synthetic fibers in the 1920s and 1930s has led to the worldwide economic demise of fibers from agaves (Smith and Cameron, 1977; Sprague, Hanna, and Chappell, 1978).

Additional uses

Although agaves have chiefly been used for the production of fiber and alcoholic beverages, they have served man in many other ways. Other food uses include the eating of flowers after boiling or scrambling them with eggs and peeling off the *cuticle* (waxy layer on the epidermis) of *Agave atrovirens* for use as a translucent wrapper for tortilla sandwiches and of *A. salmiana* to wrap meat. The Seri Indians, who inhabit the state of Sonora in northwestern Mexico, currently consume parts of nine different species of agaves; the most common food is prepared from the leaf bases, stem, or inflorescence after roasting (Felger and Moser, 1985). Aguamiel has also

Figure 1.8. (**A**) *Agave lechuguilla,* which apparently has more rosettes under natural conditions than any other agave, growing near Saltillo, Coahuila, Mexico. Leaves of the plants were marked and examined monthly as part of a productivity study with Edgar Quero. (**B**) Fibers of the harvested central spike of *A. lechuguilla* drying in the sun prior to commercial use for burlap sacks and pillow filling.

served as uncontaminated drinking water for thirsty desert travelers. Agaves were exported from Mexico to Europe as ornamentals, mainly *A. americana* (Fig. 1.9). By the eighteenth century it was established in both private and public gardens along the Mediterranean Sea and the warmer parts of Europe – the cold winters prevented the growth of *A. americana* outdoors in northern and central Europe. Indeed, *A. americana* and various other species are now horticulturally used worldwide.

Cortez in the early sixteenth century complained about the deterrent influence of agave spines on the advance of his army (Sprague et al., 1978). Indeed, agave spines were used for punishing juvenile delinquents and runaway slaves (Gentry, 1982). The stalks of the inflorescences have been used for fencing and construction. From a more industrial point of view, agaves have been used in the manufacture of paper, including that for currency, and the production of gritty soaps, shampoos, and medicines (Gentry, 1982). Indeed, many species of agaves contain by *dry weight* over 2% *sapogenins,* which are soapy compounds, widely used as shampoos, from which cortisone and sex hormones like estrogen can be synthesized (dry weight is the mass of a plant or plant part after drying to remove water, such as in an oven at 80°C). For instance, *A. schottii* can contain 2% sapogenins by dry weight and was used as a cleaning agent by Indians, and *A. vilmoriniana* can contain over 4% sapogenins (Gentry, 1982). About 6% of the world's supply of precursors for corticosteroid synthesis has come from agaves, indicating their pharmacological importance (Blunden, Culling, and Jewers, 1975; Herz, 1985). Fermentable sugars can be 50% of the dry weight of the leaves, so agaves can also be used to produce ethanol for industrial purposes and for gasahol (McDaniel, 1985; but see de Menezes and Azzini, 1985). Leaves of agaves such as *A. salmiana* have been used as cattle fodder in Mexico. The pulp of leaves of *A. fourcroydes,* a waste product of the fiber

Figure 1.9. *Agave americana,* the first-named member of the genus, used as a fence in the jungle region of northern Thailand near the border with Burma. It occurs worldwide as an ornamental plant and is used in certain parts of Mexico for the production of pulque.

industry, has been used as a source for bacterial fermentation, which can increase the protein content of the resulting meal, making it more useful as a livestock feed (Sanchez Marroquin, 1979; Blancas et al., 1982). The list of uses for agaves is indeed quite long.

CACTI

Because of their unusual stem morphology and their attractive flowers, cacti became very popular in European gardens and households beginning in the sixteenth century. Indeed, collectors, hobbyists, and landscape designers have exerted a major influence on the commercialization of hundreds of cactus species and the establishment of major cactus societies in Canada, the United States, Mexico, Great Britain, Belgium, the Netherlands, France, West Germany, Switzerland, East Germany, Czechoslovakia, Japan, Taiwan, New Zealand, Australia, Zimbabwe, South Africa, and many other countries. A cactus figures in the Aztec legend of the founding of what is now Mexico City and consequently is recognized in the national flag of Mexico,

which is embellished with an eagle perched on an opuntia, a rendition of the vision seen by the Aztecs in 1325 as a divine sign that they had reached the promised land (Sánchez-Mejorada R., 1982).

Specialized uses

Besides their worldwide use as ornamental plants, cacti became widely propagated as host plants for the cochineal scale insect (previously, *Coccus cacti*; now *Dactylopius coccus*). The Aztecs prized the rich red color of carminic acid (termed carminate for the dissociated acid) extracted from the dried bodies of the female insects, which were raised on cladodes of many species of prickly-pear cacti, including *Opuntia ficus-indica, O. littoralis, O. phaeacantha,* and *O. stricta,* among others (Evans, 1967; Donkin, 1977; Benson, 1982). Indeed, the white, cottonlike, fibrous cocoon material secreted by the female insects is still a common sight on the cladodes of platyopuntias, generally near the *areoles* (axillary buds that produce a cluster of spine primordia and which are found only on the stems of cacti).

Figure 1.10. (**A**) *Ferocactus wislizenii* leaning southward toward the equator in the Sonoran Desert near Phoenix, Arizona. (**B**) *Copiapoa cinerea* tilting northward toward the equator in the Atacama Desert near Pan de Azucar, Chile.

Montezuma annually demanded hundreds of kilograms of cochineal from the tribes under his control, each kilogram requiring about 150,000 insects to be handpicked out of their white cocoons.

Eventually, the cochineal dye of the Aztecs became highly prized in Europe. In the early sixteenth century, Cortez was instructed to send as much of the dye back to Spain as possible; he also sent opuntia plants, so cactus plantations were started in Spain and its possession, the Canary Islands, to host the insect. Similar plantations were initiated in North Africa and southern Europe along the Mediterranean Sea. The cactus plantations of the Canary Islands annually exported 1,000 metric tons (1 metric ton = 1,000 kg = 2,205 lb) of cochineal dye in the 1860s, most of which went to England for such purposes as dyeing the uniforms of the British regulars (the so-called Red Coats) and the Northwest Mounted Police (later the Royal Canadian Mounted Police), who were expanding the British Empire into the New World (Evans, 1967; Donkin, 1977; Benson, 1982).

Cacti have also been used ceremonially.

Prominent in this respect is *Lophophora williamsii* (peyote), native to northern Mexico. The active ingredient in peyote is the alkaloid mescaline (3,4,5-trimethoxyphenylethylamine), which has marked effects on visual perceptions. Such hallucinogenic powers were known to many Indian tribes in pre-Columbian times, and even today approximately 250,000 people use peyote in religious and other ceremonies (Benson, 1982). Another ancient ceremonial practice involved young men rubbing the spiny stems of *chollas* (pronounced "chóy·as," which refers to a closely related group of opuntias having jointed, rather cylindrical stems known as *cylindropuntias*) such as *Opuntia imbricata* onto their bodies during rites of initiation into manhood.

Many other specialized uses of cacti occur. Desert travelers have used barrel cacti as a source of water, either upon squeezing freshly cut stem material or after distillation from the cut stems so that bitter or noxious material could be removed. Some barrel cacti, such as *Ferocactus wislizenii* of the Sonoran Desert (Fig. 1.10A), became known as "compass plants" because they tilted toward the

Figure 1.11. *Eulychnia acida* and *Trichocereus chilensis* used as fences in central Chile.

equator. The equatorial tilt is even more pronounced for many species of *Copiapoa* (Fig.
1.10B), which is common in northern Chile.
Mucilage (a sticky, water-absorbing polysaccharide) from various species of cacti has been
used as gums, adhesives, and emulsifiers
(Mindt et al., 1975; Gibson and Nobel, 1986).
Mucilage exposed in the freshly cut surfaces
of platyopuntias can even be used to clarify
murky water by binding particulate matter. In-
flammable gases can be produced from homogenized stems of *O. ficus-indica* maintained
in closed vessels (Contreras and Toha C.,
1984). Cacti (as well as agaves) are used for
erosion control (Gatti, 1977) and windbreaks,
often to prepare sites for the planting of other
species in arid and semiarid regions. Cacti
even serve as fences (Fig. 1.11).

Food

Even though cacti had ornamental and
other uses outside of the Americas, nothing
can compare with the multiplicity of uses in
Latin America. Besides using the *cladodes*
(flattened stem segments, or pads, of platyopuntias) for food, the fruits of *Opuntia ficus-
indica* and similar platyopuntias can be eaten

raw or cooked, made into preserves, used for
a sweet drink ("miel de tuna"), and even fermented into alcoholic beverages (Bustos,
1981). Fruits of *Carnegiea gigantea* (Fig.
1.12A,B) and *Stenocereus gummosus* (Fig.
1.12C) have long been collected by Indians,
such as the Papago, Pima, and Seri, in what
is now the southwestern United States and
northwestern Mexico (Bruhn, 1971; Benson,
1982; Felger and Moser, 1985). A storable taf-
fylike product termed "queso" (cheese) can
be produced from the fruit of prickly-pear cacti
(Benson, 1982). For a long time, "cactus
candy" has been made by infiltrating exoge-
nous sugars into the stem matrix of various
species, including many barrel cacti (Shreve,
1931a). Similarly, small pieces of the stems of
Echinocactus, Ferocactus, and *Melocactus*
can be slowly cooked in water containing sodium bicarbonate, yielding an endogenous
syrup that crystallizes into the stem matrix to
form another type of cactus candy (Sánchez-
Mejorada R., 1982).

The consumption of fruits of *O. ficus-indica* and its hybrids, *O. streptacantha,* and
other species of platyopuntias is well known
in Mexico. Beginning with the introduction of

Figure 1.12. (**A** and **B**) *Carnegiea gigantea* in Pima County, southcentral Arizona. (**C**) *Stenocereus gummosus* near Ciudad Constitución, Baja California Sur, Mexico.

Figure 1.13. (A) *Opuntia ficus-indica* growing in plantations in Til Til, 50 km north-northwest of Santiago, Chile; also pictured is Ignacio Badilla, who collaborated on field studies of productivity. (B) Fruit of *O. ficus-indica*. (C) *O. ficus-indica* used as a hedge near Jericho, Israel.

O. ficus-indica (Fig. 1.13) into Spain by Columbus, its fruits have also become popular worldwide. Such fruits (Fig. 1.13B) can be seasonally purchased in many supermarkets in Chile, where fruit production from *O. ficus-indica* on steeply sloping, difficult-to-irrigate ground in semiarid regions having relatively fertile soil is favored over conventional fruit crops such as peaches, apricots, and apples. The fruit has led to the term *sabra,* meaning spiny or tough on the outside but sweet inside, which is now colloquially applied to Israelis born in Israel and to various Arab women. This borrowing of a term from a New World plant grown in an Old World setting can be traced back to the migration of the Moors from Spain to North Africa in the early part of the seventeenth century. They brought *O. ficus-indica* with them, which came to be used as fences or hedges demarking landholdings (Fig.

1.13C). The fruit from such fences is sold today throughout arid North Africa. The young cladodes are marketed as a fresh vegetable in North America, as well as other places, and recently have been sold as a processed pickle for hors d'oeuvres. Leaves of such cacti as *Austrocylindropuntia subulata* can also be purchased as vegetables in South America (Benson, 1982).

Forage and fodder

Platyopuntias have long been used as stock *forage* (eaten where grown) and *fodder* (stems harvested and then brought to the animals); barrel cacti are also often so used, although their slow growth rates limit this usage. Near the turn of the century, Luther Burbank bred and vigorously promoted various cacti for both food and fodder, such as his notorious spineless *O. ficus-indica* (Benson, 1982). In 1911 he was quoted in the *Los Angeles Examiner* as saying that the development of spineless cacti "promises to be as great or even greater value to the human race than the discovery of steam." To his credit Burbank did recognize that lack of spines would make the plants less cold tolerant (considered in Chapter 4). And at the turn of the century many Texas ranchers appreciated the value of platyopuntias, with or without spines, for consumption by range cattle.

But we should temper our expectations of the value of cactus as a forage with the Australian experience. In particular, a prickly-pear cactus, originally called *Opuntia inermis* but subsequently identified as *O. stricta,* "escaped" from household gardens in eastern Australia and hedges in southern Australia to become a pest of vast proportions in the twentieth century. One hundred years after its escape, *O. stricta* was advancing into land designated for grazing or crop use at an average rate of 100 hectares per hour, particularly in Queensland (Osmond and Monro, 1981) [1 hectare (ha) = 10,000 m^2 = 2.47 acres]. The seeds were relished by birds, and the ensuing relatively long-distance dispersal changed the original "foothold" of this species to a "stranglehold" on much of the grassland designated for use by sheep and other herbivores, especially when the relatively innocuous few-spined varieties reverted to their more spiny

ancestral forms. Chemical poisons and slaughtering of seed-disseminating birds to halt its advance had little effect. Because the introduced cochineal insect also had little effect, the moth *Cactoblastis cactorum* was introduced into eastern Australia in the 1920s. The larvae of this moth decimated the cladodes and brought the prickly-pear cactus under "control." By 1933, about 90% of the *O. stricta* in Queensland had been destroyed. This platyopuntia with its spines that deterred grazing was highly unsuitable for rangeland use in Australia (Osmond and Monro, 1981) as well as in South Africa, where again control by *C. cactorum, D. coccus,* and other insects has been attempted (Annecke and Neser, 1977; Annecke and Moran, 1978). Indeed, *Opuntia aurantiaca* and *O. ficus-indica* have infested 2 × 10^6 ha in South Africa (Zimmermann and Moran, 1982).

Despite these ecological catastrophes, both older and more recent thinking in range mangement has come to the conclusion that prickly-pear opuntias can be economically profitable, providing an important forage during drought (Thornber, 1911; Russell, 1985). Various cylindropuntias can also be used, especially for their fruit. Species or varieties with spines may be untouchable by cattle and thus may grow unhindered when water is available. When drought ensues and other herbage becomes unavailable, the spines may be inexpensively burned off and the exposed cladodes can then provide a ready supply of water and nutrients to the cattle. One person can singe 5 metric tons fresh weight of native cacti in a day; the high water content of such stems can enable cattle to go a few days without seeking additional drinking water (Thornber, 1911). Indeed, if cattle can be carried through the periods of low meat prices generally associated with drought, profits can be greatly enhanced by using prickly-pear cacti as forage, as is realized in Texas and cattle-raising regions in northern Mexico (Thornber, 1910; Benson, 1982; Russell, 1985). Yet some problems still exist. For instance, the effects of the high acidity produced during the night in CAM plants such as cacti and agaves on herbivores such as cattle and sheep have not been fully analyzed (Monjauze and Le Houérou, 1965; Samish and Ellern, 1975; Harrison, 1985).

Figure 1.14. (**A**) *Lophocereus schottii* growing near Pitiquito, Sonora, Mexico. (**B**) *Pachycereus pringlei* growing near Guaymas, Sonora; also shown is Henry J. Thompson. (**C**) *Stenocereus alamosensis* growing near Potam, Sonora. (**D**) *S. thurberi* growing near Ures, Sonora; also shown is Stanley D. Smith.

Nevertheless, the expansion of the cultivation of *O. ficus-indica,* which is well adapted to arid areas, has been highly recommended for forage and fodder to solve some of the agricultural problems of North Africa (Le Houerou, 1970).

Summary of Indian uses

The Seri Indians have many uses for cacti, which we can indicate by considering only the columnar cacti. They use the woody callus tissue lining a woodpecker nest in *Carnegiea gigantea* (saguaro, Fig. 1.12A) and *Lophocereus schottii* (senita, Fig. 1.14A) for carrying or storing food; they eat the fruit of these two species plus *Pachycereus pringlei* (cardón; Fig. 1.14B), *Stenocereus alamosensis* (sina, Fig. 1.14C), *S. gummosus* (pitaya agria, Fig. 1.12B), and *S. thurberi* (organ pipe, Fig. 1.12D); they make wine from the fruits of mainly *C. gigantea* and *S. thurberi* (the wine is sometimes kept in a container fashioned from *Ferocactus wislizenii*, Fig. 1.10A); they use ribs or wood of *C. gigantea, L. schottii, P. pringlei,* and *S. thurberi* for making drills, poles, and in construction; they fashion toys or objects for playing games from parts of *C. gigantea, L. schottii,* and *S. thurberi*; they use extracts of *C. gigantea, P. pringlei, S. gummosus,* and *S. thurberi* as a medicine or for household purposes (for example, caulking, tanning, burning); and they use *C. gigantea, L. schottii, P. pringlei,* and *S. thurberi* in religious or supernatural ceremonies (Felger and Moser, 1985). Over twenty other species, mainly in the genera *Echinocereus, Ferocactus, Mammillaria,* and *Opuntia,* are also used by the Seri Indians. The flowers and seeds of numerous species of cacti are edible. Indeed, hundreds of different uses of hundreds of species of cacti have been devised by Indian tribes in the United States and Mexico, as well as other parts of Latin America (Sánchez-Mejorada R., 1982; Davis, Kay, and Clark, 1983).

To summarize uses of cacti, let us return briefly to *C. gigantea* (Fig. 1.12A), beginning with a quotation from Lumholtz (1912) writing about the Sonoran Desert: "The . . . saguaro . . . is by far the most noteworthy representative of plant life in the desert, being, in fact, one of the most remarkable plants on the globe." Besides its use by the Seri Indians, the saguaro has been, and continues to be, used by the Pima and Papago Indians, again primarily for its fruit (Greene, 1936; Bruhn, 1971). The fruits are eaten raw or cooked, used for preserves and wine, and the seeds are dried for future consumption or as a source of oil. The ribs are used for fences, splints, canes, lances, arrows, and house construction. The spines are used for tattooing. Archeological evidence indicates that the ancestors of the Pima and Papago, the Hohokam, used saguaro wine for etching objects over 1,000 years ago (Bruhn, 1971). More recently, the saguaro has served the entertainment industry as a landscape emblem for western movies, and its flower is the state flower of Arizona. As we shall see, *C. gigantea* has also figured prominently in our understanding of the environmental biology of agaves and cacti.

Early environmental research
THE ARIZONA SCHOOL

Although early environmental research on agaves and cacti was done in many places, the most important center up to the mid-twentieth century was the Desert Botanical Laboratory of the Carnegie Institution of Washington located just outside Tucson, Arizona (Wilder, 1967; McIntosh, 1983). This is a semiarid region receiving just over 300 mm of precipitation annually, about half of which occurs in late summer. The laboratory surpassed others in sheer amount of published materials, with approximately 100 publications on some aspects of the environmental responses of cacti alone (McGinnies, 1981). Moreover, the quality of the ecological research was also high for the period, even though instrumentation was relatively crude. For instance, a potometer was used to determine water uptake by an excised shoot by following the movement of an air bubble in a glass tube, which was apparently the first measurement of *transpiration* (water loss, which was here approximately equal to water uptake) by desert plants. Transpiration was also determined from the weight losses of potted plants, and mercury thermometers were used to measure plant temperature.

The Desert Botanical Laboratory was opened in 1903 with its purpose "being to thoroughly study the relation of plants to an arid climate and to substrata of unusual composition" (Coville and MacDougal, 1903). The site was to have "a flora as rich and varied as possible, while still of a distinctly desert character," to be accessible but not depressingly hot, and to allow the researchers to get "pure water and good food." We will begin by considering some of the research of W. A. Cannon, who was the first resident investigator of the laboratory. We will then discuss the work of D. T. MacDougal, who became director of the laboratory in 1905, and his collaborators, followed by a consideration of the contributions of F. Shreve, who was put in charge of the laboratory in 1928. We will conclude with a summary of some other work emanating from the Desert Botanical Laboratory, which was closed in 1940 for reasons of finance and institutional policy (McGinnies, 1981). Our emphasis will be on the contributions that

Figure 1.15. (**A**) *Opuntia versicolor,* which has reddish stems, growing near Tucson, Arizona (Photographed by Arthur C. Gibson). (**B**) *O. phaeacantha* var. *discata* growing near Tracy, Arizona.

directly relate to topics covered in the rest of the book.

Cannon

Perhaps the greatest contribution of Cannon was his thorough study of the roots of desert plants, organs that "do not greatly excite our admiration or curiosity, and thus have received little attention in the field" (Cannon, 1916). He was particularly interested in how oxygen (O_2) gets to the roots and how temperature influences root growth (Cannon, 1925). The cylindropuntia *Opuntia versicolor* (Fig. 1.15A) had a few anchoring roots, but most of the roots were shallow when grown in native soil outdoors near Tucson. When the same species was grown outside in a cooler habitat near the Pacific coast in central California for a few years, root growth was slight unless the soil was heated; maximum root growth of 20 mm day^{-1} occurred at 33°C, decreasing 40% at 20°C and over 90% at 16°C. Sustained root growth was nearly arrested at 1% soil O_2, about 26% of maximal at 2% O_2, and 55% of maximal near 5% O_2 in the soil (Cannon, 1925).

Cannon also excavated the entire root system of a *Ferocactus wislizenii* (Fig. 1.10A) that was 0.6 m tall (Fig. 1.16). The roots were quite shallow – only about 6–10 cm below the surface (later observations found roots of *F. wislizenii* to a depth of 18 cm, which is still quite shallow; MacDougal and Spalding, 1910). The shallowness of the entire root system has important consequences for water uptake by desert succulents (discussed in detail in Chapter 3). The roots were also slender, being less than 5 mm in diameter at 1 m from the stem base, although they radiated over 3 m from the stem base (Fig. 1.16).

Cannon also made observations relevant to the *gas exchange* (transpiration and photosynthesis) of desert succulents; *photosynthesis* is the process by which CO_2 is incorporated into organic compounds in *chloroplasts,* which are the subcellular bodies or *organelles* that contain the green pigment *chlorophyll,* which absorbs light in the visible region of the electromagnetic spectrum (Nobel, 1983a; Salisbury and Ross, 1985). Even while bemoaning the lack of instruments for measuring light and the lack of techniques for determining whether net CO_2 uptake occurred, Cannon was interested in the photosynthetic apparatus of *Carnegiea gigantea* (Fig. 1.12A). He determined that the *chlorenchyma* (chlorophyll-bearing tissue) of the stem of *C. gigantea* persists throughout its life,

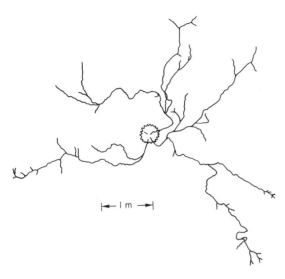

Figure 1.16. Root system of *Ferocactus wislizenii*. The plant was 0.6 m tall and 0.35 m in diameter. Modified from Cannon (1906).

which can easily be over 100 years; its chlorenchyma can be up to 7 mm thick, far greater than that of most plants (Cannon, 1908). He also noted that during the afternoon the leaves of *O. versicolor* had a threefold higher transpiration rate per unit area than did its stems (Cannon, 1906), providing some early hints on gas exchange by leafy cacti.

MacDougal and co-workers

Together with Cannon, MacDougal showed that the water content of *F. wislizenii* (Fig. 1.10A) can vary greatly, decreasing over 40% during a seven-month drought (MacDougal and Cannon, 1910). Plants survived uprooting for eighteen months outdoors, losing about 70% of their initial water content without lethal consequences, and excised cladodes of *Opuntia* survived for three years (MacDougal, 1910). Indeed, *F. wislizenii* maintained indoors under moderate conditions but without water survived for six years, losing about 30% of both its water and its dry weight over this period (MacDougal, Long, and Brown, 1915). Besides drought tolerance, growth under natural conditions was also studied. For instance, under field conditions annual height increases for *F. wislizenii* were about 3 cm for stems initially 10 or 70 cm tall (MacDougal and Spalding, 1910).

Under wet conditions the stem of *F. wis-*

lizenii could have an *osmotic pressure* of 4 bars (1 bar = 0.987 atmosphere), which in the modern unit is 0.4 megapascal (MPa) (osmotic pressure, which was determined by measuring the freezing point of the fluid mechanically expressed from the stem, is a property of a solution caused by the presence of dissolved solutes; it affects the water potential and hence water flow). After about two months of drought, the loss of water caused the osmotic pressure of *F. wislizenii* to increase to 0.6 MPa, and after four more dry months it became 1.0 MPa. Under wet conditions the osmotic pressure was 0.6 MPa for *C. gigantea,* which increased to 1.2 MPa after a few months of drought; a droughted agave had an osmotic pressure of 1.1 MPa (MacDougal and Cannon, 1910). Thus, osmotic pressures for these desert succulents ranged from about 0.4 MPa during wet periods to somewhat over 1 MPa during extended drought, which is much lower than for other desert perennials and which has important implications in their water relations.

The *xylem* serves as the plumbing for a plant, providing a conduit for water movement from the root to the stem and thence to the leaves, where water evaporation generally occurs – transpiration occurs solely through the stems for leafless cacti. MacDougal recognized the importance of the "transpiration stream" moving along the xylem. He and E. S. Spalding pointed out that the ribs of *C. gigantea* provide the stem with the accordion-like structural flexibility necessary for appreciable changes in water storage (Spalding, 1905; MacDougal and Spalding, 1910). At least 6 mm of rainfall on dry soil was necessary before morphological changes were detectable. Using the distance between rib crests as an easily measured morphological parameter, half-maximal swelling of the stem of *C. gigantea* was shown to occur eight days after a major rainfall event and maximal swelling occurred after about three weeks (Fig. 1.17), results that we will later compare with the water uptake properties of roots of *Agave deserti* and *Ferocactus acanthodes* under controlled conditions (Chapter 3). After a wet period, the dry weight of *C. gigantea* was only 9% of the *fresh weight* (the weight determined before any artificial drying). Their finding that open flowers could each lose 10 g of water per day

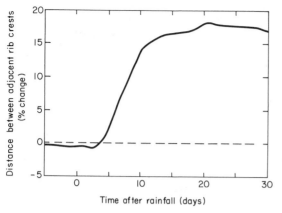

Figure 1.17. Increase in water storage by *Carnegiea gigantea* following rainfall, as monitored by changes in distance between adjacent ribs. Measurements were made 1.5 m above the ground on the north side of a stem 4 m tall (see Spalding, 1905). Data are recalculated from MacDougal and Spalding (1910) for a plant that responded to a 13 mm rainfall on 6 February 1904.

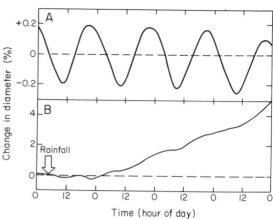

Figure 1.18. Time course of changes in the diameter of *Carnegiea gigantea*. Data were obtained during a wet period in May 1922 without appreciable net water storage (**A**) and in July 1923 following a major rainfall event at the indicated time (**B**). The dendrograph (an instrument for sensitively recording the diameter changes of trees) contacted the bases of ribs at opposite sides of a 6.5-m-tall stem at 1.2 m above the ground, where the diameter was 0.4 m. Modified from MacDougal (1924).

(MacDougal and Spalding, 1910) touches on the water costs of reproduction.

Another interesting observation had to do with the asymmetrical aspects of the ribs. The distance between rib crests tended to be less on the side where solar irradiation was greater (MacDougal and Spalding, 1910). This anticipates our discussion (Chapter 5) of the effects of *photosynthetically active radiation* (*PAR*, wavelengths of 400–700 nm that are readily absorbed by photosynthetic pigments). In any case, rib–rib spacing tended to be less on the south side of *C. gigantea* and *F. wislizenii* (in the northern hemisphere).

Daily changes in the diameter of *C. gigantea* were also observed. Diameter increased during the daytime and decreased beginning near dusk and continuing until a few hours after sunrise the next morning (Fig. 1.18A; MacDougal, 1924). This daily pattern was recognized as opposite to that of woody trees, and the changes were indicated to be the net result of water uptake from the soil minus loss by transpiration, again anticipating modern studies in water relations. The swelling following rainfall led to an increase in diameter of approximately 1% per day, measured from the base or trough of the ribs on opposite sides of the stem of *C. gigantea* (Fig. 1.18B).

Other measurements were made on gas exchange by *C. gigantea* and *O. versicolor* (MacDougal and Working, 1933). For instance, the stomata tended to be closed from late morning until dusk, and low temperatures led to more stomatal opening in the morning. Transpiration was shown to be greater when the air water vapor concentration was lower, when solar irradiation raised the stem temperature so that the stem water vapor concentration was higher, and when the wind speed was greater. Also, the stems could have an intercellular CO_2 level exceeding 1% by volume during the daytime, as could the stems of *Opuntia tomentosa* (MacDougal and Working, 1933). The cell sap of cacti had previously been found to be high in calcium and also in acidity, which was perceived as somehow being related to the photosynthetic process (MacDougal, 1910), but the reasons were not understood at such an early date. Shoots of opuntias were also found to curve toward a light source (MacDougal and Cannon, 1910).

Stems of *C. gigantea* and the platyopuntia *Opuntia phaeacantha* var. *discata* (formerly *O. engelmannii*) (Fig. 1.15B) were often 5°C to 10°C above air temperature, and soil temperatures of 64°C were recorded (Coville and MacDougal, 1903). A series of high-temperature records were also reported for a prickly-pear *Opuntia*. Cladodes continued to expand until tissue temperatures reached 51°C and could tolerate 55°C. Later, such expansion growth was shown to occur up to 55°C and then up to 58°C. Moreover, the cladodes could tolerate 1 h at a tissue temperature of 62°C and still remain alive (MacDougal, 1921; MacDougal and Working, 1921, 1922). Such a remarkable high-temperature tolerance of desert succulents will be explored in Chapter 4.

Shreve

Although perhaps best known for his contributions to the taxonomy and ecology of desert plants, Shreve also made crucial observations on the environmental responses of cacti, especially with regard to low temperature. In 1911, he wrote that "the line which marks the extreme southern limit of frost is the most important climatic boundary in restricting the northward extension of perennial tropical species," and for subtropical species like *C. gigantea* "it is obviously necessary to consider only the conditions of the coldest winters" (Shreve, 1911). Of the sixty-five species of arborescent columnar cacti, only *C. gigantea* (Fig. 1.12A), *Lophocereus schottii* (Fig. 1.14A), and *Stenocereus thurberi* (Fig. 1.14D) extend northward above the limit of frost. Shreve argued that the northern boundary of *C. gigantea* was due to low temperature, but its near absence west of the Colorado River reflected the low rainfall in eastern California. He found that *C. gigantea* could withstand 19 h of subzero air temperatures and a minimum air temperature of −8°C in Tucson. He also noted that because of transpirational cooling and radiation effects on clear nights, the surface cells of the cactus stem were below air temperature (Shreve, 1911).

Seedlings of *C. gigantea* are extremely hard to find in the field. Indeed, specimens less than 10 cm tall "are so rare, or inconspicuous, that nine botanists who have had excellent opportunities to find them report that they have never done so" (Shreve, 1910). He indicated that the drought immediately following the wet period leading to germination is the most cru-

Figure 1.19. *Coryphantha vivipara* near Mercury, Nevada, in the wintertime. This species ranges into southern Canada and is often covered with snow for extended periods. Note the 15-cm ruler in front of the cactus.

cial environmental factor affecting seedling establishment. At the other end of the lifetime, Shreve (1935) estimated the age of the tallest *C. gigantea* to be about 200 years, that of an *F. wislizenii* nearly 2 m tall about 130 years, and the lifespans of *L. schottii* and *S. thurberi* to be 100 years. In 1920, he indicated the importance of lightning in leading to the death of *C. gigantea*; for the saguaro forests near Tucson, one to two of the oldest and, hence, tallest *C. gigantea* per hectare are killed by lightning each year, the eventual bacterial decomposition of the tissue proceeding from the lightning-induced cracks (Steenbergh, 1972).

Although "the popularity of the cacti as ornamentals is partly due to the fact that they will stand more neglect than any other plants," most species "require all the light that it is possible to give them, the more direct sunshine the better" (Shreve, 1931a). Shreve also noted that greenhouse growth of desert cacti is greater if the night air temperature drops to 10°C or lower. However, temperatures below −10°C should be avoided, except for cold-hardy species such as *Coryphantha vivipara* (Fig. 1.19), *Opuntia polyacantha,* and *Pediocactus simpsonii* (Shreve, 1931a).

Others

Other researchers at the Desert Botanical Laboratory or through its auspices also made contributions to the early understanding of the responses of desert succulents to their environment. The enhanced transpiration at night was recognized by B. E. Livingston. Using the simple expedient of comparing the weight losses of *Opuntia phaeacantha, O. versicolor,* and a *Mammillaria* to that of a bowl of water under the same conditions, he showed that the relative transpiration of the cacti is much greater at night than during the daytime (Livingston, 1907). The observation of higher nocturnal transpiration by cacti was pursued by E. B. Shreve, who noted that cacti retain a higher fraction of the water taken up from the soil than do nonsucculent plants (Shreve, 1916). She also showed that water uptake from the air by spines of *Ferocactus wislizenii* was negligible and that *O. versicolor* could have a net water gain during the daytime (meaning that water uptake from the soil then exceeded transpiration) and a net loss at night (Shreve, 1915). The ideas of F. Shreve on the significance for cacti of episodic freezes were reexamined by W. V. Turnage and A. L. Hinckley, especially with regard to a cold wave in January 1937. Judging from the frost damage sustained by *Lophocereus schottii* and *Stenocereus thurberi,* the northern limits of these species as well as certain other cacti were felt to reflect their low-temperature tolerance (Turnage and Hinckley, 1938).

In 1915 H. M. Richards summarized research on the daily rise and fall of acid levels and on CO_2 exchange by cacti and other CAM plants (Richards, 1915). For instance, Th. de Saussure had noted in 1804 that the CO_2 release by an *Opuntia* sp. could be much less than that expected based on O_2 uptake accompanying *respiration* (the process taking place in organelles known as *mitochondria* whereby energy is released to the cell; if the substrate consumed is a sugar such as glucose, then the same amount of CO_2 is produced as the O_2 respired). In 1819 B. Heyne noted that the leaves of the CAM plant *Bryophyllum calycinum* tasted more acidic at the end of the night than at the end of the day; A. Mayer showed in 1875 that this was due to changes in the level of the four-carbon organic acid *malate*

$(COO^-CHOHCH_2COO^-)$. In 1883 G. Kraus demonstrated that the decrease in malate during the daytime is directly caused by light. Three years later O. Warburg indicated that the CO_2 released within CAM plants as the malate disappeared during the daytime could be fixed into organic compounds by photosynthesis. In 1890 E. Aubert reported that less acid is formed at night in CAM plants under conditions that are less favorable for photosynthesis during the daytime, such as low PAR or extreme temperatures, a conclusion supported by A. Astruc in 1903. Later, E. R. Long showed that maximum acidity in *Carnegiea gigantea* and *Ferocactus acanthodes* occurs in the morning (Long, 1915).

Using *O. versicolor* (Fig. 1.15A), Richards demonstrated that changes in tissue acidity over a 24-h period are clearly evident (Fig. 1.20). Maximum acidity occurred at dawn and then fell more than twofold by dusk. No difference in daily changes occurred from the base to the tip of its joints and the acidity changes appeared to be restricted to the chlorenchyma (Richards, 1915). Acidity increases tended to be greater for cool nights (minimum air temperatures of 10°C) than warm nights (20°C), and the daytime rate of deacidification was much less for cloudy days than for clear days. A daily pattern similar to that observed for *O. versicolor* occurred for *Mammillaria grahamii,* where a fourfold daily change in acidity level was found (Fig. 1.20).

The chemical composition of cacti was studied by H. A. Spoehr, especially *carbohydrates* (organic compounds containing C, H, and O, such as sugars and *cellulose,* a polymer of glucose). Carbohydrates represented about 20% to 30% of the dry weight, averaging 90% to 95% polysaccharides, mainly cellulose, starch, and mucilage (Spoehr, 1919). Calcium was found to be the principal inorganic element in *O. phaeacantha* (Fig. 1.15B), far exceeding all the other *cations* (positively charged ions) combined.

We should temper the impressive accomplishments of the Arizona school by indicating that the positive aspects have been emphasized here and certain erroneous ideas in the voluminous writings have not been mentioned. The research generally ignored environmental factors such as wind and PAR. Also, express-

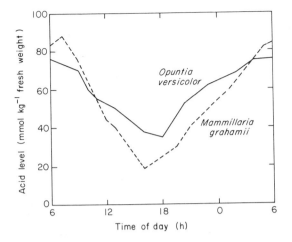

Figure 1.21. *Opuntia rastrera* planted for cattle fodder in central Coahuila, Mexico.

Figure 1.20. Changes in acidity of the chlorenchyma over 24-h periods for two species of cacti under well-watered conditions and moderate temperatures. Data are adapted from Richards (1915).

ing acidity data on a fresh-weight basis (Fig. 1.20) is less useful than on an area basis, especially when the inevitable comparisons are made to other plant groups with thin leaves or to PAR levels, which are expressed per unit area. Nevertheless, the outstanding research of the Arizona school lays a firm foundation for our consideration of the environmental responses of desert succulents in subsequent chapters.

OTHER EARLY RESEARCH

Despite the multiplicity of uses devised over the ages for agaves and cacti, prior to the twentieth century the main environmental information available on agaves and cacti related to *agronomy* (the science and economics of commercial production of crops). Species of *Opuntia* have been selected and propagated since prehistoric times for the quality of their fruit; production of thick, almost fiberless cladodes; and a high mucilaginous content of the cladodes that is suitable for soup as well as for adobe construction (Hernández Xolocotzi, 1970). The Maya in Yucatán developed highly sophisticated agronomic practices for *Agave fourcroydes* and found that annual productivity was highest for approximately 4,000 plants per hectare (Bolio, 1914). Recently, large areas of northeastern Mexico are being planted in *Opuntia phaeacantha* (Fig. 1.15B)

and *O. rastrera* (Fig. 1.21) for use as forage (Medina, Acuña, and De la Cruz, 1988). Yet there have been few quantitative studies relating the productivity of such commercially important plants to specific physical factors of the environment. Even in the husbandry of ornamental plants, particularly those specimens prized by cactus enthusiasts, little attempt has been made to relate plant growth to specific environmental factors.

Nevertheless, some of the major gaps in our knowledge up to 1940 about agaves and cacti not covered by the Arizona school were filled in by research or agronomic experiences at other locations. For instance, the worldwide cultivation of *Agave sisalana* (Fig. 1.22) provided information on its nutrient responses and its productivity (Smith, 1929; Lock, 1962; considered in Chapters 6 and 7, respectively). Information was accumulating on the productivity of opuntias, which are also cultivated worldwide, especially *O. ficus-indica* (Fig. 1.22). *Opuntia ficus-indica* has occupied about 70,000 ha in Tunisia, 100,000 ha in Sicily, and with other platyopuntias, about 300,000 ha in northeastern Brazil (Domingues, 1963; Monjauze and Le Houérou, 1965). Studies by D. Griffiths at the United States Department of Agriculture experiment station in Brownsville, Texas, indicated that annual dry-matter productivity of various platyopuntias is 3–7 metric tons of shoot dry matter ha^{-1} year^{-1}, which is 0.3–0.7 kg m^{-2} year^{-1} (Griffiths, 1905, 1906, 1908). Later studies with *O. phaeacantha* and cultivars of *O. ficus-indica* yielded productivities of over 1 kg m^{-2} year^{-1} and

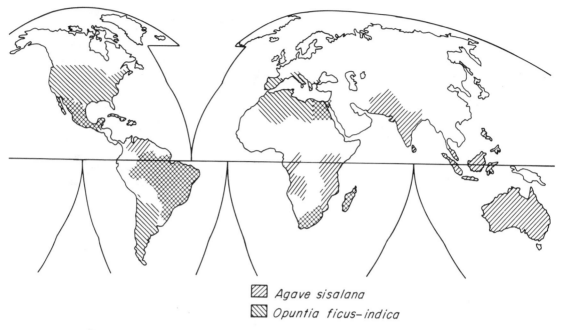

    ▨ *Agave sisalana*
    ▧ *Opuntia ficus-indica*

Figure 1.22. Countries where *Agave sisalana* or *Opuntia ficus-indica* have been or are now being commercially cultivated. Data are from Smith (1929) and Gentry (1982) for *A. sisalana* and from many sources plus personal observation for *O. ficus-indica*. The map is a modified Goode's equal-area projection and omits most small islands.

even up to 2 kg m^{-2} year^{-1} (Griffiths, 1915a). Griffiths also discussed the techniques for burning the spines off of the cladodes, experiments on feeding cladodes to cattle and other stock, and the nutritional value of the cladodes.

Certain other early studies are also worth mentioning. The state of our early knowledge on the functional morphology of cacti, especially their spines, was summarized by W. F. Ganong (1895). Also at the end of the last century, K. Goebel noted that *phototropic* responses (influences of light direction on morphology, such as curvature toward or away from a light source) occur for cacti. In particular, newly developing cladodes of the platyopuntia *Opuntia leucotricha* tended to become flattened in a direction perpendicular to a light beam, thereby intercepting more of the light compared with a random orientation of the cladodes (Goebel, 1895). At the turn of the century, C. E. Preston (1900) showed that the root systems of five species of cacti growing in the Tucson area were all shallow and radiated mainly horizontally. The freezing of cacti was studied by J. C. Th. Uphof, who noted that the ice crystals initially occurred extracellu-

larly and therefore did not injure the *protoplast* (the living part of a cell that is surrounded by the *cell wall,* the strong, nonliving cell envelope that is composed primarily of cellulose). In his experiments, certain cacti were quite cold tolerant, two *O. ficus-indica* hybrids being uninjured until $-14°C$ and $-16°C$, whereas *O. ficus-indica* was injured at $-5°C$ (Uphof, 1916). He also showed that all three of these species plus two other platyopuntias survived 6 h at 60°C, again indicating that cacti are extremely tolerant of high temperatures.

Other investigators were interested in the relationships between morphology and high temperature. For instance, F. Herzog estimated that elimination of deep ribs on cacti would increase daytime surface temperatures by 4–5°C by decreasing convective (based on air movement) heat exchange (Herzog, 1938). Convective heat exchange of cacti was measured by A. N. Watson, who also considered other thermal properties, such as heat conductance within the stems of *Carnegiea gigantea* (Watson, 1933). Also in the 1930s B. Huber reported that cladodes of *O. phaeacantha* (Fig. 1.15B) in direct sunlight could reach 65°C and those of other platyopuntias could be

34°C above air temperature without apparent injury (Huber, 1932).

This pretty much summarizes the state of our knowledge on the environmental responses of agaves and cacti prior to World War II. Technological developments reflecting the war effort led to innovations in instrumentation, which many years later were manifested in a whole series of new instruments available for the study of environmental responses of plants. Both the idea of simultaneously measuring all the environmental factors that influence leaf temperature and the sensors needed to measure them came into being in the 1960s. Photosynthesis and transpiration could be readily measured in the field thanks to the development of portable instruments in the 1970s. Sophisticated microprocessors that aided in data collection were developed along with new techniques for studying plant water relations. Computers were gaining in speed and memory while decreasing in cost, allowing the testing of more elaborate models. During the same period, similar technological breakthroughs plus the elucidation of basic metabolic pathways rapidly advanced our knowledge of plant biochemistry. Another major impetus was the dramatic increase in government financial support for scientific research, attracting a whole new set of investigators to the study of desert succulents. The discipline of plant physiological ecology came of age, as witnessed by four volumes on the topic appearing for the first time in the venerable *Encyclopedia of Plant Physiology* in the early 1980s (Lange et al., 1981–1983). Because of these developments, many of the observations and measurements made on agaves and cacti prior to the 1940s can be integrated into a much more sophisticated understanding of plant responses to the environment. The synthesis of such results is one of the objectives of this book. But first we must fill in a little more background on the whys and wherefores of agaves and cacti.

Taxonomy and morphology

Taxonomy considers the names of various organisms, including how these names are organized into an overall classification scheme. Such information is useful for knowing the correct names and the number of species of agaves and cacti and for understanding the various levels of organization and possible evolutionary relationships. We will also describe some of the morphological characteristics of the genus *Agave* and the unrelated family Cactaceae. All the sections of the genus for agaves and all the tribes (subdivisions of a family containing related genera) and many of the genera for cacti will be explicitly considered.

We should first indicate that the Plant Kingdom, which is one of the five kingdoms currently used for classifying organisms, contains over 250,000 presently existing species, most of which are seed-producing plants. Although the number of divisions into which Kingdom Plantae can be subdivided is open to interpretation and dispute, we can recognize lineages that over evolutionary time led to the algae and those that led to the land plants. The latter gave rise to the bryophytes (liverworts, hornworts, and mosses), which do not contain conducting tissues, and the plants that contain conducting or *vascular* tissue, namely xylem and *phloem* (used for conducting organic compounds throughout a plant, such as the sugars produced by photosynthesis). Vascular plants are sometimes given divisional status and called the Tracheophyta.

Early vascular plants gave rise to the seed plants, which evolved into groups we now recognize as the gymnosperms (conifers, cycads, extinct seed ferns, *Ginkgo,* and the gnetales) and the angiosperms or flowering plants (Fig. 1.23). The names for the two subclasses of angiosperms, Monocotyledoneae and Dicotyledoneae, reflect the number of *cotyledons* (initial leaves produced by the developing embryo of a seed). Monocots tend to have parallel venation in the leaves, vascular bundles scattered throughout the stem, and flower parts in threes, whereas dicots tend to have netlike venation in the leaves, a prominent vascular *cambium* (cylinder of dividing cells producing xylem on the inside and phloem on the outside) in the stem, and flower parts in fours or fives. The subclasses are divided into orders, the total number varying somewhat from expert to expert because taxonomy is a field still in flux.

The two taxa considered in this book come from different subclasses and hence from quite unrelated orders. The monocot

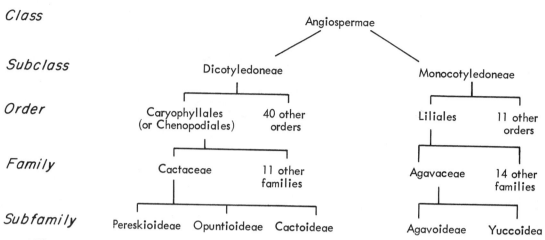

Figure 1.23. Taxonomic relations within the angiosperms. Modified from Cronquist (1981), Thorne (1981, 1983), and Gibson and Nobel (1986).

order Liliales contains fifteen families, including the Liliaceae (lily family), which contains about half of the 8,000 species in the entire order and with which the Agavaceae has many affinities (Cronquist, 1981). In turn, the Agavaceae can be divided into the subfamilies Yuccoideae, with about 40 species in two genera (*Yucca* and *Hesperaloë*), and Agavoideae, which we consider in the next section, with about 200 species in about ten genera. The dicot order Caryophyllales contains about twelve families (Fig. 1.23). Among these, the Cactaceae are probably most closely related to the Aizoaceae (ice plants) and the Didiereaceae. In turn, the Cactaceae can be divided into three subfamilies: the primitive leaf-bearing Pereskioideae with about 20 species, the Opuntioideae with slightly more than 250 species, and the large subfamily Cactoideae with nearly 1,400 species (Gibson and Nobel, 1986).

The *karyotype* (size, number, and shape of chromosomes, the bearers of genetic information) is quite different between the two taxa. Agaves have 5 large and 25 small chromosomes ($n = 30$), just over half of the species studied being *polyploid* (possessing one or more additional sets of chromosomes; McKelvey and Sax, 1933; Cronquist, 1981; Dahlgren, Clifford, and Yeo, 1985; Pinkava and Baker, 1985). Cacti generally have 11 chromosomes of similar size ($n = 11$), although polyploidy with 22, 33, 44 or more chromosomes is common for cacti, especially platyo-

puntias (Pinkava et al., 1973; Weedin and Powell, 1978; Pinkava and Parfitt, 1982). Thus the structural and physiological similarities that we will discuss for the two taxa occur against a background of major genetic differences.

AGAVES

Agaves represent a large genus of about 136 species of leaf succulents in the family Agavaceae, which has about twelve genera and 250 species. Agaves are native to North America, with their center of origin being in present-day Mexico (Gentry, 1982, 1985). Other genera in the subfamily Agavoideae that are closely related to agaves include *Furcraea*, *Manfreda*, and *Polianthes* (Dahlgren et al., 1985). The genus *Agave* was established by Linnaeus in 1753 with a brief description of *Agave americana* (Fig. 1.9), the "type" (exemplary) species for the genus.

The genus can be readily separated into two distinct subgenera (Table 1.1): Littaea with spicate or racemose flowering forms (Fig. 1.24A) and Agave with inflorescences of spreading umbellate or paniculate branches (Fig. 1.24B). As is so often true of taxonomic designations, lines between species and other levels of organization often tend to blur, making precise distinctions difficult. For instance, *Agave fourcroydes* and *A. sisalana* actually represent sterile *clones* (genetically related plants that reproduce vegetatively), not true species, because viable seeds are usually not produced (Gentry, 1985).

Table 1.1. *Summary of agave species*

Subgenus	Group or section	Number of species	Examples
Littaea	Amolae	8	*A. attenuata* Salm., *A. pedunculifera* Trel., *A. vilmoriniana* Berger
	Choritepalae	3	*A. ellemeetiana* Jacobi
	Filiferae	8	*A. filifera* Salm., *A. multifilifera* Gentry
	Marginatae	21	*A. lechuguilla* Torr., *A. xylonacantha* Salm.
	Parviflorae	4	*A. parviflora* Torr. subsp. *parviflora*, *A. schottii* Engelm., *A. toumeyana* Trel.
	Polycephalae	5	*A. celsii* Hook
	Striatae	3	*A. dasylirioides* Jacobi & Bouché, *A. stricta* Zucc.
	Urceolatae	2	*A. utahensis* var. *nevadensis* Engelm.
Agave	Americanae	6	*A. americana* L., *A. lurida* Aiton, *A. scabra* Salm-Dyck
	Campaniflorae	3	*A. promontorii* Trel.
	Crenatae	6	*A. bovicornuta* Gentry
	Deserticolae	10	*A. cerulata* Trel., *A. deserti* Engelm., *A. sobria* Brandeg.
	Ditepalae	10	*A. murpheyi* F. Gibson, *A. palmeri* Engelm.
	Hiemiflorae	12	*A. atrovirens* Karw. ex Salm, *A. lagunae* Trel.
	Marmoratae	4	*A. zebra* Gentry
	Parryanae	6	*A. parryi* Engelm., *A. parryi* var. *huachucensis* (Baker) Little ex Benson, *A. patonii* Trel.
	Rigidae	12	*A. angustifolia* Haw., *A. fourcroydes* Lem., *A. macroacantha* Zucc., *A. rhodacantha* Trel., *A. tequilana* Weber
	Salmianeae	5	*A. salmiana* Otto ex Salm ssp. *crassispina* (Trel.) Gentry
	Sisalaneae	6	*A. sisalana* Perrine, *A. weberi* Cels ex Poisson
	Umbelliforae	2	*A. shawii* Engelm.

Note: Data including authorities are from Gentry (1982, 1985). All the examples cited are pictured, discussed, or at least mentioned in the text.

Most agaves have thick short stems and thus appear as sessile rosettes (Fig. 1.25). The thick stem together with the spirally imbricating leaf bases form a large mass (Figs. 1.4 and 1.5, especially Fig. 1.5B). Subsequent leaves are produced around the rosette axis at angles averaging 137.5°, the so-called Fibonacci angle, which we will now briefly consider.

Near the beginning of the thirteenth century, Fibonacci, also known as Leonardo of Pisa, discovered a sequence of numbers formed by adding the two previous numbers in the series: 1, 1, 2, 3, 5, 8, 13, 21, 34, . . . This sequence, known as the *Fibonacci series,* has relevance to many natural patterns (including the number of cactus ribs, discussed in the next section), as well as to the angles

for agave leaves. In particular, if pairs of numbers two apart along the sequence are chosen in progression, then the following fractions can be formed: $\frac{2}{5}, \frac{3}{8}, \frac{5}{13}, \frac{8}{21}, \frac{13}{34}, \ldots$. Each fraction describes the plant's *phyllotaxy,* or arrangement of leaves along the stem. For example, a $\frac{3}{8}$ phyllotaxy means that a spiral of eight leaves would be encountered in three complete turns around the stem. The sequence of fractions rapidly converges to the value of 0.382, which when multiplied by the number of degrees in a circle (360°) gives 137.5°, the *Fibonacci angle.* Although there is admittedly an air of mystery as to how plants "obey" such a mathematical rule, the geometric patterns spatially set down early in development later become demonstrably manifested in the angles

Figure 1.24. (**A**) Spicate inflorescence of *Agave lechuguilla,* typical of the subgenus Littaea, growing near La Zarca, Durango, Mexico. (**B**) Paniculate inflorescence of *A. deserti,* typical of the subgenus Agave, growing near Anza Borrego, California; dried flowers are evident on the fruit for the current year's inflorescence in the center, whereas the inflorescence on the left is from the previous year and the basal leaves are all dead.

between successive leaves forming the characteristic agave rosette (Fig. 1.25). Moreover, the resulting leaf arrangement tends to minimize self-shading by the leaves.

Returning now to other properties of agave leaves, we note that they often live for about five years, remaining metabolically active throughout. However, gas exchange with the environment occurs predominantly when the soil is wet, a condition conducive to stomatal opening. The leaves formed just before the inflorescence emerges from the center of the rosette tend to be narrower and thinner, gradually grading into the *bracts* (reduced leaves) of the inflorescence.

The often spectacular mature inflorescences of agaves are generally 2–10 m high (Fig. 1.24). They can grow quite rapidly, daily height increments of 10–20 cm being common for *Agave deserti* and *A. lechuguilla* (Mac-

Callum, 1908; Lock, 1962; Ting, 1976a; Nobel, 1977a; Freeman and Reid, 1985). The flowers of some species are pollinated by bats – for example, those of *A. palmeri* (Howell, 1979); bees, wasps, flies, beetles, hawkmoths, other insects, hummingbirds, and other birds are also observed visiting agave flowers. Certain species, such as *A. parviflora* and *A. toumeyana,* produce nectar and pollen primarily during the daytime and tend to be pollinated by insects, while others, such as *A. palmeri* and *A. schottii,* produce nectar and pollen primarily at night, which favors bat pollination (Schaffer and Schaffer, 1977). In this regard, a thirty-year decline in seed set by *A. palmeri* in southern Arizona parallels a decline in bat numbers brought about by human disturbance of the caves inhabited by the bats (Howell and Roth, 1981). Many species of agaves primarily reproduce vegetatively, either from "off-

Figure 1.25. Rosettes of *Agave parryi* var. *huachucensis,* a multileaf variety growing near Chihuahua, Chihuahua, Mexico.

Figure 1.26. Mature stems of *Rhipsalis mesembryanthemoides* with fruit that are about 6 mm in diameter.

shoots'' (*ramets*) produced on *rhizomes* (underground stems) or by bulbils on the inflorescence. Although the rhizomes of agaves are often fleshy and thick, the roots tend to be numerous, thin, fibrous, and shallow.

As would be expected from their taxonomic diversity (Table 1.1), agaves vary considerably morphologically. For instance, *Agave dasylirioides* has thin, stiff leaves about 0.5 m long when mature and exhibits many of the evolutionarily primitive features of the genus; *A. attenuata* has wide but thin flexible leaves and is increasingly popular as a garden plant; and *A. palmeri* has stiff, thick leaves similar in length to those of the other two species. Leaf length varies from that of the small *A. utahensis*, which ranges into cold mountainous regions of Nevada and has leaves up to only about 0.2 m in length (at least for variety *nevadensis*), to the large subtropical *A. promontorii* of southern Baja California, Mexico, which can have leaves up to 1.5 m long that often arch back toward the ground, to *A. atrovirens* of the mountains of Oaxaca, which can have massive, fairly straight leaves up to 3 m long. Leaves of some species, such as *A.*

ellemeetiana, have no marginal teeth or terminal spines, leaves of other species are heavily armed, such as *A. palmeri* and some specimens of *A. deserti*, and others have a filiferous (filament-bearing) margin, such as *A. filifera*.

Most agave leaves are fairly uniformly colored, such as grayish green, but a popular horticultural variety of *A. americana* has yellow longitudinal margins, and *A. zebra* of Sonora has picturesque dark green stripes transversely across its leaves. Some species form large clusters with many rosettes, such as *A. celsii,* and others have only single isolated, nonsuckering rosettes, such as *A. bovicornuta*. Although most agaves are native to Mexico, some species occur only in Central America, such as species in section Hiemiflorae, like *A. lagunae* of Guatemala, and some only in the United States, such as *A. utahensis* (Gentry, 1982). Despite such morphological variability, most people have little difficulty in recognizing an agave when they see one.

CACTI

The Cactaceae is a family consisting of about 122 genera and just over 1,600 species (Gibson and Nobel, 1986). When Linnaeus published the first major taxonomic system for plants in 1753, he recognized 22 species of cacti. Enthusiasm of one kind or another has

Figure 1.27. Leaf-bearing cacti of Subfamily Pereskioideae: (**A**) *Pereskia aculeata* being put into gas exchange apparatus by Terry L. Hartsock and (**B**) *P. grandifolia* (photographed by David T. Tissue).

led to the proliferation of epithets. Partly because of the widespread interest cacti have engendered among hobbyists, collectors, horticulturists, and professional taxonomists, over 350 generic names and over 11,000 specific names have been used for cacti. We will take a much more manageable and conservative view here (Table 1.2).

Except for *Rhipsalis* (Fig. 1.26), whose small fleshy fruit is attractive to birds and thus could have led to widespread distribution of its seed in their feces, cacti are clearly native to the New World. Most researchers agree that the origin of the family occurred somewhere in or near present-day northern South America. The most primitive genus is *Pereskia* in Subfamily Pereskioideae (Fig. 1.27); its 15–18 species bear fairly large leaves that are not totally unlike those of ivy or other nonsucculent plants. Indeed, upon casual observation only the cognoscenti would recognize a *Pereskia* as a cactus. These species currently range from northern South America through Central America to northcentral Mexico.

Subfamily Opuntioideae was an evolutionarily early offshoot of the leafy cacti. Its *Pereskiopsis* (Fig. 1.28A) occurs natively in Mexico and Guatemala. Other leafy genera such as *Austrocylindropuntia* and *Quiabentia* (Fig. 1.28B) are more widely distributed, extending into southern areas of South America. The largest genus of the Opuntioideae is *Opuntia* (Figs. 1.2B,D, 1.13, 1.15, and 1.21), which contains at least 160 species, most with jointed stems. Moreover, one-third of the approximately 180 cactus species mentioned in this text are in the genus *Opuntia* (Table 1.2), indicating its popularity for environmental research. In this context we should point out that many genera of the Cactaceae have not been studied physiologically at all (Table 1.2); indeed, much of the data presented in this book are based on studies of two species of cacti, *Ferocactus acanthodes* and *Opuntia ficus-indica* (and one species of agave, *A. deserti*). Opuntias occur from Canada to the southern tip of South America and on nearly all islands in the New World where cacti are found. Cacti occur natively in nearly all states in the United States; in more than half of such states, the only genus represented is *Opuntia*. Platyopuntias have been introduced worldwide (Figs. 1.13 and 1.22 for *O. ficus-indica*), including such exotic places as along the coast of the Gulf of Thailand and the shores of Bali, Indonesia. The genus contains species with prominent, although ephemeral, leaves, such as *Opuntia imbricata,* and species with inconspicuous, short-lived leaves, such as the widely cultivated *O. ficus-indica* (Fig. 1.13).

Figure 1.28. Leaf-bearing cacti of Subfamily Opuntioideae: (**A**) *Pereskiopsis porteri* and (**B**) *Quiabentia chacoensis*.

Table 1.2. *Summary of cactus genera and tribes*

Subfamily	Genus or tribe	Number of genera	Number of species	Examples
Pereskioideae	*Maihuenia*	1	2–3	*M. poeppigii* (Otto) Weber
	Pereskia	1	15–18	*P. aculeata* Miller, *P. columbiana* Britt. & Rose, *P. grandifolia* Haw., *P. guamacho* Weber, *P. sacharosa* Griseb.
Opuntioideae	*Austrocylindropuntia*	1	~15	*A. cylindrica* (Lam.) Backeb., *A. subulata* (Mühlenpf.) Backeb.
	Opuntia	1	>160	
	Cylindropuntia		~40	*O. acanthocarpa* Engelm. & Bigel. var. *coloradensis* L. Bens., *O. bigelovii* Engelm., *O. echinocarpa* Engelm. & Bigel., *O. fulgida* Engelm., *O. imbricata* (Haw.) DC., *O. kelvinensis* V. & K. Grant, *O. kleiniae* DC., *O. leptocaulis* DC., *O. parryi* Engelm., *O. prolifera* Engelm., *O. ramosissima* Engelm., *O. schottii* Engelm., *O. spinosior* (Engelm.) Toumey, *O. versicolor* Engelm.
	Platyopuntia		>120	*O. amyclaea* Tenore, *O. aurantiaca* Lindl., *O. basilaris* Engelm. & Bigel., *O. bergeriana* Weber, *O. cantabrigiensis* Lynch, *O. caracasana* Salm-Dyck, *O. chlorotica* Engelm. & Bigel., *O. cochenillifera* (L.) Salm-Dyck, *O. cordobensis* Speg., *O. decumbens*

Table 1.2. *Summary of cactus genera and tribes (continued)*

Subfamily	Genus or tribe	Number of genera	Number of species	Examples
				Salm-Dyck, *O. dejecta* Salm-Dyck, *O. echios* Howell, *O. erinacea* Engelm. & Bigel., *O. fragilis* (Nutt.) Haw., *O. ficus-indica* (L.) Miller, *O. galapageia* Hensl., *O. humifusa* (Raf.) Raf., *O. hyptiacantha* Weber, *O. inamoena* K. Schum., *O. leucotricha* DC., *O. lindheimeri* Engelm., *O. littoralis* (Engelm.) Cocker. var. *littoralis, O. littoralis* var. *vaseyi* (Coult.) L. Bens. & Walkingt., *O. megasperma* Howell, *O. melanosperma* Svens., *O. microdasys* (Lehm.) Pfeiff., *O. oricola* Philbr., *O. palmadora* Britt. & Rose, *O. phaeacantha* var. *discata* (Griff.) L. Bens. & Walkingt., *O. polyacantha* Haw., *O. quimilo* K. Schum., *O. rastrera* Weber, *O. rubescens* Salm-Dyck, *O. salagria* Cast., *O. streptacantha* Lem., *O. stricta* Haw., *O. stricta* var. *dillenii* (Ker-Gawler) L. Bens., *O. sulfurea* G. Don var. *sulfurea, O. sulfurea* G. Don var. *pampeana* (Speg.) Backeb., *O. tomentosa* Salm-Dyck, *O. utkilio* Speg., *O. violacea* Engelm. var. *santa-rita* (Griff. & Hare) L. Bens., *O. vulgaris* Miller, *O. wentiana* Britt. & Rose
	Pereskiopsis	1	9–11	*P. porteri* (Brandeg.) Britt. & Rose
	Pterocactus	1	9	
	Quiabentia	1	3–4	*Q. chacoensis* Backeb.
	Tacinga	1	1–2	
	Tephrocactus	1	~50	*T. articulatus* (Pfeiff. ex Otto) Backeb. var. *inermis* (Speg.) Backeb.
Cactoideae	Browningeae	2	6–7	
	Cacteae	25	~400	*Ariocarpus fissuratus* Engelm., *Coryphantha gladiispina* (Bödek.) Berger, *C. vivipara* (Nutt.) Britt. & Rose var. *deserti* (Engelm.) W.T. Marsh., *Echinocactus grusonii* Hildm., *E. platyacanthus* Link & Otto, *E. polycephalus* Engelm. & Bigel., *Epithelantha bokei* L. Bens., *Ferocactus acanthodes* (Lem.) Britt. & Rose, *F. covillei* Britt. & Rose, *F. diguetii* (Weber) Britt. & Rose, *F. gracilis* Gates, *F. histrix* (DC.) Linds., *F. latispinus* (Haw.) Britt. & Rose, *F. viridescens* (Nutt.) Britt. & Rose, *F. wislizenii* (Engelm.) Britt. &

Table 1.2. *Summary of cactus genera and tribes (continued)*

Subfamily	Genus or tribe	Number of genera	Number of species	Examples
				Rose, *Leuchtenbergia principis* Hooker, *Lophophora williamsii* (Lem.) Coulter, *Mammillaria dioica* K. Brandeg., *M. elegans* DC., *M. elongata* DC., *M. goodrichii* Scheer, *M. grahamii* Engelm., *M. heyderi* Mühlenpf., *M. lasiacantha* Engelm., *M. longicoma* (Britt. & Rose) Berger, *M. microcarpa* Engelm., *M. rhodantha* Link & Otto, *M. tetrancistra* Engelm., *M. theresae* Cutak, *M. woodsii* Craig, *M. zeilmanniana* Bödek., *Neolloydia intertexta* (Engelm.) L. Bens., *Pediocactus sileri* (Engelm.) L. Bens., *P. simpsonii* (Engelm.) Britt. & Rose, *Turbinicarpus klinkerianus* Backeb. & Jacobs
	Cereeae	9	~110	*Cereus emoryi* Engelm., *C. pentagonus* (L.) Haw., *C. peruvianus* (L.) Miller, *C. repandus* (L.) Miller, *C. silvestrii* Speg., *C. validus* Haw., *Melocactus caesius* Wendl., *M. intortus* (Miller) Urban
	Echinocereeae	1	~50	*Echinocereus engelmannii* (Parry) Lem., *E. enneacanthus* Engelm., *E. fendleri* Engelm., *E. ledingii* Peebl., *E. maritimus* (Jones) K. Schum., *E. triglochidiatus* Engelm., *E. viridiflorus* Engelm.
	Hylocereeae	22	~140	*Acanthocereus tetragonus* (L.) Humlk., *Cryptocereus anthonyanus* Alexand., *Epiphyllum crenatum* (Lindl.) G. Don, *E. hookeri* (Link & Otto) Haw., *Harrisia tortuosa* (Forb.) Britt. & Rose, *Hylocereus costaricensis* (Weber) Britt. & Rose, *H. lemairei* (Hooker) Britt. & Rose, *Selenicereus urbanianus* (Gürke & Weing.) Britt. & Rose
	Leptocereeae	8	~30	
	Notocacteae	20	~350	*Copiapoa alticostata* Ritt., *C. applanata* Backeb., *C. cinerea* (Ritt.) Backeb., *C. ferox* Lembcke & Backeb., *C. gigantea* Backeb., *C. haseltoniana* Backeb., *C. lembckei* Backeb., *C. megarhiza* Britt. & Rose, *C. solaris* (Ritt.) Ritt., *Discocactus horstii* Buin., *Eriosyce ceratistes* (Otto) Britt. & Rose, *Eulychnia acida* Phil., *E. castanea* Phil., *E. iquiquensis* (K. Schum.) Britt. & Rose, *E. spinibarbis* (Otto) Britt. & Rose, *Gymnocalycium denudatum* (Link

Table 1.2. *Summary of cactus genera and tribes (continued)*

Subfamily	Genus or tribe	Number of genera	Number of species	Examples
				& Otto) Pfeiff., *G. mihanovichi* (Frič. & Gürke) Britt. & Rose var. *friedrichii* Werderm., *G. saglione* (Cels) Britt. & Rose, *Neoporteria chilensis* (Hildm.) Britt. & Rose, *Notocactus mammulosus* (Lem.) Berger, *N. ottonis* (Lehm.) Berger, *Parodia maassii* (Heese) Berger, *Rebutia marsoneri* Werd., *R. minuscula* K. Schum., *Rhipsalis cassutha* Gaertn., *R. fasciculata* (Willden.) Haw., *R. gaertneri* (Reg.) Vaupel, *R. mesembryanthemoides* Haw., *Schlumbergera truncata* (Haw.) Moran
Pachycereeae		13	~70	*Carnegiea gigantea* (Engelm.) Britt. & Rose, *Cephalocereus polygonus* (Lem.) Britt. & Rose., *C. royenii* (L.) Britt. & Rose, *Lophocereus schottii* (Engelm.) Britt. & Rose, *Myrtillocactus cochal* (Orcutt) Britt. & Rose, *Pachycereus pecten-aboriginum* (Engelm.) Britt. & Rose, *P. pringlei* (Berger) Britt. & Rose, *Stenocereus alamosensis* (Coult.) Gibs. & Horak, *S. eruca* (T. S. Brandeg.) Gibs. & Horak, *S. griseus* (Haw.) Buxb., *S. gummosus* (Engelm.) Gibs. & Horak, *S. montanus* (Britt. & Rose) Buxb., *S. thurberi* (Engelm.) Buxb.
Trichocereeae		13	>200	*Cleistocactus jujuyensis* (Backeb.) Backeb., *Denmoza rhodacantha* (Salm-Dyck) Britt. & Rose, *Echinopsis eyriesii* (Turp.) Zucc., *Lobivia grandis* Britt. & Rose, *L. huascha* (Weber) Marsh., *Trichocereus bridgesii* (Salm-Dyck) Britt. & Rose, *T. candicans* (Gill) Britt. & Rose, *T. chilensis* (Colla) Britt. & Rose, *T. coquimbanus* (Molina) Britt. & Rose, *T. litoralis* (Johow) Looser, *T. pachanoi* Britt. & Rose, *T. spachianus* (Lem.) Britt. & Rose

Note: Data are from Gibson and Nobel (1986), except for the authorities, which are taken from Benson (1982) for cacti from the United States and Canada, from Backeberg (1958–1962) for cacti from Latin America, or directly from modern research articles. Because of the taxonomic scheme adopted, many of the names differ from those in the papers cited; for instance, *Cereus giganteus* is here *Carnegiea gigantea*, *Lemaireocereus thurberi* is *Stenocereus thurberi*, *Opuntia compressa* is *O. humifusa*, and *O. engelmannii* is *O. phaeacantha*. All the species cited are pictured, discussed, or at least mentioned in the text, figures, or tables.

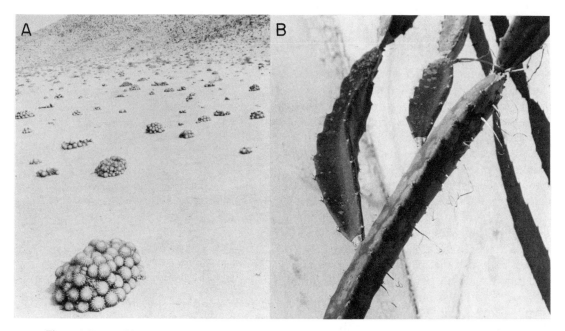

Figure 1.29. (**A**) Clumps of *Copiapoa applanata* near Pan de Azucar in northern Chile. (**B**) *Hylocereus costaricensis,* an epiphytic cactus in Tribe Hylocereeae; note the adventitious roots growing along the stem.

Opuntias can be subdivided into the cylindropuntias, with about 40 species (when the 25 species of chollas and the ribbed species are included) and the large group of platyopuntias, with over 120 species. Part of the reason for the proliferation of species of *Opuntia* is their ready hybridization. For instance, hybridization of the two sympatric cylindropuntias, *O. fulgida* and *O. spinosior,* has led to the species *O. kelvinensis* in southern Arizona (Grant and Grant, 1971; Gibson and Nobel, 1986). When the Franciscan fathers moved north into what is now California during the eighteenth century, they brought *O. ficus-indica* (the "mission fig") with them and planted it around their missions. This species readily hybridized with the local platyopuntia, *O. littoralis* var. *vaseyi,* leading to the creation of *O. "occidentalis."* Numerous other species of platyopuntias have presumably also arisen by hybridization, for instance, *O. phaeacantha* (Fig. 1.15B), which is wide-ranging in the western United States and northern Mexico (Pinkava et al., 1973; McCleod, 1975; Grant and Grant, 1980).

Let us next consider some of the characteristics of the tribes of the large subfamily Cactoideae (Table 1.2; see Gibson and Nobel,

1986). The tribe with perhaps the most primitive features is Leptocereeae, ranging from the West Indies through Central America to northwestern South America. Browingeae is a small tribe native to Peru and northern Chile that is intermediate between the Leptocereeae and the Trichocereeae, which has over 200 species in South America. Notocacteae is a large, diverse tribe also from South America that contains a few species of columnar cacti, barrel cacti such as *Copiapoa* (Figs. 1.10B and 1.29A), and *Rhipsalis* (Fig. 1.26). The Pachycereeae consists of the columnar cacti that have radiated from Mexico (six species shown in Figs. 1.12 and 1.14). Tribe Hylocereeae includes both terrestrial and *epiphytic* cacti (Fig. 1.29B; epiphytes are flowering plants, usually not rooted in the soil, that derive water and nutrients from the atmosphere and/or the plants on which they grow). Actually, 120 species of epiphytic cacti occur, mostly in the Hylocereeae or Notocacteae (for instance, *Rhipsalis*). The Cereeae are specialized columnar cacti that evolved in eastern South America, especially Brazil. Echinocereeae is a small tribe that includes the mound-forming hedgehog cacti of North America. The large tribe Cacteae contains about 400 species or one-fourth of the entire family, about half of which are in the genus *Mammillaria* (Fig. 1.2C). The Cacteae is centered in Mexico, but ranges from Canada to northern South America.

Unlike agaves, which generally produce a single inflorescence at the end of their lifetimes, mature cacti tend to produce flowers every year on already existing stems. Pollinators for cacti are similar to those for agaves – bees, hawkmoths, other insects, hummingbirds, and bats – with bees being the most common, at least in the southwestern United States (Grant and Grant, 1979).

What more can be said about the diversity in shoot morphology of cacti already presented visually in this chapter? We have seen species for which the leaves are the predominant photosynthetic organs (Figs. 1.27 and 1.28), cacti with conspicuous but drought deciduous leaves (Fig. 1.28), and others for which leaves other than the first two cotyledons are essentially absent (most of the others pictured). The variety in growth habit is exemplified by epiphytes with flattened stems (Figs. 1.26 and 1.29B), tilting barrel cacti (Fig. 1.10), erect columnar cacti (Figs. 1.12 and 1.14), platyopuntias whose cladodes can have a preferred directional orientation (Figs. 1.2D, 1.13, and 1.15B), species with very slender round stems (Figs. 1.2A and 1.27), and species with nearly spherical stems, at least when relatively young (Figs. 1.2C, 1.10, and 1.19). Some species form clumps (Fig. 1.29A) or even mats, and others have solitary stems (most of those shown from Subfamily Cactoideae). Perhaps no other family offers the morphological diversity in leaves and stems found in the Cactaceae.

Let us return for a moment to the Fibonacci series and relate it to the ribs of cacti. Ribs actually represent the coalescence or fusion of *tubercles,* the enlarged succulent leaf bases projecting from the stems of cacti. Fibonacci patterns have long been noted for the ribs of certain cacti (Troll, 1937; del Castillo Sanchez, 1982) and can be modified by ionizing radiation (Gómez-Campo and Casas-Builla, 1972) or mechanical damage to the stem apical region. For instance, using a dissecting needle to disrupt the apical region of twelve seedlings of *Ferocactus covillei* with 8 ribs

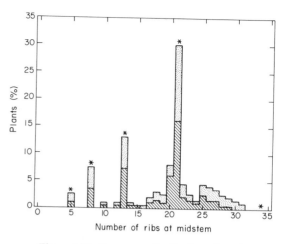

Figure 1.30. Frequency distribution of number of ribs at midstem for *Ferocactus acanthodes* at the Philip L. Boyd Deep Canyon Desert Research Center. One hundred stems were examined at each of two sites; data are from Robberecht and Nobel (1983; cross-hatched) and Gibson and Nobel (1986; stippled). Fibonacci numbers are indicated by asterisks.

caused seven of the plants to initiate new apical growth, but with 5 ribs, the number preceding 8 in the Fibonacci series (P. S. Nobel and T. L. Hartsock, unpublished observations). Also, the number of ribs at midstem for over half of the specimens of the barrel cactus *Ferocactus acanthodes* (Fig. 1.2A) examined at two sites in the western Sonoran Desert correspond to numbers in the Fibonacci series (Fig. 1.30). The number of ribs for fourteen other species of *Ferocactus* is also mainly in the Fibonacci series, in this case 8, 13, or 21 (Gibson and Nobel, 1986). However, rib number for *Echinocactus grusonii* and *E. platyacanthus,* other barrel cacti also in Tribe Cacteae, for *Lobivia grandis* and *L. huascha* in Tribe Trichocereeae, and for *Notocactus ottonis* in Tribe Notocactaceae does not tend to fall into the Fibonacci series. Likewise for columnar cacti, rib number does not conform to the Fibonacci series, at least for five species of *Stenocereus* in Tribe Pachycereeae, four species of *Cereus* in Tribe Cereeae, and sixteen other species of columnar cacti in five different tribes (Gibson and Nobel, 1986). We will reconsider rib numbers in a Fibonacci series when we discuss the productivity of *F. acanthodes* (Chapter 7).

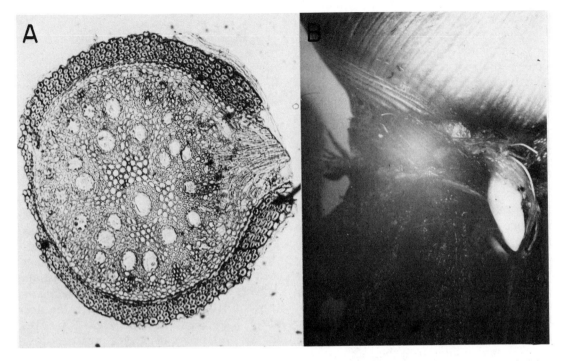

Figure 1.31. Mature and rain-induced roots. (A) Cross section of a mature root of *Agave deserti* that is 1.3 mm in diameter and with a lateral branch developing (right side). Four to six layers of cortical cells occur beneath the epidermis (which is only partially intact); the open regions in the center are the conducting cells of the xylem. (B) A rain root that is 5 mm long growing at the base of the stem of *A. deserti* only 6 h after a droughted plant had been placed in a hydroponic solution.

Organs and anatomy

We have already introduced many of the features of morphology and *anatomy* (relating to structures at a cellular or tissue level) of agaves and cacti. Our purpose here is to provide a few more details – specifically on *organs* such as roots, stems, and leaves – so that our ensuing discussion will have a proper structural basis. Emphasis will be on structures common to all seed plants as well as on features relatively special to agaves and cacti. We will draw attention to certain structural similarities between agaves and cacti, such as for the chlorenchyma and for roots.

ROOTS

As for other plants, roots of agaves and cacti anchor the plants in the ground and absorb water and nutrients from the soil. The roots of most agaves and cacti tend to be quite shallow. For instance, most of the roots of many cacti occur 3–15 cm below the soil surface (Cannon, 1911). The mean depth for roots is only 8 cm for both *Agave deserti* (Nobel, 1976) and *Ferocactus acanthodes* (Nobel, 1977b). Roots of these species tend to be rather thin (Fig. 1.31A), of fairly constant diameter along their length, and, for *A. deserti*, relatively unbranched.

Beginning at the outside, roots are covered by an epidermis generally one cell layer thick. Fine root hairs project from epidermal cells, thereby increasing the area available for water and nutrient absorption. *Mycorrhizal* associations also occur, in which the fungal hyphae penetrate into the root *cortex* just beneath the epidermis as well as extend out into the soil. These associations, which effectively increase the root surface area, have received relatively little research attention for agaves and cacti, although a number of species of *Glomus* (a fungal genus commonly involved in mycorrhizal associations) have been shown to colonize the roots of *A. deserti, F. acan-*

thodes, and other succulents in the north-western Sonoran Desert (Bethlenfalvay, Dakessian, and Pacovsky, 1984).

Conspicuous intercellular air spaces occur in the root cortex (Fig. 1.31A), facilitating the movement of O_2 and CO_2 by *diffusion* (random motion based on thermal energy, which leads to a net movement of uncharged particles toward regions of lower concentration). Water and solutes must cross the root cortex, either in cell walls (the *apoplastic* pathway) or through the protoplasts, which are interconnected by *plasmodesmata* (fine, membrane-flanked, cytoplasmic threads that pass from one protoplast through two cell walls to the protoplast of an adjacent cell). This linking of the *cytoplasm* (the part of a cell inside the cell membrane, or *plasmalemma*) of adjacent cells is termed the *symplasm;* water and solutes moving from protoplast to protoplast through the plasmodesmata are said to be moving in the *symplastic* pathway.

The region interior to the root cortex is the *stele,* which contains the vascular tissue. The vascular tissue tends to be organized into discrete bundles for the monocot *A. deserti* (Fig. 1.31A) and to consist of a central xylem with radiating xylem arms between which are regions or "poles" of phloem for the dicot *F. acanthodes* (Freeman, 1969). The vascular tissue of the root is continuous with that of the stem and the leaves, leading to the movement of water and nutrients in the xylem (part of the "transpiration stream") and the distribution of organic compounds throughout the plant in the phloem.

One of the interesting environmental responses of both agaves and cacti is the induction of *rain roots* (Fig. 1.31B) when the soil water potential in the root zone is raised by precipitation or by artificial watering. The rain roots arise both directly from the base of the stem (Fig. 1.31B) as well as from *meristematic* regions (those having cells capable of dividing, such as at the stem apex or in the vascular cambium of cacti) along the existing mature roots (Fig. 1.31A). Rain roots, whose surfaces are generally lighter in color (Fig. 1.31B) than mature roots and thus can be readily distinguished, contain cells that tend to have thinner cell walls and less intercellular air space than mature roots, especially in the cortex (water

uptake by rain roots and existing roots is discussed in Chapter 3).

In addition to mature roots and rain roots, we should also mention *adventitious* roots, which occur at abnormal locations on the stem or even from leaves. Adventitious roots can develop when a stem is not in contact with soil, as commonly occurs for epiphytes (Fig. 1.29B) and for stem segments of opuntias (Gibson and Nobel, 1986).

STEMS

Although stems are always conspicuous for cacti, especially in subfamilies Opuntioideae and Cactoideae, many agaves are called *acaulescent* – literally, "without stem" – although all agaves actually have at least a short stem. Stems can be prominent for certain agaves, such as the piña of *Agave tequilana* (Fig. 1.5B) or for *A. fourcroydes* after the older leaves have been harvested (Fig. 1.7A). Also, the inflorescences of agaves are specialized stems. Our emphasis will be on the stems of cacti and their surface appendages other than leaves.

Areoles, tubercles, and ribs

Areoles produce spines and hair (*pubescence* or *trichomes*) atop raised tubercles or on rib edges of cactus stems (Fig. 1.32). A spine contains no vascular tissue, and indeed most of its cells are not living at maturity except near the base. Spines vary considerably in length, thickness, curvature, color, arrangement, and number per areole. Such variation can even occur for a single species, reflecting environmental conditions during growth and genetic differences for different populations of that species (Gibson and Nobel, 1986). The tiny, barbed spines characteristic of Subfamily Opuntioideae are called *glochids,* which are readily dislodged from the areoles. The trichomes are usually formed as chains of cells attached end to end beginning as a projection from a single epidermal cell; at maturity most of the cells of typical cactus trichomes are dead, so the trichomes are generally quite brittle.

Tubercles can be arranged in various patterns on the stems of cacti, such as the spiral arrangements found on *Mammillaria* (Fig. 1.2C). They can also be fused into ribs (Fig.

Figure 1.32. Rib morphology for *Ferocactus acanthodes* (**A**) and *Pachycereus pringlei* (**B**) showing the characteristic spine pattern for each areole and the fusion of tubercles to form a rib.

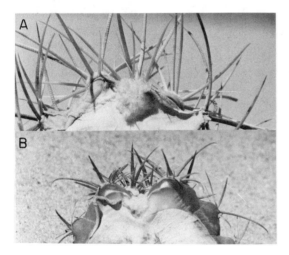

Figure 1.33. Cutaway near stem apex showing the spines and the apical pubescence: (**A**) *Ferocactus acanthodes* and (**B**) *F. viridescens*. In the apical region, spines are longer and more common and the apical pubescence is thicker for *F. acanthodes* (both stems are about 12 cm in diameter).

have meristematic regions where the dividing cells lead to new tissues. For twenty-one species of cacti in ten genera, the apical meristem at the *distal* (far) end of the stem ranged from 0.1 to 1 mm in diameter (Mauseth and Niklas, 1979; Mauseth, 1983a). They can be even larger in some other species (up to 2 mm in diameter), and indeed the Cactaceae have the largest known apical meristems among the angiosperms (Boke, 1980).

The apical regions of barrel and columnar cacti have morphological features that affect tissue temperature and hence influence the distribution of these species (Chapter 4). For *F. acanthodes,* the apical pubescence can be 7–10 mm thick, and numerous spines are present (Fig. 1.33A). On the other hand, the apical pubescence for *F. viridescens* is much thinner, and few spines occur in the apical region (Fig. 1.33B). We will quantitatively interpret the effect of such morphological features in Chapter 4, using a computer model that considers all the ways that a cactus can exchange energy with its environment.

1.32), which can either align with the stem axis or spiral around it (Gibson and Nobel, 1986). Although technically leaf bases, we will consider the tubercles together with the stem when we discuss gas exchange (Chapter 2) and thermal relations (Chapter 4) of cacti.

Meristem – apex of Cactoideae

The stems of agaves have meristematic regions where new leaves are formed, leading to the central spike of folded leaves. Cacti also

Chlorenchyma and adjacent tissues of cacti

A primary function of plant shoots is photosynthesis, those of cacti being no exception. Yet the chlorenchyma of all but the leaf-bearing cacti (to be considered shortly) is rather distinct among plants other than succulents in that it usually contains ten to twenty layers of chlorophyll-containing cells (Fig. 1.34A), in contrast with only three to five lay-

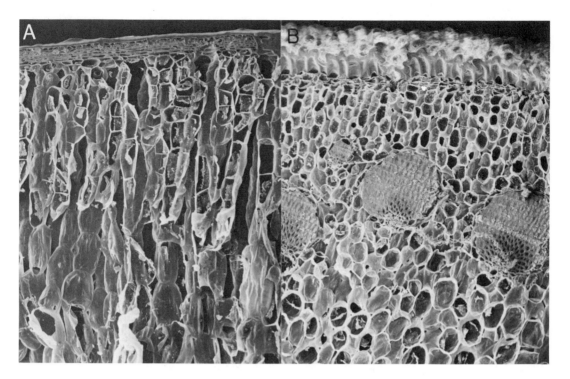

Figure 1.34. Scanning electron micrographs of the chlorenchyma in (**A**) *Opuntia ficus-indica* and (**B**) *Agave deserti*. Note the thick cuticles, the many "layers" of chlorenchyma, the conspicuous intercellular air spaces, and chloroplasts next to the broken cells (most clearly seen in upper part of the chlorenchyma of *A. deserti*). The cactus has a hypodermis composed of about four layers of thin cells immediately under a one-cell-thick epidermis, and the agave has conspicuous vascular bundles containing many fiber cells. The micrographs are from sections about 2 mm thick that were prepared by Arthur C. Gibson.

ers for most plants. Another distinctive feature of the chlorenchyma is the occurrence of aggregates of calcium oxalate crystals (druzes), leading to an abnormally high level of calcium (Ca^{2+}) in the stems of many cacti (discussed in Chapter 6).

We should also comment on the cell types that occur adjacent to the chlorenchyma. On the outside is the epidermis, which is generally one cell thick and is covered by the waterproofing cuticle that is usually 2–10 μm thick. Directly under the epidermis is a tissue of one to a few cell layers, termed the *hypodermis* (Fig. 1.34A); the hypodermis is tough but flexible, providing the cactus stem with a "skin" that can expand or contract as water is stored or released. The epidermis is interrupted by stomata, the mean stomatal frequency for five species of *Opuntia* ranging from 30 to 62 per square millimeter (Conde, 1975). The air passage through the epidermis created by a stomatal pore is continued as a substomatal canal through the hypodermis so

that exchange of gases between the chlorenchyma and the atmosphere can readily occur, at least when the stomatal pores are open. Toward the inside of the chlorenchyma is the water-storage parenchyma. Vascular tissue occurs within the chlorenchyma or just beneath it. The mucilage-containing cells are usually interspersed in the water-storage parenchyma (McGarvie and Parolis, 1981; Trachtenberg and Mayer, 1981, 1982a).

LEAVES

Agaves

The chlorenchyma of a cactus stem (Fig. 1.34A) has much in common with the chlorenchyma of an agave leaf (Fig. 1.34B). They both consist of many cell layers and hence are often about 3 mm thick, although traces of chlorophyll can be found even further away from the surface (the chlorenchyma is only 0.2–0.4 mm thick for most nonsucculent plants). The chloroplasts occur around the periphery of the cells, lying in an immobilized layer just inside the plasmalemma. For both agaves and cacti, the intercellular air spaces occupy only about 10–20% of the chlorenchyma compared with 30–50% for leaves of most nonsucculent plants.

The leaf epidermis is covered by a cuticle (Fig. 1.34B) and contains the stomatal pores. The cuticle of agaves is generally rather thick, usually 3–15 μm. The main component of the cuticle is *cutin,* a complex group of waxy polymers that waterproof the leaf surface (Wattendorff and Holloway, 1980, 1982). Indeed, the cuticles of both agaves and cacti are extremely impervious to gases like CO_2 and water vapor. The stomatal frequency is about 40 mm^{-2} for both the upper and lower sides of agave leaves – measurements on twenty-three taxa of agaves indicated that the stomatal frequency is generally $30–50 \text{ mm}^{-2}$ with the lower surface having 10 mm^{-2} fewer stomata than the upper surface (Gentry and Sauck, 1978), although *Agave sisalana* has 10 mm^{-2} more on the lower surface (Abd El-Rahman and Hassib, 1972). Such stomatal frequencies are less than the $100–300 \text{ mm}^{-2}$ for the lower surfaces of most dicot leaves and for both sides of most monocot leaves (Esau, 1977; Nobel, 1983a). The lower stomatal frequency for agaves, as well as for cacti, suggests less capacity for gas exchange with the environment. However, the real factor affecting gas exchange with the environment (Chapter 2) is the fraction of the leaf surface area occupied by stomatal pores, which open and close in response to environmental and physiological stimuli as the pair of *guard cells* surrounding the pore change their internal pressure and hence their shape.

The thick leaves of agaves also contain much water-storage *parenchyma.* Distributed among the relatively unspecialized parenchyma cells, which have fairly thin walls and are almost devoid of chloroplasts, are the vascular bundles. These bundles tend to be aligned along the leaf axis and usually contain many fiber cells (Fig. 1.34B), especially in commercially harvested species such as *A. fourcroydes, A. lechuguilla,* and *A. sisalana,* whose thin fiber cells are generally 0.5–5 mm long. Toward the center of the leaf of *A. deserti,* vascular bundles are separated from each other by about ten parenchyma cells;

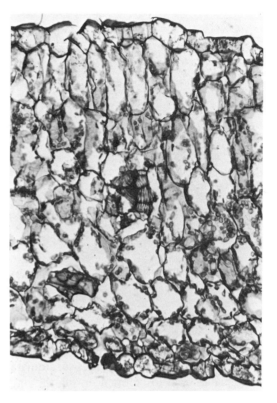

Figure 1.35. Leaf section of *Pereskia grandifolia,* which has a leaf-structure similar to that of certain C_3 plants. Note the stomatal pore near the center of the lower epidermis and the chloroplasts in the mesophyll cells. The photomicrograph, which is from a leaf 400 μm thick, was photographed by Arthur C. Gibson from a section prepared by Anne M. Sjostrom.

prominent vascular bundles also occur in the chlorenchyma (Fig. 1.34B).

Cacti

Leaves vary considerably among cacti, from the large ones of *Pereskia* of the Pereskioideae (Fig. 1.27) to relatively small, *terete* (cylindrical and generally tapering to a point) ones on *Pereskiopsis* and *Quiabentia* in the Opuntioideae (Fig. 1.28) to their essential absence for most species in Subfamily Cactoideae. For instance, leaves on *Mammillaria heyderi* are no longer than 0.05 mm (Boke, 1953).

Leaf anatomy also varies considerably among cacti (Bailey, 1968). For instance, leaves of *Pereskia grandifolia* (Fig. 1.35) are similar in cross section to leaves of most plants like ivy, lettuce, maple, oak, and spinach,

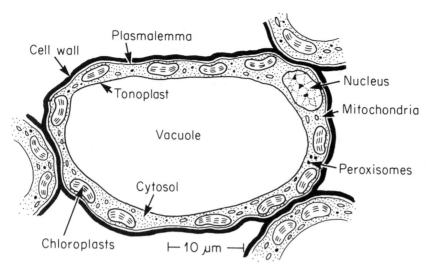

Figure 1.36. Schematic representation of a section through the center of a mesophyll cell. Modified from Nobel (1983a).

which all utilize the C_3 pathway of photosynthesis (CO_2 is fixed into a *three-carbon* molecule, 3-phosphoglycerate, which is then used in the synthesis of sugars like glucose; stomata for C_3 plants tend to be open during the day and closed at night). The leaf has an upper (*adaxial*) and a lower (*abaxial*) epidermis, with the cuticle tending to be thicker on the upper side. The intercellular air spaces in which gases can readily diffuse are fairly conspicuous for *P. grandifolia* (Fig. 1.35). The terete leaves of *Austrocylindropuntia subulata* are much thicker, and its intercellular air spaces are not as conspicuous. In the next chapter we will see that the gas-exchange characteristics of leaves of cacti tend to differ from those of their stems.

Photosynthetic cell

As our final topic in this stage-setting chapter, we briefly consider a representative photosynthetic cell of a leaf (Fig. 1.36). Such a cell occurs in the *mesophyll* – the chlorenchyma of a leaf (Figs. 1.34B and 1.35) – the basic cell characteristics being the same for a cell in the chlorenchyma of a cactus stem (Fig. 1.34A).

The cell is surrounded by a cell wall that substances cross by diffusing in aqueous solutions in the interstices between the *microfibrils* (organized arrays of about 100 cellulose molecules interspersed with other polymers; Mauseth, 1983b; Nobel, 1983a). Substances

entering the cell protoplast must next cross the plasmalemma (Fig. 1.36), either by diffusion or by *active transport*, which involves the binding of the transported molecule to some membrane component and the expenditure of metabolically produced energy for the transport process, such as *ATP* (adenosine triphosphate).

A characteristic feature of photosynthetic cells of CAM plants is a large central *vacuole*, usually comprising 90% or more of the cell volume. The vacuole is surrounded by a membrane known as the *tonoplast* (Fig. 1.36). The region between the plasmalemma and the tonoplast is the cytoplasm, a term that often excludes the *nucleus*. The increasingly used term *cytosol* (Fig. 1.36) refers to the cytoplasmic solution exterior to the nucleus and the organelles such as chloroplasts, mitochondria, and *peroxisomes* (small bodies surrounded by a single membrane and involved in metabolism). The chloroplasts in agaves and cacti occur in an immobile layer adjacent to the plasmalemma, whereas the mitochondria are mainly in that part of the cytosol that moves around the cell, this movement generally being referred to as *protoplasmic streaming*.

With this consideration of a cell plus the other introductory material, we are now armed with the definitions and much of the background necessary for our consideration of the environmental responses of agaves and cacti.

2 Gas exchange

Exchanges of CO_2 and H_2O between the atmosphere and plant tissues are among the most interesting and certainly the most investigated ways that plants interact with their environment. The exchange of O_2 also occurs, both for photosynthesis and for respiration. Taking such exchanges into consideration, we can write the net chemical reaction describing photosynthesis as follows:

$$CO_2 + H_2O \rightleftharpoons \{CH_2O\} + O_2 \qquad (2.1)$$

where $\{CH_2O\}$ represents a carbohydrate. Respiration is basically the reverse of Equation 2.1, in which a carbohydrate such as glucose interacts chemically with other compounds in the mitochondria, thereby producing CO_2 and H_2O.

Exchange of O_2 is rarely a limiting factor for plant *shoots* (aboveground stems plus leaves). Specifically, air contains 21% O_2 by volume compared with only about 0.035% CO_2, and for photosynthesis or respiration (Eq. 2.1) O_2 and CO_2 generally have a 1:1 *stoichiometry* (molar ratio of reactants and/or products in a particular chemical reaction). However, O_2 levels can fall below 10% for the air spaces in soils, especially if the volume fraction of the air spaces is reduced by soil compaction or excess water, which can cause root respiration to become O_2-limited. Indeed, one of the factors involved in the general restriction of agaves and cacti to relatively well-drained, porous soils undoubtedly has to do with belowground access to O_2. However, limitations on soil fauna and flora by the periodic drying of such porous soils may also be important, because excess water is conducive to fungal growth and root rot, matters that need further investigation. In this chapter we will restrict our attention to CO_2 and H_2O exchanges of shoots, leaving comments on the respiration of existing and rain-induced roots to Chapter 7.

Three major pathways have been identified for the fixation of atmospheric CO_2 – C_3, C_4, and CAM (Crassulacean acid metabolism). Plants from the three different pathways open their stomata at different times of the day, use two different enzymes for the initial fixation of CO_2, and exhibit different daily patterns of net CO_2 uptake (Fig. 2.1). As mentioned in Chapter 1, plants utilizing the C_3 pathway open their stomata during the daytime and initially fix CO_2 into C_3 compounds (3-phosphoglycerate) in their chloroplasts. About 90% of the Angiospermae (Fig. 1.23) use the C_3 pathway, which is considered to be the primitive (ancestral) condition. C_4 plants also open their stomata and hence take up CO_2 during the daytime (Fig. 2.1), but their initial photosynthetic products are *four-carbon* organic acids such as oxaloacetate and malate. Only about 1% of the angiosperm species use the C_4 pathway, but they include such commercially important ones as corn or maize (*Zea mays*), various species of *Sorghum*, and sugar cane (*Saccharum officinarum*), as well as eight of the ten agriculturally most noxious weeds (Holm et al., 1977; Nobel, 1983a). C_4 plants have higher average maximal rates of

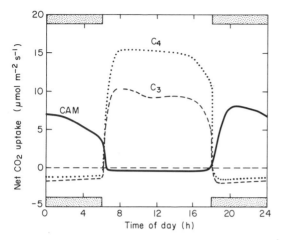

Figure 2.1. Mean net CO_2 exchange under optimal conditions of water, temperature, and PAR (assumed throughout this chapter) for plants of the three photosynthetic pathways over 24-h periods. The stippled area indicates the dark period (night). Representative data, which are adapted from Larcher (1980), Nobel (1983a), and other sources, are presented on a total leaf or stem surface area basis in all cases (total area of both sides for flat leaves) and for species with the higher observed maximum rates of net CO_2 uptake.

CO_2 uptake than do C_3 plants (Fig. 2.1) and tend to be favored in environments having relatively high temperatures, high PAR (photosynthetically active radiation), and limited water supply, habitats often occupied by C_4 tropical monocots (Nobel, 1983a).

Unlike C_3 and C_4 plants, CAM plants can open their stomata and take up CO_2 primarily at night (Fig. 2.1) when they use a biochemical pathway similar to that of C_4 plants for initially fixing CO_2. During the daytime, CAM plants tend to close their stomata and to use the C_3 pathway for fixing CO_2, whose source is not directly the atmosphere but rather from within the plants. In particular, the CO_2 fixed during the night by CAM plants is released internally and refixed during the daytime. CAM occurs in about 10% of the species of Angiospermae and in nearly thirty of its families, including the Agavaceae and the Cactaceae, in more primitive vascular plants, such as ferns, and in *Isoetes* and *Stylites* in Class Lycopodiatae (Winter, 1985). The species of angiosperms that use the CAM pathway tend to occur in arid and semiarid regions where water stress produced by extended drought is a common

occurrence. Other CAM plants (for example, many epiphytes) occur in locally dry micro-habitats.

Because gas exchange of agaves and cacti is so important, we will be treating different aspects of it throughout this book. For instance, stomatal control of gas exchange will be treated from the point of view of water relations in Chapter 3, temperature in Chapter 4, and PAR in Chapter 5. Failure to recognize PAR limitations on CO_2 uptake by agaves and cacti has led to some erroneously low estimates of the maximum net CO_2 uptake ability of agaves and cacti, especially for those studies using *environmental growth chambers* (specially designed units or even rooms in which plants are grown under artificial lighting of specified duration or *photoperiod*, controlled day and night air temperatures, and often controlled relative humidity). The effects of elevated atmospheric CO_2 levels will be discussed in Chapter 7. Here we focus on gas exchange under the optimal environmental conditions of wet soil, moderate temperatures (such as day/night air temperatures of 25°C/ 15°C), and PAR levels leading to at least 90% of maximum net CO_2 uptake. Our consideration will be for healthy plants of the species whose full identification, including authorities, is given in Tables 1.1 (agaves) or 1.2 (cacti).

Based in large measure on research at the Desert Botanical Laboratory discussed in the previous chapter, by 1940 we knew that stomatal opening and transpiration by cacti are greater at night, when the malate level in the chlorenchyma increases substantially. Our knowledge about the gas exchange of agaves and cacti did not advance substantially over the next thirty years, although the pattern of nocturnal CO_2 uptake and important details of the CAM pathway for cacti were elucidated (Thomas and Ranson, 1954; Kausch, 1965; Ting and Dugger, 1968; Holdsworth, 1971). The pathway in CAM plants whereby CO_2 is incorporated into a three-carbon compound leading to a four-carbon compound like malate was shown to occur in the cytosol, with the malate subsequently being transported into the vacuole for overnight storage. More recently, the ecological importance of CAM as a process for conserving water has been clearly recognized (Kluge and Ting, 1978; Osmond, 1978).

One of the technological advances after World War II that greatly enhanced our knowledge of the gas exchange of agaves and cacti was the development of *infrared gas analyzers* (IRGAs), which permitted accurate measurement of *net CO_2 exchange* (gross or total photosynthesis minus respiration). After illustrating representative daily patterns for net CO_2 exchange by agaves and cacti, the underlying biochemistry will be discussed. We will show how the readily observed nocturnal increases in acidity can be related to net CO_2 uptake. We will also discuss another method to assess the occurrence of CAM that utilizes the relative discrimination between atmospheric forms of CO_2 differing in *molecular weight* (a dimensionless number indicating the mass of a molecule relative to that of the common carbon isotope taken as 12). Finally, we will compare the water cost of net CO_2 uptake by agaves and cacti with that by non-CAM plants, a topic that has many ecological and agronomic ramifications.

Daily patterns for agaves

In this and the next section, we will present data on net CO_2 exchange and transpiration of agaves and cacti. We will generally express gas exchange per unit surface area, not per unit mass of plant tissue, to enable comparisons with other plant groups and to relate gas exchange to levels of incident radiation (both cases in which data are conventionally expressed per unit surface area). We will use the total surface area of the photosynthetic organs to express the *flux densities* (mass or moles of a substance moving across unit area in unit time). Although using total leaf surface area is not really necessary for the relatively thin leaves of *Pereskia* (Fig. 1.27), just as it is not necessary for leaves of most C_3 and C_4 plants, the upper surface of an agave leaf does not have the same area as the lower surface, and hence total area is then a more appropriate basis for data presentation. Moreover, the chlorenchyma on the two sides of a massive, clearly three-dimensional agave leaf is generally separated by an essentially opaque water-storage parenchyma, so the two sides act like independent surfaces with respect to PAR interception (PAR incident on either side of thin leaves can be used throughout the leaf).

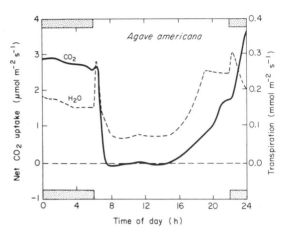

Figure 2.2. Gas exchange by *Agave americana* over a 24-h period. Net CO_2 exchange and transpiration were determined for the entire shoot of a plant maintained in an aqueous nutrient solution. Modified from Neales, Patterson, and Hartney (1968).

For cacti, the stems are not planar, the surface area is increased by ribs and tubercles, and gas exchange is again most usefully expressed on a total area basis. As for agave leaves, the stems of most cacti are opaque, so the sides act independently with respect to PAR interception.

EARLY OBSERVATIONS ON *AGAVE AMERICANA*

Independent studies by Neales, Patterson, and Hartney (1968), by Ehrler (1969, 1975), and by Kristen (1969) were the first to show that stomatal opening and net CO_2 uptake by agaves occur at night, as is characteristic of CAM plants and as had been shown for cacti in the early part of the century. All three groups worked with *Agave americana* (Fig. 1.9). Moreover, all three groups compared various plants and indicated that the agave had much lower daytime water loss per unit surface area than did the non-CAM counterparts.

Greater stomatal opening at night by *A. americana* caused a higher rate of water loss than during the day (Fig. 2.2). Moreover, net CO_2 uptake was extremely low for the 8-h period from solar times of 7 h to 15 h. Indeed, 75% of net CO_2 uptake over the 24-h period occurred at night, even though the dark period was only one-third as long as the light period

(Neales et al., 1968). Some net CO_2 uptake during the daytime occurred about 1 h after the lights went on and for the latter half of the day (Fig. 2.2). As we shall see, other species of agave also have some net CO_2 uptake at the beginning and at the end of the daytime.

OTHER SPECIES

Subgenus Littaea

Daily patterns of gas exchange have so far apparently been determined for only three species in Subgenus Littaea (Table 1.1): *Agave lechuguilla* (Fig. 1.8), *A. utahensis* (Fig. 2.3), and *A. vilmoriniana*. For these species and most of the other agaves and cacti that we will consider, measurements were made in the laboratory, as was the case for *A. americana* (Fig. 2.2), but the photoperiod was 12 h and the plants were growing in soil (compare with Fig. 2.2).

For *A. lechuguilla*, 85% of the net CO_2 uptake over the 24-h period is at night (Fig. 2.4A). Almost no CO_2 uptake occurs at the beginning of the day, and, though some does occur in the latter half of the day, the maximum daytime rate of CO_2 uptake is only about one-third of the maximum nighttime rate. The rate of transpiration is also greater at night than during most of the day, except for late in the afternoon (Fig. 2.4A) when transpiration is substantial because of appreciable stomatal opening and also the greater drop in water vapor concentration from the chlorenchyma to the ambient air, the latter caused by the higher chlorenchyma temperature during the daytime.

Similar daily patterns of gas exchange occur for *A. utahensis*, where nocturnal CO_2 uptake accounts for 78% of the total CO_2 uptake over 24 h (Fig. 2.4B). The average transpiration rate at night is about threefold higher than the morning value (Fig. 2.4B), similar to results with *A. americana* (Fig. 2.2) and *A. lechuguilla* (Fig. 2.4A). The afternoon period of net CO_2 uptake for *A. utahensis* is accompanied by appreciable transpiration (Fig. 2.4B), just as for *A. lechuguilla* (Fig. 2.4A). On the other hand, *A. vilmoriniana* has a substantial proportion of its net CO_2 uptake during the daytime (Szarek, Holthe, and Ting, 1987; Nobel and McDaniel, 1988), especially at

lower temperatures, as we will discuss in detail in Chapter 4.

Subgenus Agave

Besides *A. americana* (Fig. 2.2), gas exchange over 24-h periods has also been determined for other species in Subgenus Agave. Daily transpiration patterns parallel those of net CO_2 uptake, as above (Figs. 2.2 and 2.4), and so are omitted here.

Although the daily patterns of CO_2 exchange are somewhat different for *A. deserti*, *A. fourcroydes*, and *A. tequilana*, all have most of their net CO_2 uptake at night (Fig. 2.5). These three species have a small net CO_2 uptake in the first 1 to 2 h of the day. *Agave fourcroydes* and *A. tequilana* exhibit a moderate net CO_2 uptake in the late afternoon and a large increase followed a few hours later by a smaller decrease in the early part of the night (Fig. 2.5). In contrast, net CO_2 uptake steadily increases in late afternoon for *A. americana* (Fig. 2.2), whereas no such rise occurs for *A. deserti* (Fig. 2.5). Indeed, 99% of the net CO_2 uptake over a 24-h period occurs during the night for *A. deserti*.

What generalizations can we make about the daily net CO_2 uptake for agaves under well-watered conditions, moderate temperatures, and relatively high levels of PAR? Nearly all net CO_2 uptake occurs at night, as expected for CAM plants, ranging from 75 to 99% for the six species whose data are presented. All these species exhibit a small amount of net CO_2 uptake early in the day. More net CO_2 uptake occurs late in the afternoon for most species of both subgenera, but none occurs for *A. deserti* (Fig. 2.5). Net CO_2 uptake parallels the degree of stomatal opening, where greater opening generally leads to higher transpiration rates, although other factors such as temperature and air water vapor content are involved.

AGE EFFECTS

All of the above studies were done on mature plants, and the patterns can differ with plant age. In particular, agave seedlings can have a greater fraction of their daily CO_2 uptake during the daytime than do mature plants (Fig. 2.6 versus Fig. 2.5). This daily pattern for seedlings is reminiscent of the primitive C_3

Figure 2.3. *Agave utahensis* (**A**) growing in the Clark Mountains, California, and (**B**) placed in a gas-exchange apparatus.

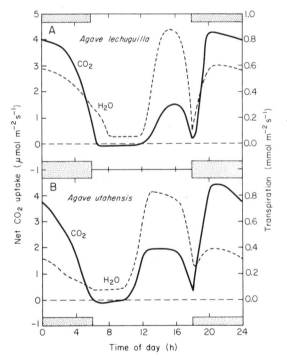

Figure 2.4. Gas exchange over a 24-h period for two species of agave in Subgenus Littaea. (**A**) *Agave lechuguilla*; CO_2 data adapted from Nobel and Quero (1986) and transpiration from unpublished observations of the same study. (**B**) *A. utahensis*; unpublished observations of P. S. Nobel and T. L. Hartsock under the same experimental conditions as for (**A**).

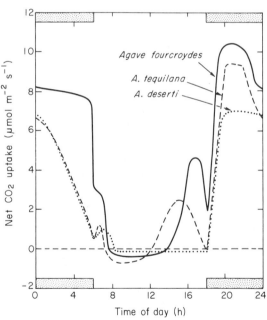

Figure 2.5. CO_2 exchange for three species of agave in Subgenus Agave over a 24-h period. Data are from Nobel (1984a) for *Agave deserti*, from Nobel (1985a) for *A. fourcroydes*, and from Nobel and Valenzuela (1987) for *A. tequilana*.

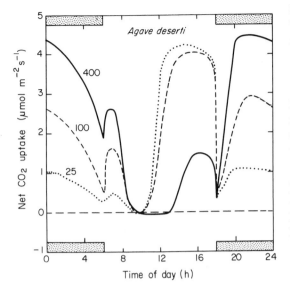

Figure 2.6. CO_2 exchange over a 24-h period by seedlings of *Agave deserti* of different ages. The age in days after germination is indicated next to the curves. The 25-day-old seedling ($\cdots\cdots$) was 14 mm tall, the 100-day-old seedling ($-----$) was 31 mm tall, and the 400-day-old seedling (———) was 63 mm tall. Unpublished observations of P. S. Nobel and T. L. Hartsock under the conditions described in Nobel (1984a).

condition, where net CO_2 uptake occurs during the daytime. In particular, the percentage of the total net CO_2 uptake over 24 h occurring at night is 27% for 25-day-old seedlings of *A. deserti*, increasing to 49% for 100-day-old seedlings and 81% for 400-day-old seedlings. Whereas the early-morning CO_2 uptake increases with seedling age, the late-afternoon uptake decreases (Fig. 2.6) and is absent for the mature plants of *A. deserti* (Fig. 2.5). We will reconsider changes in daily patterns of CO_2 uptake with age in Chapter 3 based on the concept of water-use efficiency introduced toward the end of this chapter.

Daily patterns for cacti

Compared with the limited amount of gas-exchange data for agaves, many such studies occur for cacti, especially platyopuntias. Because stomata must be open for considerable CO_2 to enter, thereby allowing water vapor to leave, daily patterns of transpiration parallel daily patterns of net CO_2 uptake and will be omitted here. We will consider all three subfamilies of cacti (Table 1.2) and again will

restrict our coverage to data expressed on a total surface area basis for plants under optimal environmental conditions.

SUBFAMILY PERESKIOIDEAE

Although Subfamily Pereskioideae is relatively small (about 20 species, Table 1.2), it is quite important in terms of the evolution of gas-exchange patterns among cacti. Nevertheless, the Pereskioideae did not receive research attention with respect to net CO_2 exchange until the 1980s (Rayder and Ting, 1981; Nobel and Hartsock, 1986a). As might be expected, the leaves turn out to be the dominant photosynthetic organs for these leafy cacti.

For two species of *Pereskia*, *P. aculeata* (Fig. 1.27A) and *P. grandifolia* (Fig. 1.27B), net CO_2 uptake over a 24-h period occurs only during the daytime and only for the leaves (Fig. 2.7A,B). Indeed, CO_2 exchange is always negative for the stems, which represent less than 10% of the shoot surface area (Nobel and Hartsock, 1986a). Rayder and Ting (1981) also reported only daytime CO_2 uptake by leaves of these two species of *Pereskia*, although their maximum rates were half of those shown in Figure 2.7, mainly because stomata were only partially open under their conditions.

Leaves of *Maihuenia poeppigii* (Fig. 2.8) are terete, not flat as are those of pereskias, and they can comprise about 80% of the shoot surface area. As for the pereskias, all net CO_2 uptake is by the leaves and during the daytime (Fig. 2.7C). We thus conclude that under well-watered conditions, moderate temperatures, and near saturating PAR levels, net CO_2 uptake by members of the Pereskioideae occurs by the leaves using the C_3 pathway, as it does for most plants. In other words, cacti in this primitive subfamily tend to use the primitive C_3 pathway.

SUBFAMILY OPUNTIOIDEAE

As for the Pereskioideae, Subfamily Opuntioideae contains many species with prominent leaves. We will begin by considering *Austrocylindropuntia subulata* (Fig. 2.9), *Pereskiopsis porteri* (Fig. 1.28A), and *Quiabentia chacoensis* (Fig. 1.28B), all of which have leaves that generally persist for many months and whose total surface area can equal or somewhat exceed that of the stem

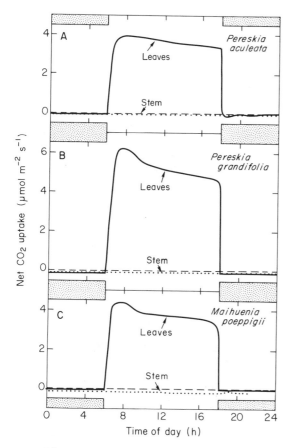

Figure 2.7. CO_2 exchange over a 24-h period by leaves and stems for three species in Subfamily Pereskioideae. **(A)** *Pereskia aculeata*; unpublished data of P. S. Nobel and T. L. Hartsock. **(B)** *Pereskia grandifolia*; modified from Nobel and Hartsock (1986a). **(C)** *Maihuenia poeppigii*; unpublished data of P. S. Nobel and T. L. Hartsock.

Figure 2.8. *Maihuenia poeppigii*, a low, leafy, freely branching cactus in Subfamily Pereskioideae found in the mountains of Chile. The terete leaves average 6 mm in length (photographed by David T. Tissue).

Figure 2.9. *Austrocylindropuntia subulata*, a leafy South American cactus in Subfamily Opuntioideae. This species, which is often included in the genus *Opuntia*, may represent a transition from the leafy cacti to the stem-succulent types.

(Nobel and Hartsock, 1986a). We will then consider other species of the subfamily, in particular platyopuntias, whose small leaves are ephemeral.

For the three leafy members of the Opuntioideae whose daily patterns of gas exchange have been determined, most of the net CO_2 uptake is again during the day and by the leaves (Fig. 2.10). However, the leaf and the stem patterns are more complicated than for the Pereskioideae (Fig. 2.7). First, leaves of *A. subulata* (Fig. 2.10A), *P. porteri* (Fig. 2.10B), and *Q. chacoensis* (Fig. 2.10C) all have some net CO_2 uptake at night. Second, some net CO_2 uptake occurs for the stems of all three species, during the daytime for *P. porteri* and *Q. chacoensis* and at night for *A. subulata* (Fig. 2.10). Considering both the daytime and the nighttime rates as well as the area of leaves and stems, 1% of the shoot net CO_2 uptake over the 24-h period is by the stems for *P. porteri*, 7% for *Q. chacoensis*, and 12% for *A. subulata* (Nobel and Hartsock, 1986a). Unlike the other leafy cacti, the relatively high nocturnal net CO_2 uptake by the stems of *A. subulata* (Fig. 2.10A) follows a pattern typical for CAM plants.

Although CO_2 exchange and transpira-

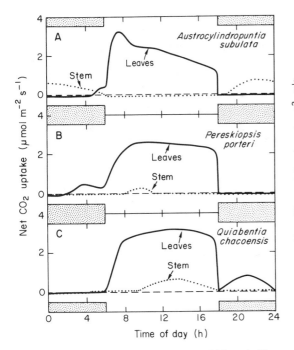

Figure 2.10. CO_2 exchange over a 24-h period by leaves and stems for three leafy species in Subfamily Opuntioideae: **(A)** *Austrocylindropuntia subulata*, **(B)** *Pereskiopsis porteri*, and **(C)** *Quiabentia chacoensis*. Modified from Nobel and Hartsock (1986a).

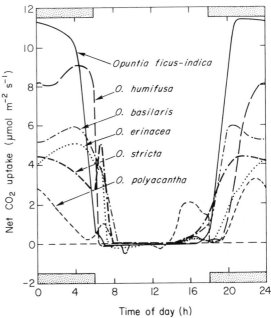

Figure 2.11. CO_2 exchange by stems of platyopuntias in Subfamily Opuntioideae. Data are adapted from Nobel and Hartsock (1983, 1984) for *Opuntia ficus-indica*; Koch and Kennedy (1980) for *O. humifusa*; Hanscom and Ting (1978a) for *O. basilaris*; Littlejohn (1983) for *O. erinacea*; Osmond, Nott, and Firth (1979b) for *O. stricta*; and Gerwick and Williams (1978) for *O. polyacantha*. Data were obtained under different conditions, so the indication of the dark period (stippled) is only approximate.

tion over 24-h periods have apparently not been published for species of *Pterocactus, Tacinga, Tephrocactus,* or cylindropuntias, many species of platyopuntias have been studied in this respect (Fig. 2.11). All of these studies examine gas exchange by the stem, and all present a similar result – namely, most of the net CO_2 uptake occurs at night. Such nocturnal net CO_2 uptake accounts for 70% to 99% of the total over 24 h (Fig. 2.11); all six species of platyopuntias considered here exhibit some net CO_2 uptake for the first 2 h of the day and in the late afternoon. Indeed, the daily patterns are very similar to those of mature leaves of agaves (Figs. 2.2, 2.4, and 2.5). We also note that *Opuntia ficus-indica* (Fig. 1.13) has the highest rate of net CO_2 uptake of those species so far examined; it is also the most widely cultivated platyopuntia (see Fig. 1.22).

SUBFAMILY CACTOIDEAE

We now turn to the largest subfamily of the Cactaceae, the Cactoideae, which contains over 80% of the species in the family (Table 1.2). Leaves tend to be small and ephemeral, and their gas-exchange characteristics have not been reported. Thus our attention is turned to the stems.

As might be expected, patterns of gas exchange have been studied for quite a few species of Subfamily Cactoideae. However, for our purposes, much of the data are of limited use because they are expressed on a fresh-weight basis (without conversion factors to relate tissue mass to stem surface area). Nevertheless, the same patterns as found for the leafless Opuntioideae prevail – namely, nearly all the net CO_2 uptake occurs at night (Fig. 2.12). Specifically, 91% to 99% of the net CO_2 uptake over 24 h occurs at night for *Cereus validus* (Tribe Cereeae), *Ferocactus acanthodes* (Tribe Cacteae), and *Schlumbergera truncata* (Tribe Notocacteae). All three species have a small net CO_2 uptake at the beginning and at the end of the day (Fig. 2.12). However, these species in Subfamily Cactoideae mainly use the CAM mode for gas exchange.

Let us next comment on the respiration rates for shoots. After 15-cm-tall plants of *Carnegiea gigantea* had been maintained in the dark for ten days, the respiration rate in air was only about 0.2 mmol CO_2 (kg fresh weight)$^{-1}$ h^{-1} (Gustafson, 1932), which is extremely low. Such a low respiration rate in part reflects the low metabolic activity of the water-storage parenchyma and the pith, which make up most of the volume of the shoot in this case. By way of comparison, for the relatively thin cylindrical stem segments of *Opuntia versicolor* at 30°C, respiration was about 30 mmol CO_2 (kg fresh weight)$^{-1}$ h^{-1} for mature stems and about 80% higher for young stems (Richards, 1915). For young cladodes of *O. basilaris*, which likewise are mainly chlorenchyma, respiration was about 20 mmol CO_2 (kg fresh weight)$^{-1}$ h^{-1} at 30°C (Szarek and Ting, 1974a).

Biochemistry

Now that we have surveyed the daily gas-exchange patterns that have been observed for agaves and cacti, we turn to the underlying biochemical pathways. In particular, we will identify the specific chemical compounds involved, the subcellular locations for their main reactions, and the daily patterns of acidity changes that can occur. After discussing carbon isotopic ratios for C_3 and CAM plants, we will summarize the existing literature on the occurrence of CAM for agaves and cacti.

C_3 AND CAM PATHWAYS

Despite the widespread occurrence of CAM, not all of its biochemical features are understood. The subcellular compartmentation of its *enzymes* (proteins that *catalyze* or speed up chemical reactions) is not known for certain and probably differs among taxa. This is not surprising, as CAM is undoubtedly of *polyphyletic* origin, having evolved independently in families that are not closely related (Kluge and Ting, 1978). Thus, Figure 2.13 summarizes only the key molecules and enzymes involved, the main biochemical cycles, and the most likely subcellular locations.

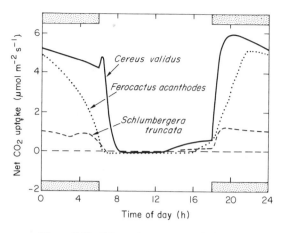

Figure 2.12. CO_2 exchange over a 24-h period by stems of three species in Subfamily Cactoideae. Data are adapted from Nobel et al. (1984) for *Cereus validus*, Nobel (1986a) for *Ferocactus acanthodes*, and Hanscom and Ting (1978b) for *Schlumbergera truncata*.

Atmospheric CO_2 diffusing toward lower concentrations eventually can enter the chloroplasts of actively photosynthesizing C_3 plants, whose stomata are open during the daytime. There ribulose-1,5-bisphosphate carboxylase/oxygenase (Fig. 2.13), or *Rubisco* for short, which is by far the most common enzyme on Earth, catalyzes the joining of a molecule of CO_2 to a five-carbon compound (ribulose-1,5-bisphosphate), leading to two three-carbon molecules (3-phosphoglycerate). These three-carbon molecules are part of a biochemical pathway leading to hexoses (six-carbon compounds) that can be incorporated into starch or other glucans or used to regenerate the CO_2 acceptor. Although CAM plants use the same pathway during the daytime as do C_3 plants, stomata of CAM plants tend to be closed during the daytime, and so the CO_2 involved must come from within the plants, as we have already indicated. In the case of agaves and cacti, the CO_2 comes from the decarboxylation of malate in the cytosol (Fig. 2.13). This malate has been sequestered overnight in the vacuole, which is ten- to twentyfold greater in volume than the cytosol (Fig. 1.36). Indeed, the large vacuoles of cells in the chlorenchyma for leaves of agaves and stems of cacti (Figs. 1.34 and 1.36) are crucial for this storage function.

Let us now turn to what happens at night,

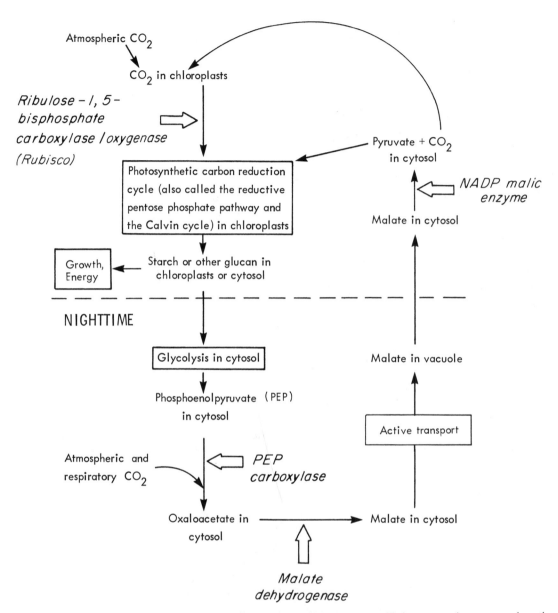

Figure 2.13. Summary of certain biochemical events and pathways most likely to occur in agaves and cacti. Biochemical cycles or processes are indicated in boxes and enzymes are indicated by italics. Data are from Dittrich (1976); Black, Carnal, and Kenyon (1982); Edwards, Foster, and Winter (1982); Sanada and Nishida (1982); Ting (1982); Winter (1985); and other sources.

when no photosynthesis can occur because of the absence of light. The polymers of hexoses accumulated during the daytime can now lead, by glycolysis (Fig. 2.13), to three-carbon compounds such as phosphoenolpyruvate (*PEP*, for short) suitable for the chemical assimila-

tion of CO_2. After passing through the four-carbon intermediate oxaloacetate, malate is produced, which is eventually stored in the vacuole (Fig. 2.13). Malate must be actively transported into the vacuole, because its increasing concentration there during the night causes its chemical potential energy to become quite high (Lüttge, Smith, and Marigo, 1982).

Now that a broad outline for the biochemical reactions has been presented, we can attempt to interpret the daily patterns of net CO_2 uptake observed for agaves and cacti in the CAM mode (Mukerji and Ting, 1968; Kluge, Fischer, and Buchanan-Bollig, 1982; Littlejohn and Ku, 1984; Pandey and Sanwal, 1984; Ting, 1985; Winter, 1985). At the beginning of the light period, net CO_2 uptake is generally observed for CAM plants and is accompanied by increased stomatal opening for up to 1 or 2 h; this CO_2 uptake may initially reflect fixation of atmospheric CO_2 by PEP carboxylase in the cytosol, but CO_2 fixation soon shifts to the chloroplasts, where Rubisco is the enzyme involved. Near midday, stomata in leaves of mature agaves and stems of cacti tend to be closed, and CO_2 exchange with the environment is very low, due only to exchange through the cuticle. Indeed, such exchange is commonly negative (outward), indicating that

daytime C_3 photosynthesis catalyzed by Rubisco (Fig. 2.13) reduces the internal CO_2 level below that of the atmosphere. At the onset of darkness, the C_3 mode of fixing atmospheric CO_2 ceases, stomata reopen, and the CAM mode using PEP to lead to the formation of malate, which is stored in the vacuole, recommences (Fig. 2.13).

DAILY ACIDITY CHANGES

As we have already noted, daily acidity changes had been reliably measured for cacti long before measurements of the same quality were done on the related issue of net CO_2 uptake. To help understand why nocturnal CO_2 uptake and tissue acidity changes are closely related, we return to a consideration of malate, including its chemical structure and the stoichiometry of its participation in the biochemical reactions of CAM plants (Fig. 2.13).

We can summarize the key chemical reactions occurring at night in the chlorenchyma of agaves and cacti as follows:

$$(2.2)$$

CO_2 is then at a higher level in the chlorenchyma than in the atmosphere (Figs. 2.2, 2.4–2.7, and 2.10–2.12). This high CO_2 level in the chlorenchyma reflects the continuing decarboxylation of malate (Ting, 1985); CO_2 levels in the intercellular air spaces of the chlorenchyma during the daytime greatly exceed the atmospheric level of about 0.035% and can reach 0.8% for *Agave deserti*, 1.3% for *Opuntia ficus-indica*, and 2.5% for *O. basilaris* (Cockburn, Ting, and Sternberg, 1979), in agreement with the values of over 1% reported much earlier for *Carnegiea gigantea* and *O. tomentosa* (MacDougal and Working, 1933). For certain agaves and cacti, stomata open partially in the afternoon, presumably as the

where we have presented substances in their undissociated forms, except for malate in the last step. We can simplify the overall reaction even further so as to focus on the relation between CO_2 fixed and hydrogen ions produced:

$$\tfrac{1}{2}\text{glucan monomer} + CO_2 \rightarrow \text{malate} + 2H^+$$

$$(2.3)$$

Equation 2.3 clearly indicates that for each CO_2 molecule fixed by a CAM plant, one molecule of malate and two hydrogen ions can be produced. This has been demonstrated for *Agave deserti* (Nobel and Hartsock, 1978) and *Opuntia ficus-indica* (Nobel and Hartsock, 1983), as well as for many other CAM plants (for instance, Lüttge and Ball, 1980). Actually, malate can be converted to other organic acids; on a mole basis, about 10% as much citrate as malate can accumulate during the night for young cladodes of *O. ficus-indica* (Teles et al., 1984).

It is really no surprise that daily acidity changes were noted in cacti as early as the nineteenth century, because they are so easy to measure. The massive increases in malate and other organic acids at night are accompanied by increases in H^+ and hence a lowering of the tissue pH ($pH = -\log[H^+]$, where $[H^+]$ indicates the concentration of hydrogen ions; using the chemically active part or chemical activity of H^+ is more appropriate, but the concentration of H^+ is very similar to its activity; see Nobel, 1983a). More useful than measuring the decrease in pH that occurs at night is the information we can obtain by restoring the pH to its initial value. In particular, we can raise the pH to a specified value by *titrating* with a base such as KOH or NaOH. The pH of the endpoint of the titration depends on the pK of the acid involved ($pK = -\log[K]$, where $[K]$ denotes the concentration of a base required to dissociate 50% of an acid).

The relationship between pH and pK can be expressed according to the Henderson–Hasselbalch equation (Lehninger, 1982):

$$pH = pK + \log \frac{[\text{dissociated acid}]}{[\text{undissociated acid}]} \quad (2.4)$$

When the pH equals the pK, half of the acid has had the H^+ removed and thus becomes dissociated (for example, from —COOH to —COO$^-$), in which case [dissociated acid] equals [undissociated acid] (see Eq. 2.4). At one pH unit above the pK, log([dissociated acid]/[undissociated acid]) equals 1 by Equation 2.4, [dissociated acid]/[undissociated acid] thus equals 10, meaning that 91% occurs in the dissociated form; at 2 pH units above the pK, 99% occurs in the dissociated form.

Equation 2.2 indicates that two H^+'s can dissociate from malic acid, and hence it has two pK's. The pK's occur at pH's of 3.4 and 5.1 for very dilute solutions at 25°C; these pK's change less than 0.1 pH unit for the entire temperature range of 0 to 50°C (Sober, 1968), but are 0.3–0.8 pH unit lower for the ionic conditions that can occur in cells (Lüttge and Smith, 1984). Hence, titration to pH 7.0 leads to removal of over 99% of the H^+'s that can dissociate (Eqs. 2.2 and 2.4). Thus, taking a piece of agave or cactus tissue of known surface area and noting how much of a base such as KOH must be added to the macerated tissue to bring it to pH 7.0 indicates the amount of organic acids and other substances from which H^+ can dissociate (such as proteins). We note that because of the *buffering* capacity of the many acidic groups on proteins and other cellular constituents, considerable amounts of base are necessary to titrate plant tissues to pH 7.0 beginning at, say, pH 5.0, which represents only 10^{-5} M H^+ (*molarity*, M, indicates moles per liter of solution, which is numerically equal to kmol m^{-3}, the unit preferred by international nomenclature agencies but seldom found in the literature).

Titration to the same final pH at dawn and at dusk can indicate the nocturnal increase in malate and hence the CO_2 fixed at night. Extending this result to net CO_2 uptake, a quantity important for understanding environmental effects on gas exchange in general and productivity in particular, is more difficult. The CO_2 in Equations 2.2 and 2.3 need not come from the atmosphere; CO_2 produced by respiration in the plant tissue at night can also be used. Under optimal environmental conditions for gas exchange, nocturnal increases in acidity represent net CO_2 uptake more accurately because stomata are open and the amount of CO_2 derived from respiration is small compared with that coming from the atmosphere. However, under conditions of water stress, stomata tend to remain closed, and a large proportion of the CO_2 then comes from respiration. Nevertheless, measurements of nocturnal increases in tissue acidity have been important for demonstrating that the CAM pathway exists in agaves and cacti. Some older studies expressed acidity changes per unit mass, which is generally less useful than a surface area basis, especially when the relative amounts of chlorenchyma (where the acidity changes actually occur) and the underlying water-storage parenchyma are unknown. Moreover, it is often desirable to relate acidity changes to gas exchange or light interception, both of which are conventionally expressed on a unit surface area basis.

By 1915, daily changes in tissue acidity per unit fresh weight were known to occur for *Mammillaria grahamii* and *Opuntia versicolor* (Fig. 1.20; Richards, 1915). Titratable acidity in the chlorenchyma expressed per unit stem surface area also increases steadily through the night and decreases during the day (Fig. 2.14) for *O. basilaris* (Fig. 1.2D) and *O. stricta* (Fig. 2.15). The peak in acidity is reached at or near dawn, and the lowest value is reached at or near dusk (Fig. 2.14). Thus, simply subtracting the dusk tissue acidity value from the dawn value shows the nocturnal increase in acidity, which can be indicative of net CO_2 uptake, at least under favorable environmental conditions. For instance, *O. basilaris* exhibits a nocturnal increase in titratable acidity of about 700 mmol H^+ m^{-2} (Fig. 2.14), which corresponds to a net CO_2 uptake of nearly 350 mmol m^{-2} in this case.

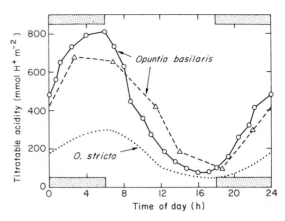

Figure 2.14. Tissue acidity based on H^+ for two species of platyopuntia over a 24-h period. Tissue samples representing only the chlorenchyma were titrated to approximately pH 7.0. Data on *Opuntia basilaris* are from Szarek and Ting (1975a; – – –△– – –) or from P. S. Nobel, N. K. Chang, and E. R. Lee (obtained in March 1979 at the Philip L. Boyd Deep Canyon Desert Research Center, unpublished; ——○——). Data on *O. stricta* are from Osmond et al. (1979b).

CARBON ISOTOPE RATIOS

As simple as it is to measure acidity levels, determining carbon isotope ratios is a simpler, though technically more sophisticated, method for indicating whether a particular species employs the CAM pathway. The method is based on how plants using the three different photosynthetic pathways – C_3, C_4, and CAM – discriminate between the two stable isotopes of carbon that occur naturally: ^{12}C and ^{13}C. The latter is 8% heavier and hence diffuses more slowly than does the former (Nobel, 1983a). Also, the enzymes fixing "CO_2" in the different pathways (Fig. 2.13) actually use different substrates: Rubisco uses CO_2, whereas PEP carboxylase uses HCO_3^- (bicarbonate), which forms upon hydration of the gaseous CO_2 in solution, a process having different rates for $^{12}CO_2$ and $^{13}CO_2$. Thus, both physical and biochemical factors play important roles in the interpretation of isotope ratios (Lerman, 1975; O'Leary and Osmond, 1980; O'Leary, 1981; Smith and Madhaven, 1982; Teeri, 1982; Farquhar, 1983; Ting, 1985; Osmond, Adams, and Smith, 1987).

Any measure of the relative frequency of ^{13}C and ^{12}C incorporated into organic compounds needs a reference, or standard, and a method for expression. The standard chosen is accessible but somewhat obscure, namely, Pee Dee belemnite limestone from South Carolina (the CO_2 is released from the carbonate shell of a Cretaceous mollusc, *Belemnitella americana*, that makes up the limestone; Craig, 1957; Lerman, 1975). The rather esoteric, but widely accepted method for expressing such data is the so-called del-13 or $\delta^{13}C$ ("delta 13-C"):

$$\delta^{13}C \ (‰) = \left(\frac{^{13}C/^{12}C_{sample}}{^{13}C/^{12}C_{standard}} - 1 \right) \times 1000$$

$$(2.5)$$

where ‰ means parts per thousand. Because C_3 and CAM plants have widely different values, $\delta^{13}C$ has become a most useful index. CAM plants with mainly nocturnal net CO_2 uptake usually have $\delta^{13}C$ values from $-11‰$ to $-18‰$ (similar to values for C_4 species, with which they are seldom confused but which also use PEP carboxylase for the initial assimilation of CO_2), whereas C_3 plants usually have $\delta^{13}C$ values from $-25‰$ to $-30‰$. The more negative $\delta^{13}C$ indicates enrichment in ^{12}C, the lighter carbon isotope. Only a relatively small sample of the chlorenchyma of a particular plant, for instance, 2 g by fresh weight, is nec-

Figure 2.15. *Opuntia stricta* growing (**A**) near Key Biscayne, Florida; (**B**) near Sydney, New South Wales, where extensive damage by *Cactoblastis cactorum* is evident (see Chapter 1); and (**C**) on the sandy shore of the Gulf of Thailand.

essary for carbon isotope ratios to be determined. The sample is dried, combusted, and the resulting CO_2 is introduced into a *mass spectrometer* to determine the ratios of ^{13}C to ^{12}C (see Eq. 2.5) and thus to indicate the metabolic pathway used to furnish the carbon that became incorporated into the tissue.

Actually, a particular CAM species can have different $\delta^{13}C$ values depending on the relative amounts of CO_2 that are fixed via PEP carboxylase ($\delta^{13}C$ near $-14\permil$) compared with Rubisco ($\delta^{13}C$ near $-27\permil$). Thus, $\delta^{13}C$ for succulents such as agaves and cacti is an integrative index of the mode of fixation of at-

mospheric CO_2. Another factor affecting $\delta^{13}C$ relates to the refixation of CO_2 released at night within the chlorenchyma by respiration. Even though PEP carboxylase and CAM are involved in this pathway, the $\delta^{13}C$ value represents that of the respiratory substrate. If the stomata are closed because of water stress, then the respiratory substrate might have been synthesized during a previous period when CO_2 uptake occurred primarily during the daytime via the C_3 pathway (consider the case for agave seedlings, Fig. 2.6). Nevertheless, assuming one has access to a ratioing mass spectrometer (or funds to pay for commercial assays), determination of $\delta^{13}C$ values can be a powerful technique for indicating whether a species employs the CAM pathway, and it has proved particularly convenient in broad surveys of plants.

OCCURRENCE OF CAM

Now that nocturnal net CO_2 uptake by the CAM mode (Fig. 2.13), the accompanying increase in tissue acidity, and the consequences for the carbon isotope ratio have been discussed, we will indicate the species of agaves and cacti that have been shown to exhibit CAM by one of these three techniques (Table 2.1). Actually, the techniques have been used with approximately equal frequency, data having been reported on over forty species using each one.

Seventeen species of agaves and over eighty species of cacti exhibit at least some of the characteristics of CAM plants (Table 2.1). Agaves on average have been studied more recently than cacti. More species in Subgenus Agave have been investigated than in Subgenus Littaea (Table 1.1), but both subgenera are composed of species whose leaves exhibit CAM. Stems of cacti in the two subfamilies for which leaves are relatively inconspicuous exhibit CAM, including examples from most of the nine tribes of Cactoideae (Table 1.2). The genus receiving the most attention has been *Opuntia*, where twenty-five species exhibit at least one of the three characteristics of CAM considered (Table 2.1). We can thus conclude that CAM is widespread among agaves and cacti.

Let us now consider leaves of pereskias, where the gas-exchange data indicate C_3 metabolism with daytime net CO_2 uptake (Fig. 2.7A,B). Nevertheless, three species of *Pereskia* can have a nocturnal increase in acidity (Table 2.1). For *Pereskia guamacho*, where the acidity data were expressed on a unit surface area basis, the nocturnal increase was only 29 mmol H^+ m^{-2}, which is over ten times lower than the nocturnal acidity increases for the stems of three other species of cacti examined in the same study (Díaz and Medina, 1984). Also for comparison, the nocturnal increase in acidity for *Opuntia basilaris* and *O. stricta* ranges from 240 to 730 mmol H^+ m^{-2} (Fig. 2.14). As indicated in the section Daily Patterns for Cacti, leaves of *Pereskia aculeata* and *P. grandfolia* exhibit daytime but not nighttime stomatal opening. Yet these two species also can have a nocturnal increase in acidity (Rayder and Ting, 1981). The CO_2 involved in this nocturnal acidity increase may be released internally by respiration and hence not represent net CO_2 uptake from the environment. The enzymes of CAM may thus be present in the relatively primitive, leaf-bearing pereskias, but not involved in net nocturnal CO_2 uptake from the environment, which, as we shall see in the next section, is the key to the water-conserving attributes of CAM.

Gas-exchange equations

We will conclude this chapter by developing an analytical framework for discussing gas exchange. We can then quantitatively interpret the importance of microclimatic and plant factors for controlling the exchange of CO_2 and H_2O with the environment. First we will introduce some equations describing gas exchange, then we will present a ratio that helps assess the advantage of the CAM pathway over the C_3 or C_4 pathway with respect to water conservation.

TRANSPIRATION AND CO_2 UPTAKE

Transpiration

Transpiration requires a higher concentration of water vapor in the intercellular air spaces adjacent to the cell walls of chlorenchyma cells (Figs. 1.34–1.36) than in the atmosphere surrounding the shoot. The actual rate of transpiration is controlled by the degree of stomatal opening. By Fick's first law, diffusion depends linearly on the concentration difference (Nobel, 1983a); the air in the intercellular air spaces is approximately saturated with water vapor (nearly 100% relative humidity), whereas the surrounding atmosphere generally has a lower relative humidity and lower water vapor concentration. We can represent the flux density of water vapor loss (J_{wv}) as follows:

Transpiration per unit area and per unit time
$$= J_{wv} = g_{wv}\,\Delta c_{wv} \quad (2.6a)$$

Table 2.1. *Summary of species of agaves and cacti that have been demonstrated to exhibit CAM*

Species	Nocturnal net CO_2 uptake	Nocturnal acid accumulation	CAM-like carbon isotope ratio
Agaves			
Agave americana	Neales et al. (1968)	Ramachandra Reddy and Rama Das (1978)	Troughton, Card, and Hendy (1974) (**A**)
A. deserti	Nobel (1976a)	Hartsock and Nobel (1976)	
A. fourcroydes	Nobel (1985a) (**B**)	B	
A. lechuguilla	Eickmeier and Adams (1978) (**C**)	C	Eickmeier and Bender (1976) (**D**)
A. lurida		Thomas and Ranson (1954) (**E**)	
A. murpheyi		Nobel and McDaniel (**F**)	
A. parryi		Nobel and Smith (1983) (**G**)	
A. salmiana		Nobel and Meyer (1985)	
A. scabra			D
A. schottii			D
A. shawii	Mooney, Troughton, and Berry (1974) (**H**)		H
A. sisalana		F	
A. tequilana	Nobel and Valenzuela (1987) (**I**)	I	
A. utahensis	Nobel and Hartsock (1981)		
A. vilmoriniana	Szarek et al. (1987)	F	
A. virginica L.[a]	Martin, Lubbers, and Teeri (1982) (**J**)	J	J
A. weberi		F	
Cacti			
Acanthocereus tetragonus		Díaz and Medina (1984) (**K**)	
Austrocylindropuntia cylindrica		E	
A. subulata	Nobel and Hartsock (1986a)		
Carnegiea gigantea	Despain, Bliss, and Boyer (1970)	Richards (1915) (**L**) Long (1915) (**M**)	
Cephalocereus royenii		Ting (1976b) (**N**)	
Cereus emoryi			H
C. peruvianus			Bender et al. (1973) (**O**)
C. repandus		Díaz (1983) (**P**)	
C. sylvestrii		Seeni and Gnanam (1980)	A
C. validus	Nobel et al. (1984) (**Q**)	Q	
Copiapoa cinerea			H
C. ferox			H
C. haseltoniana			H
C. lembckei			H
C. megarhiza			H
C. solaris			H
Coryphantha vivipara	Nobel (1981a)	Littlejohn, Green, and Williams (1982) (**R**)	
Cryptocereus anthonyanus			A

Table 2.1. *Summary of species of agaves and cacti that have been demonstrated to exhibit CAM (continued)*

Species	Nocturnal net CO$_2$ uptake	Nocturnal acid accumulation	CAM-like carbon isotope ratio
Echinocereus engelmannii		Ting and Dugger (1968) (S)	H
E. fendleri	Dinger and Patten (1972) (T)		
E. ledingii	T		
E. maritimus			H
E. triglochidiatus	T		
E. viridiflorus	Green and Williams (1982)	R	
Echinopsis eyriesii	Nuernbergk (1961) (U)		
Epiphyllum hookeri		Smith et al. (1985) (V)	V
Eulychnia acida			H
E. castanea			H
E. iquiquensis			H
E. spinibarbis			H
Ferocactus acanthodes	Patten and Dinger (1969) (W)	Jordan and Nobel (1981)	
F. gracilis	H		H
F. viridescens	H		H
F. wislizenii		M	
Hylocereus lemairei			V
Lophocereus schottii	H	Smith, Didden-Zopfy, and Nobel (1984)	H
Mammillaria dioica	Nobel (1978)		H
M. goodrichii			H
M. grahamii		L	
M. rhodantha	U		
M. tetrancistra		S	
M. woodsii	Winter, Schröppel-Meier, and Caldwell (1986) (X)	X	
Melocactus caesius		P	
M. intortus		N	
Myrtillocactus cochal	H		H
Neolloydia intertexta		Oeschager and Lerman (1970)	
Neoporteria chilensis			H
Notocactus mammulosus			A
Opuntia acanthocarpa	W	N	Sutton, Ting, and Troughton (1976) (Y)
O. aurantiaca	Whiting and Campbell (1984)	Whiting, van de Venter, and Small (1979)	
O. basilaris	Szarek and Ting (1975a)	S	Y
O. bigelovii	Y	N	Y
O. chlorotica	Nobel (1980a) (Z)		
O. cochinellifera		Master (1959)	
O. decumbens	Kausch (1965)		
O. dejecta		Mukerji, Sanwal, and Krishnan (1964)	
O. echinocarpa	S	S	
O. echios		Nobel (1981b) (AA)	
O. erinacea	Littlejohn (1983)	R	

Table 2.1. *Summary of species of agaves and cacti that have been demonstrated to exhibit CAM (continued)*

Species	Nocturnal net CO_2 uptake	Nocturnal acid accumulation	CAM-like carbon isotope ratio
O. ficus-indica	Nobel and Hartsock (1983)	Samish and Ellern (1975)	
O. fulgida		N	H
O. humifusa	Koch and Kennedy (1980) (**AB**)	**AB**	O
O. leptocaulis		L	
O. oricola	H		H
O. phaeacantha	Nisbet and Patten (1974)	L	D; Szarek and Troughton (1976)
O. polyacantha	Gerwick and Williams (1978)		Troughton, Wells, and Mooney (1974)
O. prolifera	H		H
O. ramosissima	S		
O. rubescens		N	
O. stricta	Osmond et al. (1979b) (**AC**)	**AC**	Osmond et al. (1973)
O. versicolor		L	
O. vulgaris	X	X	
O. wentiana		K	
Pachycereus pringlei			A
Pereskia aculeata		Rayder and Ting (1981) (**AD**)	
P. grandifolia		**AD**	
P. guamacho		K	
Rhipsalis cassutha			V
Schlumbergera truncata	Hanscom and Ting (1978b)	U	O
Stenocereus griseus		K	
S. gummosus	H	Z	H
S. thurberi		N	
Tephrocactus articulatus			O
Trichocereus chilensis		AA	H
T. coquimbanus			H

Note: Nocturnal CO_2 uptake and nocturnal acid accumulation expressed on dry-weight, fresh-weight, and surface-area bases are included. Authorities are given in Tables 1.1 and 1.2 (the taxonomic scheme adopted there leads to names that may differ from those in the papers cited). To avoid repetition, studies cited more than once are assigned a boldface letter the first time they occur, this letter being used for subsequent citations. This list expands upon those developed by Black and Williams (1976), Szarek and Ting (1977), and Szarek (1979), but indicates only those references with actual data; when multiple studies are on the same species, the earliest one is indicated.

[a] This species, which is not recognized as an agave by Gentry (1982), is often referred to as *Manfreda virginica* (L.) Salisb. and sometimes as *Polianthes virginica* (L.) Shinners. It occurs in many states in the midwestern and eastern United States.

where Δc_{wv} is the drop in water vapor concentration from the chlorenchyma to the atmosphere surrounding the shoot and g_{wv} is the *water vapor conductance*, sometimes referred to as the *stomatal conductance* for water vapor because it is approximately proportional to the open area of the stomatal pores (Nobel, 1983a). A conductance such as g_{wv} is useful because it directly shows how the flux density will change for a given change in the driving force.

Instead of using the drop in concentration as the driving force for water loss, the drop in the *mole fraction* of water vapor from the chlorenchyma to the atmosphere (ΔN_{wv}) is increasingly being used:

$$J_{wv} = g'_{wv}\,\Delta N_{wv} \qquad (2.6b)$$

where the dimensionless quantity N_{wv} is the ratio of the number of moles of water vapor to the total number of moles of all molecular species in that volume, and g'_{wv} is a water vapor conductance having different units than g_{wv}. Long ago, J_{wv} could have been determined by periodically weighing cacti (Livingston, 1907) and dividing the water loss by total shoot surface area and the time involved, whereas instruments are now available that directly measure J_{wv} or g_{wv}.

The *saturation* or maximum values of water vapor, both c^*_{wv} and N^*_{wv}, increase rapidly with temperature (Fig. 2.16). Specifically, warmer air can hold substantially more water vapor at saturation than can cooler air; for example, heating air at 20°C that is saturated with water vapor (100% relative humidity) to 32°C (at constant pressure or at constant volume) causes the relative humidity to drop to 50%. Therefore, synchronizing the timing of stomatal opening with the lowest tissue temperatures and hence lowest chlorenchyma water vapor levels is the key to the water-conserving aspects of CAM. We also note that c^*_{wv} or N^*_{wv} (Fig. 2.16) times the relative himidity in percent divided by 100 equals c_{wv} or N_{wv}, so the content of water vapor in the atmosphere can be readily expressed if we know the air temperature and the ambient relative humidity.

Agaves and cacti can have extremely low values for minimal water vapor conductance, about 10 to 100 times lower than for leaves of

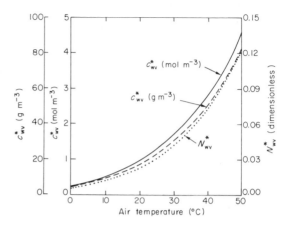

Figure 2.16. Dependence of saturation values of water vapor concentration (c^*_{wv}) and mole fraction (N^*_{wv}) on air temperature. An ambient pressure (needed to determine N^*_{wv}) of one atmosphere was assumed. Modified from Nobel (1983a).

representative C_3 plants (Szareк, Johnson, and Ting, 1973). For instance, g_{wv} can be as low as 0.005 mm s^{-1} (g'_{wv} of 0.2 mmol m^{-2} s^{-1}) for *Agave deserti* and *Ferocactus acanthodes* (P. S. Nobel, unpublished observations). The low g_{wv} mainly reflects the difficulty for water vapor to diffuse across the thick, nearly impervious cuticle, although the low frequency and tight closure of stomata also help reduce g_{wv}. Such a low water vapor conductance decreases water loss during extended periods of drought. Also, CO_2 released within the photosynthetic organs cannot readily diffuse out across the epidermis when the stomata are closed, an important aspect when daytime malate decarboxylation raises the CO_2 level in the intercellular air spaces of the chlorenchyma to 1 or 2%.

CO_2 uptake

Let us now turn to the CO_2 flux density, J_{CO_2}. CO_2 diffusing into a leaf encounters the same constrictions resulting from stomatal pores (Fig. 1.35) and the same tortuosity of the intercellular air spaces as does water vapor diffusing out. Thus, the CO_2 conductance for the gas phase, $g^{gas}_{CO_2}$, is proportional to the water vapor conductance (g_{wv} or g'_{wv} in Eq. 2.6; because the appropriate symbol is clear from whether Δc_{wv} or ΔN_{wv} is used as the driving force, we will henceforth omit the prime on g_{wv}). The difference between $g^{gas}_{CO_2}$ and g_{wv} is related to molecular weight, which is 18 for H_2O and 44 for CO_2. The heavier CO_2 molecules diffuse more slowly, which lowers the conductance of CO_2 relative to that of H_2O (Nobel, 1983a):

$$g_{CO_2}^{gas} = \frac{1}{1.6} g_{wv} \qquad (2.7)$$

We will next present an equation describing net CO_2 uptake using parameters for the gas phase:

Net CO_2 uptake per unit
area and per unit time

$$= J_{CO_2} = g_{CO_2}^{gas}(N_{CO_2}^{ta} - N_{CO_2}^{ias}) \qquad (2.8)$$

where the superscript ta refers to the *turbulent air* outside the shoot, and ias refers to the *intercellular air spaces* in the chlorenchyma.

We can carry our analysis one step further to indicate what happens in the cells of the chlorenchyma. Namely, the same flux density of CO_2, which is expressed per unit total surface area of the shoot, continues undiminished into the cytosol, where PEP carboxylase can bind the CO_2 in the CAM mode (primarily at night), or into the chloroplasts, where Rubisco can bind CO_2 in the C_3 mode during the daytime. Because the pathway across the cell wall, the plasmalemma, and through the cytosol (Fig. 1.36), including the path into the chloroplasts in the case of Rubisco, is in a liquid phase, we will represent its CO_2 conductance by $g_{CO_2}^{liq}$:

$$J_{CO_2} = g_{CO_2}^{liq}(N_{CO_2}^{ias} - N_{CO_2}^{chl}) \qquad (2.9)$$

where the superscript chl indicates the location in the *chlorenchyma* where the CO_2 is enzymatically fixed (Fig. 2.13). We note that $g_{CO_2}^{liq}$ includes both physical factors, such as diffusional barriers, and biochemical factors, such as glucan levels.

Chlorenchyma cells

The surfaces of the chlorenchyma cells of agaves (Fig. 1.34B) and cacti (Figs. 1.34A and 1.35) represent an extremely large area across which CO_2 can diffuse, which effectively increases the liquid-phase conductance. In particular, we can relate the liquid-phase CO_2 conductance per unit shoot surface area to that per unit cell area ($g_{CO_2}^{cell}$) by using the

ratio of chlorenchyma cell wall area to shoot surface area, A^{chl}/A:

$$g_{CO_2}^{liq} = \frac{A^{chl}}{A} g_{CO_2}^{cell} \qquad (2.10)$$

Thus, for a particular cellular CO_2 conductance, the greater A^{chl}/A is, the higher $g_{CO_2}^{liq}$ is, and hence the greater the net CO_2 uptake is (Eq. 2.9).

Although A^{chl}/A is much larger for CAM plants with their many layers of cells in the chlorenchyma than for the typically thin leaves of representative C_3 and C_4 plants, few data exist on actual values for agaves and cacti. For *Agave deserti* (Fig. 1.24B), A^{chl}/A is 82 (Nobel and Hartsock, 1978), and for *Ferocactus acanthodes* (Fig. 1.2A) it is 137 (Nobel, 1977b). These values should be compared with an A^{chl}/A of only 6–20 for most C_3 and C_4 plants when expressed on the basis of total leaf area (both sides of the leaves for A; Nobel, Zaragoza, and Smith, 1975; Nobel, 1980b; Björkman, 1981). The value of $g_{CO_2}^{cell}$ for *A. deserti* is comparable to that for various C_3 plants (Nobel and Hartsock, 1978), so the relatively large area available for CO_2 diffusion into its chlorenchyma cells per unit leaf area can enhance net CO_2 uptake (Eqs. 2.9 and 2.10). Yet there is still no fully satisfactory explanation as to why the chlorenchyma of agaves and cacti generally tends to be 2–3 mm thick with an A^{chl}/A of about 100 compared with values up to tenfold lower for leaves of C_3 and C_4 plants.

WATER-USE EFFICIENCY

We now turn to a direct assessment of the water costs involved in acquiring carbon. Stomatal opening represents a compromise between the need for net CO_2 uptake and the inevitable diffusion of water vapor out the open pores in the shoot surface. Transpiration can play an important role in reducing leaf temperature via evaporative cooling. In addition, some transpiration is necessary for moving the nutrients obtained from the soil to all parts of the shoot via the xylem. Flow in the transpiration stream is also essential for providing the water that will subsequently move in the phloem. Nevertheless, an advantage can be gained in arid and semiarid environments by limiting water loss.

Basic equation

The criterion that has been developed over the years for evaluating the "cost" of transpiration with respect to carbon gain employs the ratio of net CO_2 uptake to water loss and is called the *water-use efficiency*, WUE (Nobel, 1983a):

$$\text{WUE} = \frac{\text{net } CO_2 \text{ uptake}}{\text{transpiration}}$$

$$= \frac{J_{CO_2}}{J_{wv}} = \frac{g_{CO_2}^{gas}(N_{CO_2}^{ta} - N_{CO_2}^{ias})}{g_{wv}(N_{wv}^* - N_{wv}^{ias})}$$

$$(2.11)$$

To obtain Equation 2.11, we used Equations 2.6 and 2.8 and assumed the intercellular air spaces in the chlorenchyma to be saturated with water vapor (relaxing this latter assumption to a more realistic water vapor content in the chlorenchyma would mean replacing N_{wv}^* in Eq. 2.11 by about $0.99 N_{wv}^*$, which is on the border of measurement sensitivity).

Although the concept of WUE has found widespread application, data for gas flux densities have been expressed in many ways: on an instantaneous basis, integrated for the whole night, integrated over an entire 24-h period, and on a seasonal or an annual basis. Data on a seasonal or annual basis have the most agronomic and ecological relevance, especially if expressed for the whole shoot. For practical purposes it is often convenient to express net CO_2 uptake on the basis of dry-weight gain (determined upon harvesting), where 1 mole CO_2 incorporated into plant tissue leads to an average dry-weight gain of 27 g (Lieth and Whittaker, 1975; Salisbury and Ross, 1985). We note that the reciprocal of WUE, termed the *transpiration ratio*, is also widely used, especially in the older literature, to describe the relation between water vapor loss and CO_2 gain.

Examples

Water-use efficiency can be calculated from Figures 2.2 and 2.4, where both net CO_2 uptake and transpiration are presented. Near midnight, the instantaneous WUE for *Agave americana* (Fig. 2.2) is about (2.9 μmol CO_2 m^{-2} s^{-1})/(0.18 mmol H_2O m^{-2} s^{-1}) or 16 μmol CO_2/mmol H_2O, which is 0.016 CO_2/

H_2O (Table 2.2). For the whole night, the WUE is also 0.016 CO_2/H_2O, whereas the higher temperatures during the daytime increased transpiration relative to net CO_2 uptake; over a 24-h period, WUE is 0.008 CO_2/H_2O (Fig. 2.2 and Table 2.2). For *A. lechuguilla* and *A. utahensis* (Fig. 2.4), WUE is 56% and 30% lower, respectively, than for *A. americana*; for *A. deserti* it is 173% higher (Table 2.2).

The water-use efficiencies just indicated are fairly typical for agaves and cacti but are much higher than values observed for C_3 and C_4 plants. For instance, the WUE of *A. americana* over a growing season can be nearly double that of *Zea mays* (corn; Ehrler, 1969, 1983). Over a 24-h cycle, *A. americana* has a WUE similar to that of another CAM plant, *Ananas comosus* (pineapple), but threefold higher than *Helianthus annuus* (sunflower) and *Nicotiana glutinosa* (tobacco; Neales et al., 1968). Over a 24-h period, the WUE of many agaves and cacti can be 0.0040 to 0.0100 CO_2/H_2O (Table 2.2), whereas daily WUE under optimal environmental conditions averages about 0.0009 CO_2/H_2O for C_3 plants and 0.0016 CO_2/H_2O for C_4 plants (Szarek and Ting, 1975b; Osmond, Björkman, and Anderson, 1980; Nobel, 1983a). Although the actual values of WUE depend on temperature and air water vapor content (see Eq. 2.11), CAM plants have higher values than do C_3 or C_4 plants. Indeed, even over an entire year (Table 2.2), the WUE of agaves and cacti can be substantially higher than the values for C_3 or C_4 plants on favorable days.

The high WUE of CAM plants is not attributable to the enzymology of the CAM pathway – C_4 plants also use PEP carboxylase for the initial CO_2 fixation – but rather to the temporal pattern of stomatal opening. In particular, stomata of CAM plants tend to open at night when air and tissue temperatures are lower, so the driving force on water loss, $c_{wv}^* - c_{wv}^{ta}$ (Eq. 2.6a) or $N_{wv}^* - N_{wv}^{ta}$ (Eq. 2.6b), is less. Let us assume a water vapor content of air of 5 g m^{-3}, which would not tend to change appreciably over the course of a 24-h period unless there were a change in weather, such as the arrival of a storm front. Then, for a given degree of stomatal opening (a specific value for g_{wv}, Eq. 2.6), transpiration would be

Table 2.2. *Summary of water-use efficiencies for various agaves and cacti*

| Species (source of data) | WUE (mol CO_2/mol H_2O) | | | |
	Instantaneous value at midnight	Entire nighttime	24-h period	Annual
Agave americana (Fig. 2.1)	0.0161	0.0155	0.0083	—
A. deserti (Nobel, 1976)	0.0251	0.0260	0.0227	0.0164
A. lechuguilla (Fig. 2.3A)	0.0070	0.0064	0.0045	—
(Eickmeier and Adams, 1978)	—	0.0133	0.0053	—
A. utahensis (Fig. 2.3B)	0.0121	0.0105	0.0058	—
Ferocactus acanthodes (Nobel, 1977b)	0.0049	0.0050	0.0041	0.0058
Opuntia basilaris (Szarek and Ting, 1974a, 1975b; Hanscom and Ting, 1978a)	0.0070	0.0062	0.0053	0.0051
O. ficus-indica (Acevedo, Badilla, and Nobel, 1983; Nobel and Hartsock, 1983)	0.0131	0.0119	0.0101	0.0059

three times greater for a chlorenchyma temperature of 30°C during the daytime than for 15°C at night. Specifically, c^*_{wv} is 30 g m^{-3} at 30°C and 13 g m^{-3} at 15°C (Fig. 2.16), so Δc_{wv} would be $30 - 5$ or 25 g m^{-3} at 30°C and only 8 g m^{-3}, or one-third as large, at 15°C. If both daytime and nighttime temperatures were 10°C less, then the WUE in the present case would be seven times greater at night than during the daytime. The very strong dependence of the saturation water vapor content in air on temperature (Fig. 2.16), coupled with the fact that the air in intercellular spaces in the chlorenchyma is nearly saturated with water vapor, ensures that nocturnal stomatal opening, which occurs when tissue temperatures are lower, will conserve water and hence tend to increase the water-use efficiency (Eq. 2.11).

Stomatal effects

We can summarize our discussion on how WUE varies among taxa by saying that it tends to be two to six times higher for agaves and cacti than for C_3 and C_4 plants. But before we conclude that a higher water-use efficiency is exclusively beneficial, we should examine the effect of partial stomatal closure on WUE.

It is useful to reexpress the net CO_2 uptake, J_{CO_2}, in terms of the gas-phase and liquid-phase conductances, which we can do by eliminating $N^{ias}_{CO_2}$ between Equations 2.8 and 2.9:

$$J_{CO_2} = \frac{g^{gas}_{CO_2} g^{liq}_{CO_2}}{g^{gas}_{CO_2} + g^{liq}_{CO_2}} (N^{ta}_{CO_2} - N^{chl}_{CO_2})$$

(2.12)

Partial stomatal closure causes g_{wv} to decrease with a proportional decrease in transpiration (Eq. 2.6); for net CO_2 uptake, two conductances in *series* (in sequence), $g^{gas}_{CO_2}$ and $g^{liq}_{CO_2}$, are involved (Eq. 2.12), so a smaller decrease occurs during partial stomatal closure. Moreover, water-use efficiency increases with partial stomatal closure, even though CO_2 uptake is decreased, which we can see by combining Equations 2.6, 2.7, 2.11, and 2.12:

$$\text{WUE} = \frac{g^{liq}_{CO_2}}{1.6(g^{gas}_{CO_2} + g^{liq}_{CO_2})} \frac{(N^{ta}_{CO_2} - N^{chl}_{CO_2})}{(N^*_{wv} - N^{ta}_{wv})}$$

(2.13)

Equation 2.13 indicates that as stomata

close and, hence, $g_{CO_2}^{gas}$ decreases, WUE increases. This helps account for the high daily water-use efficiencies for *A. deserti* in Table 2.2, because the stomata never opened fully on the day considered. Also, the maximum water vapor conductance is never very high for *A. deserti* (Nobel, 1976), which helps it to conserve water and have a high annual WUE (Table 2.2) but at the cost of decreased CO_2 uptake. Analytic relations can be developed between $g_{CO_2}^{gas}$ and $g_{CO_2}^{liq}$ for CAM plants that predict the maximization of WUE for a given amount of water transpired per day, which help interpret the gas-exchange patterns ob-

served for agaves and cacti (Comins and Farquhar, 1982). During the night, glucan levels decrease, and those of malate increase, leading to decreases in $g_{CO_2}^{liq}$ and increases in $N_{CO_2}^{chl}$. Thus, partial stomatal closure can reduce transpiration, as shown for agaves in Figures 2.2 and 2.4, without greatly diminishing net CO_2 uptake beyond that already caused by the increased $N_{CO_2}^{chl}$ and the lowered $g_{CO_2}^{liq}$ (see Eq. 2.12). The stomatal control of gas exchange is indeed a fascinating topic that we will return to in subsequent chapters, where influences of individual environmental factors on gas exchange are discussed.

3 Water relations

Water is the key environmental variable limiting CO_2 uptake by desert succulents, as is true for many other groups of plants. Indeed, even at hot times of the year when little gas exchange might be expected, rainfall can induce stomatal opening and consequently net CO_2 uptake by agaves and cacti (Nobel, 1976; Ting, 1976b). The particular water-conserving feature of agaves and cacti results from the daily timing of stomatal opening that is characteristic of CAM plants. As we discussed in the previous chapter, stomata of CAM plants tend to open at night when temperatures are lower, which generally leads to a twofold to sixfold lower transpiration rate compared with the same degree of stomatal opening during the daytime. Thus, CAM plants tend to be favored in habitats with a restricted water supply or where long periods of drought can occur, during which the plants cannot obtain any water from the soil or other substrate (Teeri, Stowe, and Murawski, 1978). For instance, over 80% of the aboveground biomass in parts of the northwestern Sonoran Desert can be cacti (Ting, 1976a), and epiphytic CAM species occupying arid microhabitats can be common in the humid tropics (Smith, 1984).

In the first chapter we mentioned that the water potential Ψ indicates the energy of water and that the osmotic pressure contributes to Ψ, important aspects of plant water relations. We will now present such topics in ways that can be applied to the environmental biology of agaves and cacti, beginning with a formal definition of Ψ (Nobel, 1983a):

$$\Psi = P - \pi + \rho_w gh \qquad (3.1)$$

where P is the *hydrostatic pressure*, or *turgor pressure* (unidirectional force per unit area exerted in an aqueous solution), π is the osmotic pressure, and $\rho_w gh$ is the *gravitational energy* per unit volume of liquid water (ρ_w is the density of water, g is the gravitational acceleration, and h is the height above the arbitrary zero level for gravitational energy); $\rho_w g$ equals 0.0098 MPa m^{-1}, so the gravitational term is usually important only when water moves over large vertical distances and thus will be ignored for most applications in this chapter. For pure water at atmospheric pressure and where h equals zero, Ψ is zero, whereas Ψ is generally negative in the soil and in plants. Equation 3.1 also can be used to describe the driving force for the percolation of water down into a soil following rainfall and for the movement of water in the xylem from the root to the transpiring surfaces, although differences in π often have little effect in such cases. We also note that water spontaneously tends to move toward regions of lower Ψ, and therefore values of Ψ indicate directions for water movement.

By the van't Hoff relation, π is proportional to the concentration c of osmotically active solutes:

$$\pi = cRT \qquad (3.2)$$

where RT (the gas constant times the absolute

temperature) is 2.44 MPa M^{-1} at 20°C (Nobel, 1983a). Thus, osmotic pressure tends to increase when plant tissues lose water during drought, because approximately the same amount of solute is dissolved in less water.

In this chapter we consider the influence of rainfall on the *soil water potential*, Ψ^{soil}. When Ψ^{soil} is higher (less negative) than Ψ^{plant}, water movement into the plant can occur. An intriguing aspect of agaves and cacti is their ability to store water in their tissues, which helps buffer their physiological responses during the initial phases of subsequent drought. We discuss responses of mature agaves and cacti to the periodic availability of water, such as stomatal opening, C_3/CAM shifts in gas-exchange patterns, and induction of rain roots. Drought considerations are crucial for seedling establishment. We also consider some of the important consequences that water relations portend for the reproductive process of agaves and cacti.

Water uptake from the soil

The water transpired by a plant or stored in its tissues generally comes from the soil. Although there have been reports of some water uptake from the air by the spines of *Discocactus horstii*, *Mammillaria theresae*, and *Turbinicarpus klinkerianus* (Schill and Barthlott, 1973; Schill, Barthlott, and Ehler, 1973; Barthlott and Capesius, 1974), Shreve long ago showed that no water uptake occurs via the spines of *Ferocactus wislizenii* (Shreve, 1916). Also, water loss through the spines at midstem on mature plants of *Ferocactus acanthodes*, *Carnegiea gigantea*, and *Stenocereus thurberi* is immeasurably small (less than $\frac{1}{1000}$ of the water loss for the same area of stem over a 24-h period for well-watered plants; P. S. Nobel, unpublished observations). Indeed, mature spines generally contain no functioning vascular tissue and thus would not conduct water to the plant surface. Although dew can condense on the spines of *Copiapoa haseltoniana*, a cactus native to the Atacama Desert of Chile, and flow down to the areoles, such water does not enter the chlorenchyma (Mooney, Weisser, and Gulmon, 1977). Moreover, if water uptake could occur through the spines or from

the areoles, water loss from these structures would also be expected, which would be detrimental to cacti in arid and semiarid regions. Therefore, we will focus exclusively on water uptake from the soil.

RAINFALL AND SOIL WATER POTENTIAL

To permit water uptake by agaves and cacti, rainfall or supplemental watering must raise the water potential of the soil in the root zone above the water potential of the plant. The amount of water available in a soil is in part dependent on particle sizes in the soil, as we will discuss in Chapter 6. For soil of the northwestern Sonoran Desert, 77% of the non-gravel soil fraction has particle sizes of 0.05–2 mm (Nobel, 1976), which is classified as sand. When such a soil is dry, a rainfall of 7 mm can raise Ψ^{soil} to -0.5 MPa at 10 cm below the soil surface (Fig. 3.1A) and induce stomatal opening by *Agave deserti* (Fig. 3.1B). Similarly, 6 mm of rainfall are needed to initiate changes in the distance between rib crests of *Carnegiea gigantea* (Fig. 1.17; MacDougal and Spalding, 1910) and to result in water uptake by *Opuntia basilaris* (Szarek and Ting, 1975b). In actuality, the amount of rainfall required to bring about a physiological response depends on the existing soil water content, the ambient air temperature, the ambient air water vapor content, and the timing of the rainfall during a particular storm – light rainfall during the first few hours will lead to a gradual wetting of the soil crust, permitting a much higher rate of subsequent water infiltration into the soil (Nobel, 1983a).

A computer model has been developed to predict daily variations in Ψ^{soil} at various depths in the soil (Young and Nobel, 1986; Hunt and Nobel, 1987a). A particular rainfall event is assumed to raise successive soil layers, beginning at the surface, to their water-holding capacity (*field capacity*). The temperature of the soil surface is assumed to vary sinusoidally over a day, as is approximately the case at the field site (Figs. 1.1 and 1.2) in the northwestern Sonoran Desert used for validation of the model. The number of days per year when the soil is wet enough for water uptake by agaves and cacti increases approx-

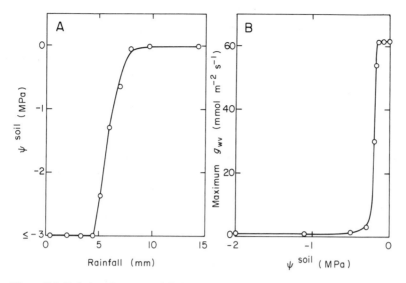

Figure 3.1. Relations between rainfall, soil water potential, and stomatal opening for *Agave deserti*. **(A)** Influence of single rainfall events occurring for dry soil on the maximum soil water potential in the root zone 10 cm below the soil surface. **(B)** Influence of Ψ^{soil} maintained for five days for plants in initially dry soil on the maximum water vapor conductance observed during the fifth day. Modified from Nobel (1976).

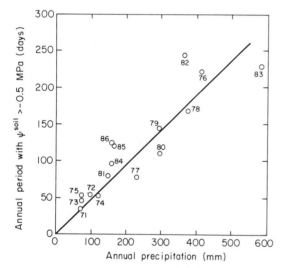

Figure 3.2. Relation between annual precipitation at a site in the northwestern Sonoran Desert (Figs. 1.1 and 1.2) and the number of days when the soil water potential 10 cm below the soil surface was above −0.5 MPa. Data indicated for specific years were obtained by simulation together with field measurements for a few years that were used for validation (Nobel, 1977b, 1985b, 1987a; Young and Nobel, 1986; Hunt and Nobel, 1987a). Used by permission from Nobel (1987a), with additional measurements for 1986. © 1987 by The University of Chicago. All rights reserved.

imately linearly with the annual precipitation (Fig. 3.2).

Precipitation in deserts can differ measurably over short distances (Fogel, 1981) and generally varies considerably from year to year (Figs. 3.2 and 3.3). Although the precipitation in the northwestern Sonoran Desert tends to be annually bimodal (Ting and Jennings, 1976), occurring in both late summer and winter, the pattern is variable. For instance, essentially no winter rainfall occurred in 1971/72 and 1974/75, and no summer rainfall occurred in 1971, 1975, and 1980 (Fig. 3.3). Such rainfall vagaries have important consequences for year-to-year variations in biomass productivities of mature plants (Chapter 7) as well as for the establishment of seedlings (discussed later in this chapter).

PLANT WATER POTENTIAL

We expect water uptake from the soil when Ψ^{soil} exceeds the plant water potential, Ψ^{plant}. As we indicated in Chapter 1, the osmotic pressure, which is a major contributor to Ψ^{plant} (Eq. 3.1), is often about 0.5 MPa for agaves and cacti from the Sonoran Desert. Consistent with this, π under wet conditions

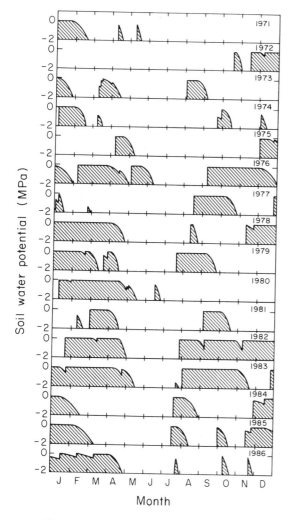

Figure 3.3. Annual variations in soil water potential at a site in the northwestern Sonoran Desert, emphasizing periods when Ψ^{soil} was above -2 MPa. Data were obtained and permissions are as described for Figure 3.2.

sure the relative humidity of air in equilibrium with plant tissue in a sealed chamber and, with proper calibration, can indicate the water potential of plant tissues. Alternatively, a shoot can be placed in a pressure chamber, or "pressure bomb," with the cut surface protruding and internal air pressure gradually increased until water is just forced out the cut surface, at which time the applied pressure equals $-\Psi$. Largely because of instrumentation, measurements of π for agaves and cacti tend to predate measurements of Ψ (Table 3.1). Also, values of Ψ and π for a particular species change appreciably depending on the water status of the plant.

The highest values for Ψ occur under wet conditions, when cells of the chlorenchyma and water storage parenchyma are turgid, meaning that they have a positive hydrostatic pressure. For example, for mean values of Ψ and π determined for a particular species under wet conditions (Table 3.1), P calculated using Equation 3.1 ($\Psi = P - \pi$ when the gravitational term is ignored) ranges from 0.17 to 0.23 MPa. During drought, Ψ^{plant} decreases as water is lost. Moreover, P then tends to decrease, so Ψ becomes similar to $-\pi$. Using mean values of Ψ and π determined under drought conditions (Table 3.1), we find that the calculated P is 0.00 to 0.02 MPa, indicating an essentially complete loss of turgor.

The values of π in CAM plants can vary considerably over a daily cycle because of the substantial daily changes in malate concentration (Fig. 3.4). For instance, the average stem malate concentration of *Cereus validus* (Fig. 3.5) increases from about 55 mM at dusk to 155 mM at dawn (Fig. 3.4B), and the osmotic pressure increases from 0.37 to 0.61 MPa (Fig. 3.4C; Lüttge and Nobel, 1984). Based on the van't Hoff relation (Eq. 3.2), we predict an increase in the osmotic contribution from malate of 0.24 MPa, in complete agreement with the observed increase in π of 0.24 MPa.

As the tissues lose water during drought, the osmotic pressure increases (Table 3.1). For the examples presented, about half of the water is lost over a four- to five-month drought, leading to an approximate doubling of the solute concentration and a consequent doubling of π (see Eq. 3.2). Similarly, severed cladodes of two species of platyopuntia lose

is 0.5–0.7 MPa for seven species of cacti from Jamaican coastal deserts (Harris and Lawrence, 1917). In this subsection we summarize values for plant water potentials and osmotic pressures for agaves and cacti, especially under wet and drought conditions, which will also indicate when water uptake from the soil is expected.

Instruments for measuring osmotic pressure have been available for a long time, but those for measuring water potential did not come into widespread use until the 1970s (Slavik, 1974; Turner, 1981). Psychrometers mea-

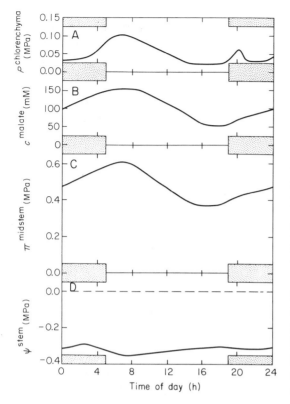

Figure 3.4. Variations in several water relations parameters for *Cereus validus* over a 24-h period. (A) Hydrostatic pressure in uppermost chlorenchyma cells measured with a pressure probe placed into individual cells. (B) Malate concentration determined for an extract from chlorenchyma cells at midstem. (C) Osmotic pressure of the extract used in (B). (D) Water potential measured for the whole stem placed in a pressure bomb. Data are adapted from Lüttge and Nobel (1984). Although Ψ^{stem} varied little over the 24-h period, the increase in malate concentration at night caused $\pi^{chlorenchyma}$ to increase, leading to water uptake by the cells and, hence, to an increase in their turgor pressure.

about half of their fresh weight in five months (Chow, Burnside, and Lavy, 1966). For *Carnegiea gigantea* in southern Arizona, π decreases with increasing elevation, reflecting the increase in precipitation with elevation as well as the decrease in temperature (Soule and Lowe, 1970), which reduce ΔN_{wv} and hence transpirational water losses ($J_{wv} = g'_{wv} \Delta N_{wv}$, Eq. 2.6b). In addition, osmotic pressure tends to be higher on the south side than on the north side of the stems, especially under drought conditions when the difference generally av-

erages 0.03–0.07 MPa. On occasion, the south side can have a π that is 0.2 MPa higher than the north side for *C. gigantea*, and similar π asymmetries are found for *Ferocactus wislizenii* (Walter, 1931). The higher π on the south side was attributed to greater transpirational water loss there because of higher daytime stem temperatures (Soule and Lowe, 1970). More likely, greater PAR interception on the south side leads to greater nocturnal stomatal opening reflecting the enhanced capability for net CO_2 uptake (Chapter 5) and leading to more malate accumulation.

Under wet conditions, π is about 0.5 MPa and Ψ is about -0.3 MPa for agaves and cacti (Table 3.1). For water uptake to occur under such conditions, Ψ^{soil} must be above -0.3 MPa. As the tissues lose water and Ψ^{plant} falls, the Ψ^{soil} leading to water uptake does not have to be as high. For example, after a four-month drought, representative desert succulents can take up water when Ψ^{soil} is above approximately -1.0 MPa (Table 3.1). Averaged over most conditions, water uptake into agaves and cacti generally occurs when Ψ^{soil} exceeds about -0.5 MPa.

ROOT PROPERTIES

A striking characteristic of the roots of agaves and cacti is their shallow habit. Among other investigators, Cannon (1911) long ago clearly recognized this and noted that most of the roots of cacti are 3–15 cm below the soil surface. Two of the three species featured throughout this book, *Agave deserti* (Fig. 1.24B) and *Ferocactus acanthodes* (Fig. 1.2A), have mean root depths of only 8 cm (Nobel, 1976, 1977b), allowing physiological responses to rainfalls of only 6–7 mm. Even though shallow, roots of desert succulents can laterally extend over considerable distances (Fig. 1.16), which has important consequences for competition between plants for water and nutrients.

Rain roots

One of the most interesting environmental responses of the root systems of agaves and cacti is the rapid induction of rain roots (Fig. 1.31) caused by precipitation or supplemental watering. Kausch (1965) noted that rain roots formed on established roots of *Opuntia de-*

Figure 3.5. *Cereus validus* used for the data in Figure 3.4 as well as for salinity studies (Chapter 6). The plants were nine months old, and the stems averaged 10 cm in height (Nobel et al., 1984a) (photographed by J. Andrew C. Smith).

Table 3.1. *Examples of water potentials and osmotic pressures for leaves of agaves and stems of cacti*

Species	Wet		Drought		Reference
	Ψ (MPa)	π (MPa)	Ψ (MPa)	π (MPa)	
Agave deserti	−0.30−−0.45	0.52–0.68	−1.13−−1.63	1.18–1.57	Nobel (1976)
A. fourcroydes	−0.31−−0.45	0.47–0.71	−0.96−−1.24	0.99–1.22	Nobel (1985a); P.S. Nobel and T.L. Hartsock (unpublished observations)
A. sisalana	—	0.8–1.0	—	1.4–1.9	Abd El Raman, Gamassy, and Mandour (1968)
Carnegiea gigantea	—	0.4–0.6	—	1.2	Livingston (1906); MacDougal and Cannon (1910); Walter (1931)
	—	0.35–0.43	—	0.73–0.96	Soule and Lowe (1970)
Cereus validus	−0.28−−0.34	0.36–0.62	—	—	Lüttge and Nobel (1984)
Copiapoa cinerea	—	—	−1.24	1.34	Rundel et al. (1980
Ferocactus acanthodes	−0.09−−0.20	0.28–0.35	−0.45−−0.80	0.59–0.69	Nobel (1977b)
F. wislizenii	—	0.4	—	1.0	MacDougal and Cannon (1910)
Opuntia phaeacantha	—	0.7	—	1.5	Walter (1931)
Selenicereus urbanianus	—	0.37	—	0.51	Hadač and Hadačová (1974)

Note: Wet generally refers to a fully hydrated plant, whereas *drought* generally means a four- to five-month period without uptake of water from the soil. The ranges presented encompass 90% of at least ten measurements on a population, or the daily variation in the case of *Cereus validus*.

cumbens within a few hours after the soil is moistened and occur within 24 h after rewetting for *Trichocereus bridgesii*, *T. pachanoi*, *T. spachianus*, and *Opuntia ficus-indica*. A 5-mm-long rain root develops on *Agave deserti* only 6 h after rewetting (Fig. 1.31B), and for *Ferocactus acanthodes* rain roots are visible in 8 h, averaging over 2 mm in length (some 6 mm long) in 24 h (Nobel and Sanderson, 1984). Rain roots tend to shrivel and die once drought recommences, usually in a matter of weeks, but some survive and become "established" roots after a year or so, although much remains to be learned about the time course of the anatomical and physiological changes involved.

Significant water uptake and consequent physiological responses can also occur shortly after watering droughted plants. Stomatal opening is detectable 12 h after a droughted *A. deserti* receives water and full stomatal opening occurs in about 48 h, although it is not sustained for the whole night (Nobel, 1976). Water uptake causes swelling of the stems of *Carnegiea gigantea* (Fig. 1.18B) and induces net CO_2 uptake by *Echinocereus viridiflorus* (Green and Williams, 1982) 24 h after rainfall. Substantial water uptake occurs for *Opuntia basilaris* and *O. stricta* 24–36 h after rainfall (Szarek et al., 1973; Szarek and Ting, 1975b; Osmond et al., 1979b). Based on the morphological and physiological evidence, it is tempting to speculate that the rain roots lead to the rapid uptake of water that occurs for the first day or so after drought is interrupted by a sizable rainfall. However, most of the initial water uptake is by the established roots (Fig. 3.6). Four days after the rewetting, water uptake by the new rain roots equals that taken up by the previously existing established roots for *A. deserti* (Fig. 3.6), with similar results for *F. acanthodes* (Nobel and Sanderson, 1984).

To help analyze water uptake by rain roots and established roots, let us represent the volume flux density of water into the roots by J_V and the surface area of a particular group of roots by A:

Plant water uptake
per unit time

$$= J_V A^{\text{total}}$$
$$= J_V^{\text{est}} A^{\text{est}} + J_V^{\text{rain}} A^{\text{rain}} \qquad (3.3)$$
$$= L_P^{\text{est}} A^{\text{est}} \, \Delta\Psi + L_P^{\text{rain}} A^{\text{rain}} \, \Delta\Psi$$

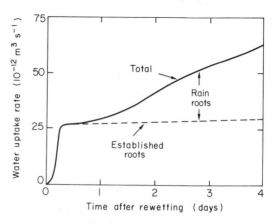

Figure 3.6. Estimated rate of water uptake by droughted *Agave deserti* at various times after being placed in a nutrient-containing solution. The five-year-old plant had nine leaves and was 20 cm tall. A $\Delta\Psi$ of 0.1 MPa was assumed to occur from the solution to the root xylem. Modified from Nobel and Sanderson (1984).

where the superscripts est and rain refer to established roots and rain roots, respectively; L_P is the *root hydraulic conductivity*, which indicates the volume flux density of water flowing radially into the roots at their surface for a particular drop in water potential, $\Delta\Psi$, from the soil adjacent to the roots to the root xylem; that is, J_V equals $L_P \Delta\Psi$ (the small $\Delta\Psi$ along the root xylem is ignored here, and the same $\Delta\Psi$ is assumed to occur for both rain roots and established roots). Even though L_P^{rain} is higher than L_P^{est}, presumably because of the greater water permeability of the epidermis and the cortical region of rain roots, A^{rain} is initially zero. Four days are required for A^{rain} to increase sufficiently so that $L_P^{\text{est}} A^{\text{est}}$ becomes equal to $L_P^{\text{rain}} A^{\text{rain}}$ for *A. deserti*, meaning that water uptake by the rain roots then becomes equal to water uptake by the established roots (Eq. 3.3 and Fig. 3.6).

Rectifierlike activities

A perplexing question regarding the roots of desert succulents has always been, "Why isn't the water in the fleshy leaves of agaves and fleshy stems of cacti lost to the soil during drought?" Indeed, during drought a wet tissue can be situated near a very dry soil, and Equation 3.3 predicts that water should flow toward the region of lower Ψ. Equation 3.3 also contains the resolution of the apparent ilemma; namely, L_P of the root changes de-

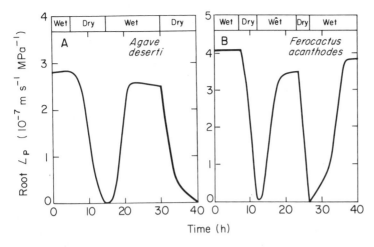

Figure 3.7. Time courses for changes in hydraulic conductivity of individual roots. (A) An established root 35 mm long with a rain root 8 mm long from *Agave deserti*. (B) A young lateral root 26 mm long from *Ferocactus acanthodes*. Water uptake was determined for excised roots to which a $\Delta\Psi$ of 0.04 MPa was applied. Data are from Nobel and Sanderson (1984) and unpublished observations.

pending on whether the soil is wet or dry. Thus, the roots of agaves and cacti act like electronic rectifiers, passing water (current) readily from the soil to the plant but not in the reverse direction.

To measure L_P, roots of *Agave deserti* and *Ferocactus acanthodes* were excised and small capillaries were carefully attached (Russel and Sanderson, 1967). By placing the root in an aqueous solution and applying a partial vacuum to the capillary (see Gibson and Nobel, 1986, for a drawing of the device used), one can induce water flow and L_P can be determined from the relation $L_P = J_V/\Delta\Psi$. The excised root can be removed from the solution to simulate the effect of soil drying.

For both *A. deserti* (Fig. 3.7A) and *F. acanthodes* (Fig. 3.7B), L_P decreases more than tenfold after about 5 h of drying and could decrease more than 1,000-fold in a few days (Nobel and Sanderson, 1984). Yet when the dried roots are rewetted, L_P rapidly increases; for instance, L_P increases more than tenfold within 5 h of rewetting for both species (Fig. 3.7). Although the anatomical details are not fully understood, rewetting allows rehydration of regions external to the stele, including the root cortex (see Fig. 1.31A). The rehydration of the extrastelar region facilitates water entry into the stele, where cells are alive and never become dehydrated, and thus into the root xylem. Such quick rehydration allows water

uptake by established roots to resume in a matter of hours (Fig. 3.6) after the soil becomes wet following rainfall. Only after several days do the rain roots, which have higher hydraulic conductivities than established roots, develop enough surface area to contribute appreciably to water uptake by the plant (Eq. 3.3 and Fig. 3.6). When drought inevitably sets in, L_P decreases dramatically, beginning when Ψ^{soil} falls below about -1.2 MPa (Jordan and Nobel, 1984). In this way, the water contained in these desert succulents is not lost back to a drying soil, which can rapidly develop a considerably lower water potential than can be endured by the tissues of agaves and cacti.

Distribution in soil

We have already indicated that the roots of desert succulents are shallow, as is clear for established and rain roots of *Agave deserti* and *Ferocactus acanthodes* (Fig. 3.8). For *A. deserti*, the rain roots tend to occur as groups of small lateral rootlets projecting from the sides of established roots, whereas for *F. acanthodes* they arise primarily by renewed elongation of root tips, although lateral roots are also initiated. In both cases, rain roots, which can be induced when Ψ^{soil} rises above about -1 MPa (Jordan and Nobel, 1984), have approximately the same distribution with depth as do the established roots (Fig. 3.8). The mean root depths for the various plants ex-

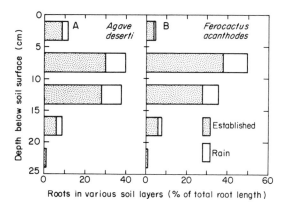

Figure 3.8. Distribution of established (stippled) and rain (open) roots in the soil for (**A**) *A. deserti* and (**B**) *F. acanthodes*. Data on total root length (rain plus established) are for medium-sized plants averaged for 5-cm-thick soil layers and were obtained about two weeks after a major rainfall. Modified from Jordan and Nobel (1984) and Hunt and Nobel (1987a).

cavated range from 7 to 12 cm. No roots occur deeper than 25 cm, and except directly under the shoots essentially no roots occur in the upper 3 cm of the soil, an aspect that is largely controlled by temperature (Chapter 4).

The shallow root distributions of agaves and cacti allow them to take up a sizable fraction of the annual rainfall in a desert. Ignoring runoff and "runon" for a particular area, both of which depend on topography, simulation modeling has shown that 60% to 70% of rainfall in the northwestern Sonoran Desert evaporates from the soil surface (Young and Nobel, 1986). Field measurements indicate that *A. deserti* can take up about 35% of the water incident on the soil surface within the region of its root extension (Nobel, 1976) and *F. acanthodes* can take up about 22% (Nobel, 1977b). The values of Ψ^{plant} for such desert succulents (Table 3.1) average about -0.5 MPa. For the sandy soils involved, just over 75% of the non-evaporated water is stored between field capacity ($\Psi^{soil} \cong 0.0$ MPa) and Ψ^{soil} of -0.5 MPa (Young and Nobel, 1986; discussed in more detail in Chapter 6), most of which is taken up by the species considered. Thus, the water-uptake characteristics of *A. deserti* and *F. acanthodes* are very well matched to the water availability in the sandy desert soils of their native habitats.

Simulation modeling also indicates that

the observed root distributions of *A. deserti* and *F. acanthodes* (Fig. 3.8) are highly advantageous for water uptake. A two-dimensional model incorporating root distributions, with both depth in the soil and radial distance from the plant base, shows that observed root distributions lead to more water uptake than those in which all the roots are in a single, 5-cm-thick layer at any depth (except at the soil surface, where thermal considerations become of overriding importance; Hunt and Nobel, 1987a). For the root distributions observed in the field, water uptake by *A. deserti* and *F. acanthodes* is predicted to be 85% of the maximum possible (Hunt and Nobel, 1987b). Water uptake can be enhanced by a higher proportion of rain roots with their higher L_P's than for established roots, but the "costs" in terms of energy, carbon, and nutrients for producing such generally ephemeral roots must be considered, a topic that we will return to in Chapter 7.

Water storage – capacitance

Agaves and cacti are termed succulents because of their thick, water-storage tissues. Water storage is important for drought tolerance, and it also affects their thermal relations (Chapter 4). We will first consider the structural aspects of agaves and cacti pertaining to water storage and then develop a formal framework for analyzing the water relations of succulence.

MORPHOLOGICAL ASPECTS

When rainfall interrupts a period of prolonged drought, net water uptake into the stems of cacti can be readily demonstrated. For *Carnegiea gigantea*, the distance between adjacent rib crests increases about 7% one week after rainfall and 15% in two weeks (Fig. 1.17), indicating major changes in stem morphology. Changes in its stem diameter are detectable about one day after rainfall, with the diameter increasing about 2% per day for the next few days (Fig. 1.18B). Small daily changes in diameter are superimposed on the stem swelling following rainfall. In particular, at the beginning of the night, the diameter of *C. gigantea* is about 0.2% greater than the daily mean; nocturnal stomatal opening leads to a net loss of water through the night, and

Table 3.2. *Ratios of volume/surface area for leaves of agaves and stems of cacti*

Species	V/A (cm)	Reference
Agave deserti	0.97	Nobel (1976)
A. fourcroydes	2.02 ⎫	P.S. Nobel (unpublished
A. lechuguilla	0.32 ⎭	observations)
Ferocactus acanthodes	6.3	Nobel (1977b)
Lophocereus schottii	0.7–1.5	Felger and Lowe (1967)
Opuntia imbricata	0.40 ⎫	
O. kleiniae	0.23 ⎪	
O. leptocaulis	0.14 ⎬	Conde (1975)
O. lindheimeri	0.45 ⎪	
O. phaeacantha	0.53 ⎭	

the diameter becomes about 0.2% less than the daily mean by the next morning (Fig. 1.18A). Likewise, cladodes of the hybrid *Opuntia "occidentalis"* are thickest at the beginning of the night and thinnest near the beginning of the day (Schroeder, 1975). These patterns of daily changes in stem thickness are opposite in phase to those of C_3 and C_4 plants, which have daytime stomatal opening.

The decreases in stem diameter at night primarily reflect a net loss of water. When the stomata open and water is lost, the water potentials of the chlorenchyma and the xylem decrease. Because Ψ^{plant} is negative and the solution in the xylem has a very small osmotic pressure (Nobel, 1983a), a lower Ψ means that P^{xylem} becomes more negative ($\Psi = P - \pi$, Eq. 3.1). This negative P, or tension, can be transmitted throughout the stem by the bonding together of the water molecules in the apoplast. As the tension of water in the xylem and the cell walls increases at night, the whole stem can contract. Such contractions during periods of high transpiration can be readily measured for trees and cactus stems using a dendrograph (Fig. 1.18).

A simple, but useful, index for indicating water storage is the volume of the organ (V) divided by the surface area (A) across which water can be lost by transpiration, V/A. The ratio V/A is generally similar to the mean cross-sectional area of a stem divided by its mean circumference, especially if the stem is long relative to its diameter. A long, thin cylinder has a low V/A relative to a short, wide cylinder. We also note that V/A represents the average depth of tissue in which water can be stored per unit surface area where water can be lost. Instead of using tissue volume, we could more appropriately use water volume or even fresh weight, measurements that would, however, require harvesting of the tissue.

For the thin pencil chollas, *Opuntia kleiniae* and *O. leptocaulis*, V/A is about 0.2 cm, whereas for the massive barrel cactus *Ferocactus acanthodes*, V/A can exceed 6 cm (Table 3.2), indicating much more water storage capability per unit surface area in the latter case. Indeed, substantial stomatal opening by *F. acanthodes* can occur for about 40 days after water uptake from the soil no longer occurs (Nobel, 1977b); *Agave deserti*, with a V/A near 1 cm (Table 3.2), can maintain nocturnal stomatal opening for only up to about 8 days after Ψ^{soil} falls below Ψ^{plant} (Nobel, 1976). For *Lophocereus schottii* and other species in Tribe Pachycereeae, V/A tends to increase with increasing latitude, a pattern discussed in more detail in the next chapter with respect to tissue temperatures; such increases in V/A are correlated with lower annual precipitation (Felger and Lowe, 1967). Similarly, for *Agave cerulata*, *A. deserti*, and *A. sobria*, V/A increases as the probable drought length in their native habitats increases (Burgess, 1985). Thus, V/A can be a useful index for indicating the water storage capacity of agaves and cacti and for correlations between water storage and environmental factors.

ANATOMICAL ASPECTS

For leaves of agaves and stems of cacti, the water-storage parenchyma is usually anatomically distinct from the chlorenchyma (Fig. 1.34). We will consider some of the cellular differences in these tissues that affect water retention during desiccation. Also, we will consider the pathway for water flow from the water storage tissue to the xylem, from which the water can readily move to the chlorenchyma.

During a drought lasting four to six months, agaves and cacti can lose 50% of their stored water (MacDougal et al., 1915; Nobel, 1976, 1977b). However, the water does not come equally from all cells. For well-hydrated plants of *Carnegiea gigantea*, *Ferocactus acanthodes*, and *Opuntia basilaris*, 95% or more of the plant water is in the stem (Barcikowski and Nobel, 1984). Of the stem water, most is in the water-storage parenchyma located in the *cortex* (region internal to the chlorenchyma and external to the main vascular region) and in the *pith* (region internal to the main vascular system and therefore in the center of the stem). Such water-storage regions tend to lose a greater fraction of their water than does the chlorenchyma, a process that we can analyze after suitably modifying the van't Hoff relation (Eq. 3.2, $\pi = cRT$).

We begin by noting that the concentration c indicates the number of moles of solute, n, divided by the volume of water in a cell of volume V. We also note that because part of V is made up of membranes, proteins, and other nonwater components having a non-aqueous volume b, the water volume is more accurately presented as $V - b$. We can then rewrite Equation 3.2 as follows:

$$\pi = \frac{n}{V - b} RT \tag{3.4a}$$

or

$$V = \frac{nRT}{\pi} + b \tag{3.4b}$$

which is known as the Boyle–van't Hoff relation (Nobel, 1983a).

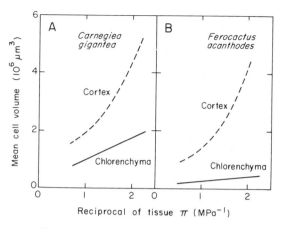

Figure 3.9. Volumes of cells in the water-storage cortex and in the chlorenchyma of *Carnegiea gigantea* (**A**) and *Ferocactus acanthodes* (**B**). The osmotic pressure of the tissues was determined about 3 h into the daytime after various periods of drought; cell volume was calculated from cell dimensions measured on living cells. Modified from Barcikowski and Nobel (1984).

As the osmotic pressure in stem cells increases during drought and hence $1/\pi$ decreases, cells in the chlorenchyma of *C. gigantea* and *F. acanthodes* shrink (Eq. 3.4b and Fig. 3.9). The thin-walled, water-storage cells in the cortex have even greater fractional volume losses than do the thick-walled chlorenchyma cells as $1/\pi$ decreases for both species (Fig. 3.9). Although the types of molecules have not been identified, n apparently decreases for the cortex, and thus less water remains there at a given π (Eq. 3.4). Because n does not decrease for the chlorenchyma, a larger fraction of water is lost from cells of the cortex during drought. For *O. basilaris*, the osmotic responses of the two cell types are similar, but the initial osmotic pressure is 0.13 MPa higher in the chlorenchyma (0.60 vs. 0.47 MPa; Barcikowski and Nobel, 1984). As water is lost and P goes to zero, Ψ and $-\pi$ eventually become equal in the pith and in the chlorenchyma ($\Psi = P - \pi$, Eq. 3.1). However, a larger fraction of the water in the water-storage cells must be lost to get to the same π as in the chlorenchyma, because the latter initially has a higher π (Fig. 3.9). The chlorenchyma also has a higher π than the thin-walled water-storage parenchyma for *Epiphyllum crenatum* and *Opuntia ficus-indica* (Wiebe and Al-Saadi, 1976). The preferential loss of water from the storage tissue of all three species allows the photosynthetic tissue to remain active for a longer period during drought. We also note that in the hydrated state, the water-storage cells are twofold larger in volume than the chlorenchyma cells for *O. basilaris*, nearly threefold larger for *C. gigantea* (Fig. 3.9A), and tenfold larger for *F. acanthodes* (Fig. 3.9B).

Water stored in the thin-walled parenchyma cells can move to the xylem in the cell walls – the apoplastic pathway – or from cell to cell by going through plasmodesmata and/or the plasmalemmas – the symplastic pathway. This has been investigated in detail for the leaves of *Agave deserti*, where an average of eight water-storage parenchyma cells occur between adjacent vascular bundles (Fig. 1.34B). Simulations based on a model incorporating both cell wall and plasmalemma properties (Molz and Ferrier, 1982; Boyer, 1985) show that most of the movement of water from the storage cells to the vascular bundles is via a symplastic pathway. The hydraulic conductivity L_P (see Eq. 3.3, which can be simplified to $J_V = L_P \Delta\Psi$) here equals 2×10^{-6} m s^{-1} MPa^{-1} for a cell along this pathway (Smith and Nobel, 1986). Compared with movement along the apoplastic and symplastic pathways from the water-storage cells to the xylem, movement along the xylem is easier (Smith and Nobel, 1986). For instance, the hydrostatic pressure gradient $\Delta P/\Delta x$ leading to flow in the xylem axially along a leaf of *A. deserti* supporting maximum rates of transpiration is about 0.05 MPa m^{-1}. To obtain the measured fluxes of water from the water-storage cells to the xylem supporting the maximum transpiration rates, we need a gradient of about 20 MPa m^{-1}, but because of the short distance involved the drop in water potential is only 0.009 MPa (Smith and Nobel, 1986). Even though the $\Delta\Psi$ required is small, the process is fairly slow, the *half-time* (time to go halfway from some initial to some final state, here measured for rehydration of desiccated leaf slices) being about 1 h. We will return to such considerations after we formally introduce the concept of capacitance in the next subsection.

Mucilage, which can bind considerable amounts of water, has sometimes been implicated in the water storage ability of desert succulents (Mindt et al., 1975; Esau, 1977). How-

ever, mucilage cells occupy only about 3% of the volume of the parenchyma of *O. ficus-indica* and may be more involved in its calcium relations (Trachtenberg and Mayer, 1982b) or possibly as a carbohydrate reserve (Sutton et al., 1981). Other studies on *O. ficus-indica* similarly indicate that only 3% of the water stored in the stem is associated with mucilage and other polymers (Wiebe and Al-Saadi, 1976). Indeed, nearly all the water that is stored or released for *A. deserti* (Smith and Nobel, 1986) and various cacti (Barcikowski and Nobel, 1984) comes from the thin-walled, water-storage parenchyma cells.

ELECTRICAL CIRCUIT
REPRESENTATION

Considerable progress in analyzing the water relations of plants has been achieved by representing the various parts of the pathway making up the transpiration stream as if they were part of an electrical circuit (Nobel, 1983a). Water movement is analogous to the flow of charge (that is, the electrical current), water potential is analogous to electrical potential, and each cellular structure can have a resistance to water flow that is analogous to electrical resistance. Moreover, water storage in agaves and cacti is analogous to the capacitance of electrical circuits.

Resistance

For *steady-state conditions* in which water flows and water potentials are not changing with time, we can represent the entire pathway of the transpiration stream by a series of resistances. Indeed, resistances are implicit in Equation 3.3, which we can rewrite as follows:

$$J_V A = L_P A \, \Delta\Psi = \frac{\Delta\Psi}{R} \qquad (3.5)$$

where R is the resistance and has units of MPa s m^{-3}.

For water flow in a pipe of lumen radius r, an equation known as the Hagen–Poiseuille relation was determined experimentally over 100 years ago (see Nobel, 1983a):

$$J_V A = \frac{\pi r^4}{8\eta} \frac{\Delta P}{\Delta x} \qquad (3.6)$$

where η is the viscosity of the solution (1.31 mPa s at 10°C and 0.80 mPa s at 30°C for water), and ΔP is the drop in hydrostatic pressure that occurs over distance Δx. Assuming the conducting elements in the xylem behave as pipes and using Equations 3.5 and 3.6, we can represent the resistance for flow in the xylem as

$$R = \frac{\Delta \Psi}{J_V A} = \frac{8\eta \, \Delta x}{\pi} \sum_i \frac{1}{r_i^4} \qquad (3.7)$$

where the summation is over all i conducting elements in the xylem and we have equated $\Delta \Psi$ to ΔP for flow along the xylem; for vertical movement we will have to include gravitational effects (Eq. 3.1), where $\Delta \Psi / \Delta z$ equals $\Delta P / \Delta z + \rho_w g$ ($\Delta \pi = 0$ along the xylem).

Capacitance

As the plant water potential decreases during transpiration at night or during an extended drought, a net movement of water occurs out of the water-storage tissue and the chlorenchyma. We can define *capacitance*, a capacity for the tissues to hold water, as follows (Jarvis, 1975; Nobel, 1983a, 1987b):

Capacitance
of plant $= C^j$
component j

$\quad$ change in water
$= \dfrac{\text{volume in component } j}{\text{change in average water}}$
$\qquad\qquad$ potential of component j

$= \dfrac{\Delta V_{\text{water}}^j}{\Delta \overline{\Psi}^j} \qquad (3.8)$

Capacitances can be estimated from pressure–volume or P–V curves, which relate Ψ to V_{water}^j or, more frequently, to *relative water content* (Tyree and Hammel, 1972):

Relative
water $= $ RWC
content

$= \dfrac{\text{fresh weight at a given } \Psi - \text{dry weight}}{\text{fresh weight at a } \Psi \text{ of } 0 - \text{dry weight}}$

$\qquad\qquad\qquad\qquad\qquad (3.9)$

Because ρ_w is a constant, RWC numerically equals the relative water content on a water volume basis (Nobel and Jordan, 1983). Thus, a change in RWC equals ΔV_{water} divided by the water volume present at full hydration.

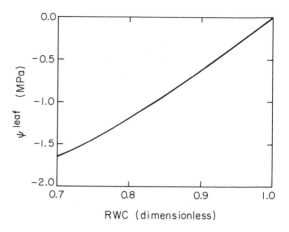

Figure 3.10. Response of leaf water potential to relative water content for *Agave deserti*. Data from pressure–volume curves are adapted from Nobel and Jordan (1983) and Calkin and Nobel (1986).

P–V curves have been determined for the leaves of *Agave deserti* (Fig. 3.10). The capacitance per unit volume of leaf water present at full hydration (the reciprocal of the slope in Fig. 3.10) is about 0.18 MPa^{-1} over a wide range in Ψ. This specific capacitance times the water volume in a leaf indicates the capacitance of that leaf in m^3 MPa^{-1}, the conventional unit for C^j (Eq. 3.8).

Storage resistance

Before assembling the various components into an electrical circuit and presenting specific values, we need to consider the resistance for water moving to or from the storage capacitance. We first calculate this storage resistance R_s^j using a relation implicit in Equation 3.5 [$R = 1/(L_P A)$], and then we introduce another method for calculating such resistances.

We will assume that, on average, water moves from the water-storage parenchyma to a vascular bundle of *Agave deserti* by crossing three cells and hence six membranes. Because the L_P of 2×10^{-6} m s^{-1} MPa^{-1} for *A. deserti* (Smith and Nobel, 1986) refers to the entry or the exit of water from an individual water-storage cell, the hydraulic conductivity of the overall pathway is about $L_P/6$ or 0.3×10^{-6} m s^{-1} MPa^{-1} (geometric complications are ignored in this calculation). Cells in the water-storage parenchyma of *A. deserti* contact each

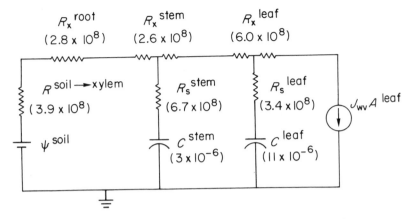

Figure 3.11. Organization of the soil-to-root xylem resistance ($R^{\text{soil} \to \text{xylem}}$), organ xylem resistances (R_x^j), organ storage resistances (R_s^j), and organ capacitances (C^j) into an analog electrical circuit for *Agave deserti*. Each organ capacitance is depicted as connected through its storage resistance to the center of a particular xylem resistance, indicating that C^j's represent average values. Numerical values for a plant 0.2 m tall with nine leaves, indicated in MPa s m^{-3} for resistances and m^3 MPa^{-1} for capacitances, are from Calkin and Nobel (1986).

other over approximately 25% of their surface area in the direction radial to the xylem axis (Smith and Nobel, 1986). Considering this contact area and the sizes of the cells involved, the mean area across which water flows from the water-storage parenchyma into the vascular bundles is about 270 m^2 per m^3 of leaf tissue for *A. deserti*. Hence, the storage resistance on a unit volume basis is about 1/[(0.3 × 10^{-6} m s^{-1} MPa^{-1})(270 m^2 m^{-3})], or 1.3 × 10^4 s MPa.

We can also calculate the storage resistance by examining the time course involved in a particular change, such as the swelling of a previously desiccated tissue as water moves back into the water-storage parenchyma or the change in Ψ^j that follows upon a rapid change of air pressure in a pressure bomb (Nobel and Jordan, 1983). By analogy with electrical circuits, the time constant τ^j for component j can be represented as (Jarvis, 1975; Nobel, 1983a)

$$\tau^j = R_s^j \, C^j \qquad (3.10)$$

The time constant (or *lifetime*) for a process that varies exponentially with time equals 1/ln 2 or 1.44 times the half-time. Using the C^j per unit volume of 0.30 MPa^{-1} appropriate to the L_P just used to calculate R_s (Smith and Nobel, 1986) and the observed half-time of about 1 h (3600 s), we find by Equation 3.10 that the storage resistance on a unit volume

basis is (3600 s)(1.44)/(0.30 MPa^{-1}) or 1.7 × 10^4 s MPa, similar to the value of 1.3 × 10^4 s MPa just estimated by an entirely different technique. We are now in a position to discuss the total transpiration stream for *A. deserti*.

Electrical circuit

The various resistances and capacitances for water movement through *Agave deserti* can be arranged into an electrical circuit (Fig. 3.11). We have ignored root capacitance, because for *A. deserti* it is usually small relative to that of the stem or the leaves (Nobel and Jordan, 1983).

As transpiration for *A. deserti* increases at the beginning of the night (Fig. 2.5), water moves into the plant from the soil, out of storage in the stem, and out of storage in the leaves (Fig. 3.12). Water release from the shoot causes a lag of about 4 h between peak transpiration and the subsequent peak flow from the soil into the root. During the initial part of this lag period, most of the transpired water comes from the leaves (Fig. 3.12). During the night about 24% more water is transpired than is taken up from the soil. When transpiration decreases after dawn (Fig. 2.5), water continues to flow from the soil into the roots; water then flows mainly into storage in the stem and the leaves (Fig. 3.12). As drought begins, less water tends to be transpired each day because

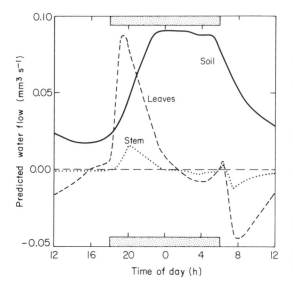

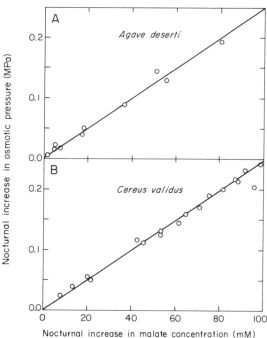

Figure 3.12. Predicted water flow in the transpiration stream of a well-watered plant of *A. deserti* whose resistances and capacitances are presented in Figure 3.11. Water flow is from the soil (———), stem capacitance (· · · · ·), or leaf capacitance (– – – – –). Flow is positive when coming out of a compartment. Modified from Calkin and Nobel (1986), using the simulation program SPICE (Nagel, 1975) to solve the electrical circuit.

Figure 3.13. Correlations between nocturnal increases in malate concentration and in osmotic pressure for **(A)** *Agave deserti* and **(B)** *Cereus validus*. The solid lines represent predictions by the van't Hoff relation ($\pi = cRT$, Eq. 3.2) at 25°C. Data (open circles) for *A. deserti* were obtained by assuming that the changes in the measured titratable H^+ concentration were twice those of malate concentration and are adapted from Smith, Schulte, and Nobel (1987). Data for *C. validus* are from Lüttge and Nobel (1984).

the stomata do not open fully; most of the net loss of water from storage in the leaves during the early part of the night is replaced later in the night by water from the soil (Calkin and Nobel, 1986).

Similar daily patterns in the direction of water movement from the soil to the shoot, to and from storage in the shoot, and to the atmosphere are indicated by simulations for *Ferocactus acanthodes* (Hunt and Nobel, 1987c). Its larger capacitance than that of *A. deserti* increases the time lag between peak transpiration and the subsequent peak flow into the roots from the soil to 12 h. In the field, *F. acanthodes* would never reach steady state because of the capacitance effects associated with its massive stem (Fig. 1.2A). In particular, its τ^{stem} (Eq. 3.10) is 19 h, so changes in environmental conditions, and hence in stomatal opening as part of the day/night cycle, occur faster than a steady state can be achieved (Hunt and Nobel, 1987c).

Osmotic effects

We have already indicated that nocturnal CO_2 uptake by agaves and cacti is accompanied by the accumulation of malate in the vacuoles of the chlorenchyma cells (Fig. 2.13). By the van't Hoff relation ($\pi = cRT$, Eq. 3.2), we expect that as the concentration of malate increases the osmotic pressure will increase. Indeed, this is the case for *Agave deserti* (Fig. 3.13A) and *Cereus validus* (Fig. 3.13B; note the close agreement in Fig. 3.13 between measured π and π predicted from the van't Hoff relation). Because the value of π at dusk was about 0.5 MPa for both *A. deserti* and *C. validus*, the nocturnal increases in malate typically represent a 40% increase in osmotic pressure (Lüttge and Nobel, 1984; Smith, Schulte, and Nobel, 1987). Such a large increase in π

can have a major effect on the shoot water potential or its hydrostatic pressure ($\Psi = P - \pi$, Eq. 3.1), which in turn means that the daily malate oscillations could have a substantial effect on water movement. For instance, an increase in shoot osmotic pressure during the night in CAM plants could decrease shoot water potential, thereby favoring water uptake from the soil.

To examine the influence of changes in osmotic pressure in the chlorenchyma on water fluxes for *A. deserti*, we must refine the electrical circuit model (Fig. 3.11) from an organ level to a tissue level (Fig. 3.14). In particular, no daily changes in malate or π are expected for the water-storage parenchyma, which constitutes 64% of the leaf by volume. The chlorenchyma is represented by a 1-mm-thick outer layer, where 68% of the nocturnal acid accumulation per unit leaf area occurs, and a 2-mm-thick inner layer, where the rest of the nocturnal acid accumulation takes place (Fig. 3.14). Also, the electrical circuit can be modified to represent the daily cycling of osmotic pressure by voltage sources in series with the storage capacitors (Smith et al., 1987). Simulations for *A. deserti* indicate that as π in the chlorenchyma increases during the night, the turgor pressure increases to nearly the same extent, in agreement with measurements of *P* for *C. validus* (Lüttge and Nobel, 1984). For the first hour of the night, one-third of the water transpired by the leaf comes from the outer chlorenchyma, but the net water loss from this tissue soon ceases, and indeed it is gradually recharged with water during the second half of the night when both its *P* and its π increase. Thus, the acid accumulation during the night for agaves and cacti can have a major influence on cell turgor, which in turn has many influences on metabolism and gas exchange (Schulze, 1986). Understanding the consequences of such *P* and π changes is an important area for future research on the water relations of CAM plants.

Temporal variations in water availability

The various physiological responses of agaves and cacti to the availability of water are realized over a wide range of time scales. We have already discussed the induction of

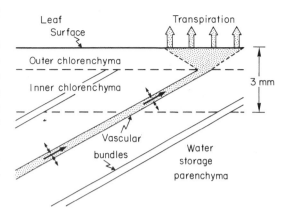

Figure 3.14. Schematic illustration of the leaf anatomy of *Agave deserti* used for modeling its water relations. Each vascular bundle is represented as supplying a particular part of the leaf surface (Smith and Nobel, 1986). The vascular bundles do not extend across the chlorenchyma, so water in the transpiration stream apparently follows a nonxylary pathway across the outer chlorenchyma. Most of the leaf water is stored in the water-storage parenchyma, which represents 81% of the leaf capacitance. Modified from Smith et al. (1987).

rain roots that occurs a few hours to a few days after rainfall, and in the following section we will discuss seedling survival, which requires endurance of droughts that last many months. Here we focus on responses of adult plants to water status over both the short term and the long term.

SHORT-TERM STOMATAL RESPONSES

Stomata of agaves and cacti can open during the day, as occurs for C_3 and C_4 plants, and can open at night, which characterizes the CAM mode. Daytime opening can predominate for seedlings of *Agave deserti* (Fig. 2.6) and for the leaves of cacti (Figs. 2.7 and 2.10); however, most stomatal opening for agaves and cacti occurs at night. The anatomy of the guard cells is apparently similar for C_3, C_4, and CAM plants (Faraday, Thomson, and Platt-Aloia, 1982), and the guard cells may also respond to the same physiological factors.

There are two principles that we would expect to be relevant to the control of stomatal opening (Jarvis and Mansfield, 1981; Zeiger, 1983). First, the major function of the chlorenchyma is the fixation of CO_2 into carbo-

Table 3.3. *Changes leading to nocturnal stomatal closure for agaves and cacti*

Change	Species	Reference
Increased ΔN_{wv} (at constant temperature)	*Opuntia humifusa*	Conde and Kramer (1975)
	O. stricta	Osmond et al. (1979a)
Increased ΔN_{wv} induced by increases in tissue temperature	*Agave deserti*	Nobel and Hartsock (1979)
	Carnegiea gigantea	Nobel and Hartsock (1981)
	Ferocactus acanthodes	Nobel (1977b)
Raised internal CO_2 level	*Agave americana*	Neales (1970)
	A. deserti	
	Opuntia ficus-indica	Cockburn et al. (1979)

hydrates and other products. Thus, if the CO_2 concentration in the intercellular air spaces in the chlorenchyma were to become lower than some specific threshold level, such as the concentration in the ambient air surrounding the shoot, it would be physiologically advantageous for stomata to open and allow CO_2 to diffuse into the shoot. The internal CO_2 level could be lowered by light-mediated C_3 photosynthesis during the day as well as by binding of CO_2 by PEP carboxylase at night. Second, the opening of the stomata leads to an inevitable water loss from the plant. If such a loss were to be "excessive," it would be advantageous to initiate some stomatal closure. Thus, partial stomatal closure would be a useful response to drastic lowering of the water vapor concentration or mole fraction in the ambient air, because such an atmospheric change by itself would increase transpiration ($J_{wv} = g_{wv} \, \Delta c_{wv} = g'_{wv} \, \Delta N_{wv}$, Eq. 2.6). Similarly, raising the tissue water vapor level by increasing the temperature could induce stomatal closure, because the saturation level of water vapor in the intercellular air spaces increases nearly exponentially with temperature (Fig. 2.16), again creating a greater driving force for water loss.

Table 3.3 summarizes the stomatal response of agaves and cacti to changes in CO_2 and water vapor levels. The two principles outlined above are supported: (1) stomata tend to close when the internal CO_2 level increases, which could be caused by decarboxylation of malate during the daytime (Fig. 2.13); and (2) stomata tend to close when the drop in water vapor mole fraction from the intercellular air spaces to the ambient air increases (Table 3.3).

Although the physiological consequences of such stomatal movements are clear, understanding exactly how stomata sense levels of CO_2 and water vapor is currently an area of active research.

INTERMEDIATE-TIME RESPONSES

By "intermediate-time" responses we mean changes that occur over a period of weeks. Such responses can be induced in the laboratory – for instance, by withholding water for one week from leafy cacti or by daily watering of *Agave deserti* (a daily watering protocol designed for a desert fern was mistakenly applied to *A. deserti*, with interesting results). We will also consider the drought responses of the stems of cacti and the leaves of agaves.

Overwatering an agave

After ten weeks of daily watering, a mature plant of *A. deserti* can shift from the typical CAM mode of stomatal opening and net CO_2 uptake to a C_3 mode with predominantly daytime CO_2 uptake (Fig. 3.15). Also, the nocturnal increase in titratable acidity changes from 400 mmol m^{-2} in the CAM pattern to less than 40 mmol m^{-2} in the C_3 mode (Hartsock and Nobel, 1976). Thus, leaf succulent agaves may have considerable flexibility in their CO_2 uptake pathway.

The shift from a CAM to a C_3 pattern (Fig. 3.15) occurred under the long nights characteristic of winter. During the summer the higher PAR levels and the longer daytime should lead to substantially more net CO_2 uptake in the C_3 mode than when the shift occurs during the winter. As we indicated at the end

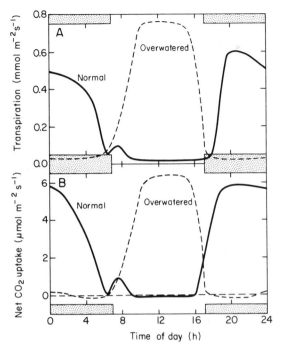

Figure 3.15. Influence of water status on (**A**) the daily pattern of transpiration and (**B**) net CO_2 uptake for *Agave deserti*. "Normal" refers to a plant receiving daily watering for two weeks, and "overwatered" refers to the same plant after an additional ten weeks of daily watering. Modified from Hartsock and Nobel (1976).

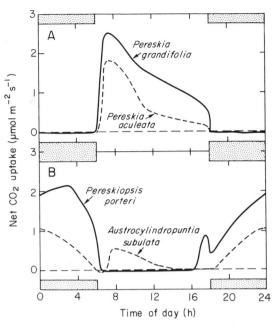

Figure 3.16. CO_2 exchange over a 24-h period by leaves on plants droughted seven days for species in (**A**) Subfamily Péreskioideae and (**B**) Subfamily Opuntioideae. Data are adapted from Nobel and Hartsock (1987) and unpublished observations.

of the previous chapter, predominantly daytime stomatal opening will lead to a lower water-use efficiency and thus would be expected to occur only under conditions when water is not limiting. Interestingly, overwatering cacti such as *Opuntia basilaris* does not lead to a shift to the C_3 mode with its daytime net CO_2 uptake (Hanscom and Ting, 1977, 1978a).

Droughting leafy cacti

Net CO_2 uptake by leafy cacti is primarily by the leaves and during the daytime for species in subfamilies Pereskioideae and Opuntioideae. But what happens to them during drought? The already small net CO_2 uptake by the stems of leafy cacti decreases during drought, both in absolute amount and as a fraction of total plant uptake (Nobel and Hartsock, 1987). Therefore, we will focus on net CO_2 uptake by the leaves.

For *Pereskia aculeata* and *P. grandifolia*

(Subfamily Pereskioideae), a drought of 7 days decreases net CO_2 uptake by the leaves over a 24-h period by 75–80% (Fig. 3.16A compared with Fig. 2.7A,B), but net CO_2 uptake still occurs in the daytime. Similarly, 10 days of drought for these two species led to stomatal opening only in the early part of the day (Rayder and Ting, 1981). For *Austrocylindropuntia subulata* and *Pereskiopsis porteri*, whose leaves are more succulent than the leaves of the two species of *Pereskia* (Figs. 2.9 and 1.28A compared with Figs. 1.27 and 1.35), a drought of 7 days reduces net CO_2 uptake over a 24-h period by 40% to 60% (Fig. 3.16B compared with Fig. 2.10A,B). However, in the case of these members of Subfamily Opuntioideae, the drought shifts the net CO_2 uptake pattern of the leaves from primarily daytime to primarily nighttime. Although the underlying biochemical events have not been investigated, the drought is apparently inducing the plants to shift from the conventional C_3 pattern to the water-conserving CAM pattern. Indeed, the water-use efficiency (Eq. 2.11) over the 24-h period increases fivefold to sixfold in response to the 7 days of drought (from 0.003 to

0.015 CO_2/H_2O for *A. subulata* and from 0.002 to 0.012 CO_2/H_2O for *P. porteri*).

Droughting plants in CAM mode

We will next consider what happens to gas exchange over 24-h periods when mature agaves and cacti having predominantly nocturnal CO_2 uptake are exposed to drought for a period of weeks. We might expect that the small amount of stomatal opening and net CO_2 uptake during the daytime observed for leaves of agaves (Figs. 2.2, 2.4, and 2.5) and for stems of cacti (Figs. 2.11 and 2.12), with its inherently lower water-use efficiency (WUE), might be more susceptible to drought than nocturnal gas exchange. Also, tissue temperatures tend to decrease through the night in the field, so ΔN_{wv} from the tissue to the air would decrease, and hence WUE would tend to increase for a given degree of stomatal opening (Eq. 2.11). Thus, we might expect maximal stomatal opening to occur later in the night as drought proceeds, at least under field conditions.

For the three species of agaves examined, drought reduces both daytime and nighttime net CO_2 uptake, but the greater fractional decrease occurs for the daytime uptake (Fig. 3.17). Similar results occur for *Agave americana* (Neales, 1975). The length of drought required to halve net CO_2 uptake over a 24-h period is about 11 days for *A. deserti* and 8 days for both *A. fourcroydes* and *A. lechuguilla*. Maximal nocturnal stomatal opening tends to occur later in the night as drought proceeds; for *A. deserti* droughted for 9 days (Fig. 3.17A), most of the transpiration occurs in the latter half of the night.

Drought has similar influences on the pattern of gas exchange by cacti. Daytime net CO_2 uptake for *Opuntia auriantiaca* is also more sensitive to drought than is nighttime net CO_2 uptake; total uptake over a 24-h period is halved after 6 days of drought, such a short time reflecting its relatively thin cladodes, which have a low V/A (Whiting and Campbell, 1984). Under field conditions, maximal stomatal opening for *Ferocactus acanthodes* occurs later in the night as drought progresses. Under wet conditions, g_{wv} (Eq. 2.6) is maximal 2–3 h into the night; after 10 and 50 days of drought, g_{wv} is maximal after about 4 h and

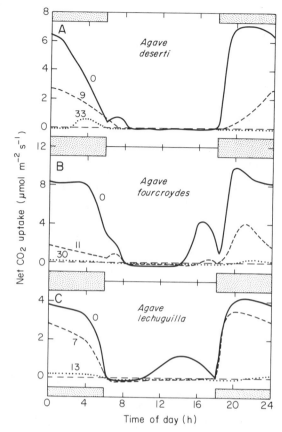

Figure 3.17. CO_2 exchange over a 24-h period by leaves of agaves after various periods of drought, indicated in days next to the curves. Data for (**A**) *Agave deserti* are adapted from Nobel (1984a), for (**B**) *A. fourcroydes* from Nobel (1985a), and for (**C**) *A. lechuguilla* from Nobel and Quero (1986). Gas exchange was measured in the laboratory under nearly saturating PAR, and the night temperature was constant at about the mean annual value of 15°C for *A. deserti*, 20°C for *A. fourcroydes*, and 9°C for *A. lechuguilla*.

8 h into the night, respectively (Nobel, 1977b). Similarly, the greatest degree of stomatal opening and the maximal transpiration occurs later in the night for *Opuntia ficus-indica* as drought increases from 0 to 20 to 35 days under field conditions; however, in this case the stomatal opening at the beginning of the day is not substantially affected (Acevedo, Badilla, and Nobel, 1983). For *O. polyacantha*, nighttime net CO_2 uptake is more sensitive to drought than is the small amount of net CO_2 uptake occurring at the end of the afternoon (Gerwick and Williams, 1978). For *O. basilaris*, watering does not cause a shift from

CAM to C_3, but drought causes maximal g_{wv} and maximal transpiration to occur later in the night (Hanscom and Ting, 1977, 1978a, 1978b), which is also the case for *O. stricta* (Osmond et al., 1979b). Therefore, maximizing WUE by reducing reliance on daytime net CO_2 uptake and delaying nocturnal stomatal opening as drought proceeds generally, but not always, occurs for agaves and cacti.

We can also indicate how drought length affects the net CO_2 uptake over 24-h periods for cacti, anticipating our introduction of a "water index" in Chapter 7. For *F. acanthodes*, two weeks of drought are necessary before net CO_2 uptake begins to decrease, and nearly one month is required to halve it (Fig. 3.18). For *O. ficus-indica*, drought-induced decreases become apparent in about one week, and three weeks are required for halving the net CO_2 uptake over 24-h periods (Fig. 3.18). The shorter drought periods for halving net CO_2 uptake by agaves (Fig. 3.17) mainly reflect their lower V/A compared with the cacti considered here. Even though net CO_2 uptake over a 24-h period is zero after about 50 days of drought for *O. ficus-indica* (Fig. 3.18), nocturnal acidity increases can still be 40% of maximal (Acevedo et al., 1983). Nocturnal acidity increases that occur during drought in the absence of net CO_2 uptake represent incorporation of CO_2 released within the tissue by respiration, as was first clearly recognized for *O. basilaris* (Szarek et al., 1973).

Watering droughted plants

Now that we have presented the responses of agaves and cacti to drought, we will briefly comment on the responses of droughted plants to watering (see also Figs. 1.17 and 1.18). As we have already indicated, rainfall or supplemental watering can trigger stomatal opening at any time of the year, the details depending on the rehydration of existing roots and the induction of rain roots (Fig. 3.6), among other factors such as temperature. After *Agave fourcroydes* is droughted for 30 days, rewatering causes a return to 48% of the maximal net CO_2 uptake rate in 3 days and 91% in 7 days (Nobel, 1985a). Watering droughted *Opuntia decumbens* causes transpiration to increase fivefold and net CO_2 uptake to double within 1 day (Kausch, 1965). For *O. auran-*

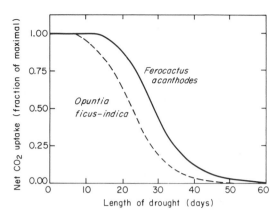

Figure 3.18. Influence of drought duration on net CO_2 uptake over 24-h periods for two species of cacti. Data are from Nobel (1977b), Acevedo et al. (1983), Nobel and Hartsock (1984), and unpublished observations.

tiaca, 77% of the maximal net CO_2 uptake over a 24-h period occurs 2 days after watering a droughted plant (Whiting and Campbell, 1984). For *O. basilaris*, over 90% of the maximal nocturnal acidity increase occurs 5 days after watering a droughted plant (Szarek and Ting, 1975a). For *O. stricta*, nearly maximal nocturnal acidity increases occur 3 days after watering (Osmond et al., 1979b). In summary, return to nearly maximal gas exchange occurs within one week after droughted agaves and cacti receive appreciable water.

LONG-TERM RESPONSES

"Long-term" responses may occur over seasonal, yearly, or even longer time periods. The WUE (Eq. 2.11) over an entire year was 0.016 CO_2/H_2O for *Agave deserti* (Nobel, 1976), 0.006 CO_2/H_2O for *Ferocactus acanthodes* (Nobel, 1977b), and 0.005 CO_2/H_2O for *Opuntia basilaris* (Szarek and Ting, 1974b). WUE over even longer periods can affect the type of plants occurring in a particular locality. For example, CAM plants like *A. lechuguilla* tend to dominate arid, low-elevation habitats in the Chihuahuan Desert, which would be expected from considerations of water conservation, but the sympatric CAM plant, *Opuntia phaeacantha*, dominates at high elevations that undoubtedly represent different habitats, indicating that other long-term factors are also involved (Eickmeier, 1978, 1979).

Many studies have indicated that values

of $\delta^{13}C$ (Eq. 2.5) for agaves and cacti do not change appreciably on a seasonal basis (Eickmeier and Bender, 1976; Sutton, Ting, and Troughton, 1976; Szarek and Troughton, 1976; Hanscom and Ting, 1978a). The simplest interpretation is that in the field these plants do not appreciably shift from the CAM to the C_3 mode, or vice versa. Also, the shift in $\delta^{13}C$ is not substantial upon watering *Opuntia basilaris* or *O. littoralis* or in depriving *O. acanthocarpa*, *O. basilaris*, or *O. bigelovii* of water for up to three years (Smith and Madhaven, 1982). Although major shifts in $\delta^{13}C$ appear not to be the case for mature agaves under field conditions and for nonleafy cacti, agave seedlings and leafy cacti in Subfamily Opuntioideae may well have variable $\delta^{13}C$'s (see Figs. 2.6 and 3.16B).

For *Opuntia humifusa* in Iowa, net CO_2 uptake under field conditions occurs from April through September, a period that is generally wet (Koch and Kennedy, 1980). Even though it is also wet during the winter, no net CO_2 uptake occurs then, presumably due to low temperatures (Chapter 4). Similarly, for *O. erinacea* in eastern Washington, nocturnal increases in acidity could occur throughout the year when water is available, again except during the winter (Littlejohn and Williams, 1983).

Agaves and cacti supposedly occur where there are two rainy periods per year (Walter and Stadelmann, 1974) or where the annual precipitation is low (100 to 200 mm) but annually regular (Ellenberg, 1981). Yet agaves occur in regions with a mean annual rainfall below 100 mm and even in regions with rainless years (Gentry, 1972). Species of *Copiapoa* are the dominant plants in regions of northern Chile, where the mean annual precipitation is only 25 mm and a few years without rainfall are not uncommon (Gulmon et al., 1979) – indeed, *C. cinerea* in Figure 1.10B was photographed after a period of six years without rainfall. Such long-term responses are possible because of tissue tolerances of water loss, which are undeniably high for agaves and cacti. For instance, *Carnegiea gigantea*, *F. acanthodes*, and *O. basilaris* can lose 81% of their stem water and still survive (Barcikowski and Nobel, 1984) and *Coryphantha vivipara* (Fig. 1.19) can lose up to 91% of its stem water and survive (Nobel, 1981a). Also, as just indicated, *Opuntia acanthocarpa*, *O. basilaris*,

and *O. bigelovii* can survive up to three years without water (Szarek and Ting, 1975b; Smith and Madhaven, 1982).

The preceding examples of long-term effects of water availability can involve other environmental factors, such as temperature and radiation, and sometimes morphological studies can be useful for sorting out the effects. For instance, much-branched, narrow-stemmed species such as *Opuntia leptocaulis* and *O. ramosissima* with a low V/A tend to have deep roots extending to 1.5 m below the soil surface and are restricted to flat sandy areas; but wide-stemmed, mostly unbranched species with a high V/A, such as the barrel cactus *F. acanthodes* and the hedgehog cactus *Echinocereus engelmannii*, have shallow root systems that are above 0.2 m and occur in adjacent, steeper, rockier sites (Cody, 1986a). This pairing between root morphology and the shoot volume per unit surface area may be related to the water availability in different habitats.

Because soil water availability generally controls plant growth in deserts (Noy-Meir, 1973), competition for soil water is often suspected for neighboring plants. Specifically, many studies have indicated that cacti compete with conspecifics or other species for water (Pieper, Rea, and Fraser, 1974; Rundel and Mahu, 1976; Yeaton, Travis, and Gilinsky, 1977; Yeaton, 1978; Vandermeer, 1980; McAuliffe, 1984a; McAuliffe and Janzen, 1986). Root distribution with depth has seldom been measured, so the degree of root overlap with neighboring plants is generally not known, and Ψ^{plant} with and without potential competitors has essentially never been measured, although such measurements are necessary for quantitatively assessing competition for water. Also, the many effects of nurse plants (Fig. 1.1) have not been fully analyzed (but see Chapters 4 and 5). We will reconsider long-term responses in the context of annual productivity of agaves and cacti and the interaction of environmental factors in Chapter 7.

Seedling establishment

The key to the presence of a particular plant species in a specific locality is often seedling establishment, and the key to seedling establishment is often water relations, especially for desert succulents. We will leave questions of seed germination to the next chapter. Here we will focus on the water relations of germinated seeds with respect to the establishment of the seedlings under natural conditions.

AGAVES

As indicated in the first chapter, seedlings of desert succulents are often extremely difficult to find in the field. This is due both to the small size of seedlings and to the rarity of their establishment.

Seedlings of *Agave deserti* can withstand desiccation leading to plant water potentials of −1.6 MPa, at which stage about 70% of the tissue water is lost (Jordan and Nobel, 1979). As the seedlings grow in size, they can tolerate longer periods of drought (Fig. 3.19). For instance, the tolerable period of drought increases from 2 weeks for a 4-week-old seedling to 12 weeks for a 15-week-old seedling. This change primarily reflects increases in V/A, which is 0.07 cm for a 10-day-old seedling, 0.18 cm at 120 days, 0.29 cm at 445 days (Jordan and Nobel, 1979), and about 0.91 cm for leaves of an adult (Nobel, 1976). Therefore, as the seedlings become older, they have more stored water per unit surface area, which helps them tolerate longer droughts.

Considering both seed availability and temperature requirements, germination of seeds of *A. deserti* would be expected to occur during a summer wet period. Using the observed rainfall events, the duration of the summer wet period and the subsequent drought can be predicted for any particular year (see Fig. 3.3). For the seventeen-year period from 1961 to 1977, only 1967 was deemed to have a sufficiently long wet period followed by a sufficiently short drought to permit establishment of *A. deserti* at the field site involved (Fig. 3.19). At the end of this period, an area of the field site containing over 1,000 rosettes of *A. deserti* was carefully examined and the six smallest plants were identified (Jordan and Nobel, 1979). By monitoring the annual growth of these "seedlings" (Fig. 3.20), such as by counting the number of new leaves unfolding, and by noting the dead leaves and leaf scars at the stem base, one can estimate the age of the plants. These estimates of plant age considered the different rainfall patterns in each year (Fig. 3.3), which affects the annual

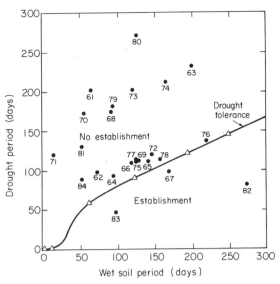

Figure 3.19. Relation between yearly variations in water availability and seedling establishment for *Agave deserti*. To determine the lengths of drought that could be tolerated, seedlings were raised in the laboratory and then subjected to drought (here, $\Psi^{soil} < -1.6$ MPa, because of the high π^{plant} occurring during desiccation) for various periods ($\triangle$). Field measurements at a site in the northwestern Sonoran Desert (Fig. 1.1) plus simulations (Young and Nobel, 1986) were used to determine the longest wet period (here, $\Psi^{soil} \geq -1.6$ MPa) beginning in a particular year ($\bullet$) and the subsequent period of continuous drought ($\Psi^{soil} < -1.6$ MPa) for 1961 to 1984 (wet and/or drought periods could extend into the next calendar year). Used by permission from Nobel (1985c).

Figure 3.20. A thirteen-year-old "seedling" of *Agave deserti*. The plant, which was 16 cm tall (note the 15-cm ruler in the picture) and was estimated to have become established in 1967, was photographed in 1980 at the same field site as for Figure 1.1. The nurse plant is again *Hilaria rigida*, as commonly occurs for *A. deserti* at this site (Jordan and Nobel, 1979).

growth, and effects of the nurse plant (Fig. 3.20), which alter the radiation environment of the seedling, again influencing growth. The mean estimated year of germination for the six smallest rosettes is 1967 (Jordan and Nobel, 1979). This is in striking agreement with the prediction that 1967 is the only year suitable for establishment of *A. deserti* from 1961 to 1977 based on simulations of field Ψ^{soil} and laboratory measurements of drought tolerances (Fig. 3.19).

The above example underscores the importance of the first year of life for the establishment of *A. deserti* from seed. Since 1977, when the detailed study ended, two additional years, 1982 and 1983, have been deemed suitable for establishment (Fig. 3.19). Indeed, seedlings that became established in 1982 can be identified in the field (Fig. 1.1).

Seeds of *A. deserti* germinate and seedlings become established during wet periods. Initially, net CO_2 uptake is primarily during the daytime, as we have already discussed (see Fig. 2.6). For 25-day-old seedlings, the WUE (Eq. 2.11) is 0.0011 CO_2/H_2O, which increases to 0.0029 for 100-day-old seedlings and to 0.0061 for 400-day-old seedlings. Thus, nearly six times as much CO_2 is fixed into organic products per unit of water lost as seedling age increases from 25 to 400 days, indicating the water-saving consequences of shifting from primarily daytime to primarily nighttime net CO_2 uptake (Fig. 2.6). Such a C_3/CAM shift in their natural habitat for seedlings of *A. deserti* and other agaves would be advantageous in environments that on average become more arid shortly after the wet periods favoring seedling establishment.

CACTI

Most of the principles for seedling establishment of agaves also apply to seedling establishment of cacti. The ratio *V/A* is again initially very small and it remains low, because initial growth is slow, especially in the field. For instance, one-year-old seedlings of *Carnegiea gigantea* may be only 0.3 cm tall in the field (Steenbergh and Lowe, 1969), although after one year they can be 1.3 cm tall in a lath house (Turner et al., 1966) and 2.5 cm tall when grown in a commercial nursery (Despain, 1974). In a wetter natural habitat, *Opun-*

tia humifusa can also be 2.5 cm tall after one year (Baskin and Baskin, 1977), whereas *Echinocactus platyacanthus* requires two years to reach this height (Trujillo Argueta, 1982). The significance of less water stored in smaller stems has also been recognized for *Copiapoa*, where seedling establishment in a region with an average annual precipitation of only 25 mm is extremely difficult (Gulmon et al., 1979). Because they already exhibit the CAM pattern, even under wet conditions, the water-conserving shift from C_3 to CAM does not occur for 2-cm-tall seedlings of *C. gigantea* (Despain, Bliss, and Boyer, 1970) or *Ferocactus acanthodes* (Jordan and Nobel, 1981). On the other hand, two-week-old cladodes developing on established cladodes of *Opuntia ficus-indica* have predominantly daytime stomatal opening, which switches to the CAM pattern by two months of age (Acevedo et al., 1983). Most of the research on seedling establishment for cacti is apparently on *C. gigantea* (Fig. 1.12A) and *F. acanthodes* (Fig. 1.2A), so we consider these two species here.

Carnegiea gigantea

Fully hydrated, seven-month-old seedlings of *C. gigantea* can lose 60% of their water content and still survive (MacDougal, 1912). Such desiccation tolerance is crucial for the establishment of this species. As for *Agave deserti* (Jordan and Nobel, 1979), most successful seedlings of *C. gigantea* are associated with nurse plants, which affect water relations, temperature, and radiation (Steenbergh and Lowe, 1969). In addition to these environmental factors, seed and seedling predation by herbivores, such as rodents and rabbits, also affect seedling establishment (Niering, Whittaker, and Lowe, 1963; Turner, Alcorn, and Olin, 1969).

Based on the response of net CO_2 uptake in the laboratory to various environmental factors and the annual height increments observed for *C. gigantea* in the field, growth in height can be predicted for a series of years (Fig. 3.21). A plant 0.1 m tall is estimated to be 8 years old and one 0.3 m tall about 13 years old for a particular site in northern Arizona. By comparing the heights measured at the site with the expected heights for various ages (Fig. 3.21), the approximate years for seedling

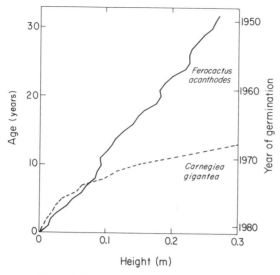

Figure 3.21. Estimated age and year of germination for *Carnegiea gigantea* at Wikieup, Arizona, and for *Ferocactus acanthodes* near Palm Desert, California, as a function of height in 1981. Growth was estimated from field measurements of height changes (Hastings and Alcorn, 1961; Steenbergh and Lowe, 1977; Jordan and Nobel, 1979) together with laboratory measurements of net CO_2 uptake under various conditions (Jordan and Nobel, 1981; Nobel and Hartsock, 1981). The negative growth occasionally noted for *F. acanthodes* represents stem shrinkage during extended drought. Used by permission from Jordan and Nobel (1982). © 1982 by the University of Chicago. All rights reserved.

where less than 10% of the years are suitable (Jordan and Nobel, 1982). The western distributional boundary of this species apparently reflects the local precipitation patterns and water relations of seedling establishment (Shreve, 1911; Brum, 1973; Steenbergh and Lowe, 1977; Jordan and Nobel, 1982), underscoring the critical importance of this phase in the environmental biology of *C. gigantea*.

Ferocactus acanthodes

As for *C. gigantea*, the growth of *F. acanthodes* has been predicted at a specific site for the years 1950 to 1981. Even though the curve shapes are different, the mean rate of height increase over nearly the first ten years of life is similar for the two species (Fig. 3.21); after ten years, annual height increments for *C. gigantea* become much greater than for *F. acanthodes*, because the former changes from *globose* (nearly spherical) to club-shaped, whereas *F. acanthodes* continues to be globose for essentially thirty years, its height and diameter being quite similar throughout this period.

For a given surface area, a sphere can store more water than any other shape, indicating that a globose form has advantages in terms of water storage. However, for a given volume of stored water, a sphere has the least surface area across which net CO_2 uptake can occur; therefore, the survival demands based on water storage can be in conflict with the growth demands based on CO_2 uptake. For a sphere, V/A equals $\frac{4}{3}\pi r^3/4\pi r^2$ or $r/3$, where r is the radius. Thus, as *F. acanthodes* remains fairly globose over its initial thirty years, the water-storage tissue per unit of surface area across which transpiration can occur increases approximately proportionally to the radius. For a small seedling seven days after germination, V/A is about 0.02 cm, which under wet conditions in the laboratory increases to 0.06 cm at two months and 0.17 cm after one year, when the stem is about 1.5 cm tall (surface irregularities caused by ribbing are included in the estimates of V/A). Such young seedlings can lose 84% of their stored water and still survive, which affects the length of droughts they can tolerate (Jordan and Nobel, 1981).

To determine possible establishment years, let us next consider the stem heights

establishment can be deduced. For *C. gigantea*, the measured heights closely match those expected from the years predicted to be suitable for establishment (Jordan and Nobel, 1982). In particular, using an analysis similar to that for *A. deserti* (Fig. 3.19), one can predict that nine years from 1950 to 1981 are suitable for seedling establishment of *C. gigantea* near Wikieup in northern Arizona, whereas only one year during this period (1952) is deemed to be suitable near the Colorado River in southeastern California. Indeed, the population of *C. gigantea* near Parker Dam, California, seems to be declining in numbers, and no seedling establishment is apparent from 1953 to 1981 (Jordan and Nobel, 1982).

At a series of sites along the Colorado River in Arizona and California, *C. gigantea* occurs where 10% or more of the years are suitable for seedling establishment but not observed for *F. acanthodes* at a site in the northwestern Sonoran Desert (Figs. 1.1, 1.2, and 3.20). The discrete peaks in its height distribution plot (stem height versus number of individuals in a particular height interval) suggest that seedling establishment is intermittent, occurring only in favorable years. Individuals with heights slightly greater or less than those at a peak most likely reflect different growth rates for seedlings established in a favorable year. Because the annual increase in stem height for *F. acanthodes* in the field is approximately constant at 9 mm per year (Fig. 3.21), a height distribution plot can be readily converted to an age distribution plot, which can be related to years favorable for seedling establishment (Fig. 3.22). Indeed, the years estimated for germination during the period 1950 to 1981 based on observed plant height (Fig. 3.22A) coincide fairly well with the years in which tolerable drought exceeds actual drought, the latter indicating when seedling establishment would be expected (Fig. 3.22B).

For a series of sites within the distributional boundaries of *F. acanthodes* in southern California, this species occurs in regions where at least 10% of the years are suitable for seedling establishment and it is absent in areas where less than 10% of the years are suitable (Jordan and Nobel, 1982). Although 10% should not be construed as a magic number, populations of *F. acanthodes* (and *C. gigantea*) clearly do not become established where seedling establishment is very infrequent. Mountain ranges, which influence precipitation patterns by increasing rainfall on the windward side (the direction approached by the prevailing wind) and by decreasing it on the leeward side (a "rain shadow" effect), affect the locations where *F. acanthodes* can be found. In particular, *F. acanthodes* occurs at lower elevations on the windward side of a mountain than on the leeward side due to the elevational effects on precipitation for cases in which storms tend to come from a particular direction. Just as for *C. gigantea*, and presumably other desert succulents, the water relations governing seedling establishment of *F. acanthodes* can be the most important water relations factor determining its distribution (temperature effects will be treated in the next chapter).

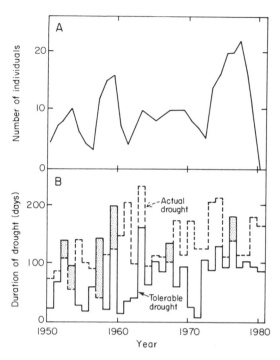

Figure 3.22. Comparison of the number of individuals of *Ferocactus acanthodes* estimated to have germinated in a particular year (**A**) with length of droughts faced by seedlings germinating in the summer wet period (**B**). Individual age classes are based on heights measured in the field and the estimated growth rate of *F. acanthodes* (Fig. 3.21). Lengths of tolerable drought are based on the desiccation tolerance of *F. acanthodes* determined in the laboratory, and the lengths of actual drought are based on predicted Ψ^{soil}, as for *Agave deserti* (Fig. 3.19). Stippled areas indicate periods when tolerable drought exceeds actual drought and thus when establishment would be expected. Modified from Jordan and Nobel (1982).

Reproduction

The water relations of agaves and cacti affect reproduction in many ways. For instance, drought reduces the number of flower buds for *Schlumbergera truncata* (Heins, Armitage, and Carlson, 1981). *Carnegiea gigantea* tends to produce fewer flower buds and fewer flowers in very dry years or at dry sites (Brum, 1973; Steenbergh and Lowe, 1977), although a decrease in Ψ^{stem} is supposed to favor flower bud formation in this species (Walter and Stadelmann, 1974). Young flower buds of *Opuntia ficus-indica* have mainly daytime stomatal opening, whereas gas exchange by fruit utilizes the CAM pattern, although CO_2 and

Figure 3.23. Inflorescence of *Agave deserti* emerging from a basal rosette of leaves in the early spring (photographed at Agave Hill near Palm Desert, California; see Fig. 1.1).

H_2O exchange with the environment by the fruit become very low after three months (Acevedo et al., 1983). In this section we discuss various aspects of the water relations of reproduction for *Agave deserti*, which has been studied in considerable detail, and then summarize the water costs for reproduction of certain cacti.

AGAVES

As we indicated in the first chapter, the massive inflorescence of *A. deserti* (Fig. 3.23) can have daily height increments of over 10 cm. Such growth requires considerable amounts of water, which as we shall see comes mainly from the leaves. Vegetative reproduction also occurs for *A. deserti*, and again the parent rosette plays an important role in the water relations of the daughter ramets (an individual member of a clone, here connected by a rhizome to the parent).

Inflorescence requirements

For the approximately five-month period from the emergence of an inflorescence (Fig. 3.23) to the production of viable seed, 18 kg

of water enters the inflorescence and lateral branches of *A. deserti* (Nobel, 1977a). During this period, 3.1 kg of water is stored in the stalk, 4.3 kg is transpired from its surface, and 10.8 kg goes into the lateral floral branches (Fig. 1.24B). Of the last, half is transpired for the two months that fruits are present, although because of the copious nectar produced, a fourfold higher daily water loss occurs during the flowering stage.

The water required by the inflorescence can be fully provided by the water stored in the leaves (Nobel, 1977a). This can be deduced from leaf morphological changes during the five-month flowering period, a time when water uptake from the soil is generally slight; in particular, the thickness at midleaf decreases from 4.1 to 1.4 cm, representing a net loss of 18 kg of water. Diversion of water from the leaves to the inflorescence has also been noted for *Agave americana* (MacCallum, 1908). If the emerging inflorescence is severed, either by an investigator or by the local bighorn sheep (*Ovis canadensis*) that often feed on the young inflorescences, the sizable decrease in leaf thickness does not occur. In these cases, the rosette does not die, and indeed inflorescences of agaves are often cut off to prolong the life of the plant or to divert to the stem base the dry matter that would have gone to the inflorescence (Fig. 1.4A).

Usually only the largest rosettes of *A. deserti* produce an inflorescence, presumably in direct response to a hormonal stimulus linked to plant age. Soil water potential also affects the annual percent of these monocarpic perennials that flower, which can vary fiftyfold from year to year (Nobel, 1987a). There is little correlation between the percent flowering in a particular year and the number of wet days in that year, the year before, or three years before. However, the percent flowering is positively correlated with the annual number of wet days occurring two years previously ($r^2 =$ 0.33, $P \leq 0.10$, meaning that such an observation accounts for 33% of the annual variation and has at most only a 10% probability of occurring by chance). Another factor, apparently biological, has even more of an impact – years of copious flowering tend to alternate with years of sparse flowering ($r^2 = 0.64$, $P \leq 0.05$). Taken together, the year-to-year alternations and the annual number of wet days two years before flowering account for 95% of the annual variation in percent flowering ($r^2 = 0.95$, $P \leq 0.01$; Nobel, 1987a). This suggests that differentiation leading to the inflorescence occurs about two years before the inflorescence of *A. deserti* appears, and rainfall patterns (Fig. 3.3) that favor an above-average percentage of flowering in a particular year may leave a lower-than-average number of the older rosettes available for flowering in the next year. We also note that based on the mean fraction of plants that flowered annually for the eight years considered (0.018), the average age at flowering is about 1/0.018 or 56 years for *A. deserti* (Nobel, 1987a).

Ramets

Even though *Agave deserti* produces inflorescences and the seed is viable, about 95% of its reproduction at the field site in the northwestern Sonoran Desert is vegetative by ramets (Nobel, 1977a). By severing the rhizome between parent rosette and daughter ramets, one can show that the ramets depend on the parent for survival, such dependence decreas-

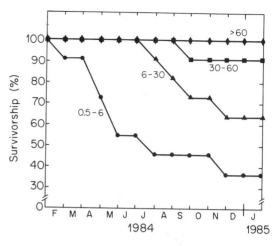

Figure 3.24. Survivorship of ramets of *Agave deserti* over a one-year period. At the beginning of February 1984, eleven ramets in each of various size classes (indicated as dry weight in grams next to the curves) were severed from their parents. Plants were defined as dead when less than 10% of their surface area was green. Used by permission from Raphael and Nobel (1986). © 1986 by The University of Chicago. All rights reserved.

ing with ramet size (Fig. 3.24). For instance, only 37% of the small, severed ramets (0.5 to 6 g in dry weight) survived for one year, whereas all the large, severed ramets (> 60 g) survived. Ramets apparently depend on the parent for water, because unsevered ramets have the same water potential as their parent but the water potential of severed ramets averages 0.26 MPa less after five months of drought (Raphael and Nobel, 1986).

The ability of ramets to utilize the water and also the carbohydrate reserves of their parents allows them to grow much faster than seedlings. Indeed, a thirteen-year-old seedling growing under a nurse plant (Fig. 3.20) has about the same dry weight as a three-year-old ramet of *A. deserti* (approximately 8 g; Nobel, 1984a; Raphael and Nobel, 1986). Other chemical signals may also move from parent to daughter ramet. For instance, even though inflorescences are chiefly produced by large rosettes, occasionally small rosettes will also produce inflorescences (Nobel, 1987a). In such cases, an intact rhizome occurs between the parent and the daughter ramet, which allows the passage of a hormone or set of chemical conditions involved in triggering flowering. Such movement along the rhizome can lead to the precocious formation of an inflorescence in the still-attached ramets, which are fiftyfold smaller than typical adult rosettes of *A. deserti* that flower (Nobel, 1987a).

CACTI

In contrast to *Agave deserti*, an individual cactus tends to flower every year once it reaches maturity, which can be at a height of 2.2 m and an age of 33 years for *Carnegiea gigantea* (Steenbergh and Lowe, 1977). During the single day of its opening, each flower of *C. gigantea* can secrete 0.5 g (Schmidt and Buchmann, 1986), about 5 g (McGregor et al., 1959), or up to 11 g (MacDougal and Spalding, 1910) of nectar. Considerable water is lost during the bud stage as well as during the fruit stage of *C. gigantea*, although transpiration per unit area by these floral structures has not been reported.

Transpiration has been measured for the buds, flowers, and fruits of *Ferocactus acanthodes*, so its total water cost for reproduction can be estimated (Nobel, 1977b). Stomata are present on the outer flower parts and the fruit, but they appear to be nonfunctional, because the water vapor conductance (Eq. 2.6) does not change appreciably from day to night. About 44 g of water is transpired for each reproductive structure during the three months that it is present; just over half of the water loss comes during the seven days of flowering, which have a twelvefold higher daily water loss rate than during the bud stage and sixteenfold higher than during the fruit stage (Nobel, 1977b). Summed for all reproductive structures, transpirational water loss and water stored in the reproductive structures represent about 4% of the water taken up from the soil during the year. Such a water requirement for reproduction represents about 6% of the stem water at the time of flowering, which for *F. acanthodes* generally occurs during the drought of late spring. Thus, although only a small fraction of the water stored in the stem or annual transpiration is diverted to the reproductive structures for *F. acanthodes*, such diversion comes at a time of the year when there is generally no water uptake from the soil (Nobel, 1977b). Much remains to be learned about the water costs of reproduction for agaves and cacti.

4 Temperature

We have already discussed several phenomena that are influenced by temperature, the most commonly measured environmental variable. For example, CO_2 and H_2O diffusion during gas exchange (Chapter 2) is based on the random thermal motion of molecules. Cells are mostly water (Chapter 3), which has important thermal properties: a high *heat of vaporization* (energy necessary to cause unit amount of a liquid to become a gas), a high *specific heat* on a mass basis (energy needed to raise the temperature of unit mass by 1°C), and a high *thermal conductivity* for a liquid (energy flux density per unit temperature gradient). The properties of water dictate the temperature range over which physiological activities such as CO_2 fixation can occur, usually from 0°C to 50°C; essentially all biochemical reactions exhibit optimal temperatures within this range. The optimal temperatures of a process can shift in response to changing ambient temperatures, indicating that the plant can *acclimate*; we also note that shifts in physiological responses occurring over evolutionary time scales and which are fixed genetically are termed *adaptations*. Subzero temperatures can cause detrimental ice formation in plants, and temperatures above 50°C or so can *denature* proteins (inactivation that is usually caused by a change in their three-dimensional molecular structure). Actually, the temperature of an object is the net result of all the ways that energy can enter it, leave it, or be distributed within it, processes described by an *energy budget* (the analysis of energy reception, loss, and storage).

Many of the studies on gas exchange for agaves and cacti have focused on the influence of temperature, so we will begin this chapter by examining their results. We will then consider the influence of extreme temperatures outside the physiological range, both low and high temperatures. Because energy budgets can be very useful for predicting plant temperature, we will develop analytic expressions for the different energy terms and indicate how we can incorporate various morphological features into energy budgets. Finally, we will relate various aspects of plant temperature to the ecology of agaves and cacti, emphasizing the effects of extreme temperatures.

Gas exchange

Nearly all of the studies on the effects of temperature on gas exchange for agaves and cacti have been performed when the plants are operating predominantly in the CAM mode, meaning that most of the net CO_2 uptake is occurring at night. Many of the studies are done in the field, where concomitant variations in water status and radiation complicate the interpretation of temperature effects. Also, gas-exchange data have often been expressed on a unit mass basis (both fresh weight and dry weight), instead of a surface-area basis, which as we previously argued (Chapter 2) is more useful when relating gas exchange to radiation interception or when comparing CO_2 uptake per unit area among species.

A common feature of nocturnal CO_2 uptake by CAM plants is the low optimal temperatures of about 10°C to 22°C (Kluge and

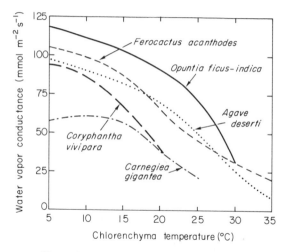

Figure 4.1. Influence of nighttime chlorenchyma temperature on water vapor conductance for an agave and various cacti. The drop in water vapor mole fraction from the chlorenchyma to the ambient air was approximately constant in each case. Data, which are for slightly different temperature ranges for the different species, are from Nobel (1985b) for *Agave deserti*, Nobel and Hartsock (1981) for *Carnegiea gigantea*, Nobel (1981a) for *Coryphantha vivipara*, Nobel (1977b) for *Ferocactus acanthodes*, and Nobel and Hartsock (1984) for *Opuntia ficus-indica*.

Ting, 1978; Winter, 1985). Higher temperatures generally lead to appreciable stomatal closure. These responses lead to limited transpiration and a relatively high water-use efficiency (Eq. 2.11), both of which are consistent with the water-conserving attributes of CAM. Here we will review the temperature effects on stomatal opening (as indicated by changes in water vapor conductance), nocturnal acid accumulation, and net CO_2 uptake for agaves and cacti (influences of temperature on their productivity will be discussed from a geographical perspective in Chapter 7).

WATER VAPOR CONDUCTANCE AT NIGHT

As nighttime temperatures increase, the water vapor conductance decreases for an agave and four species of cacti (Fig. 4.1), indicating substantial stomatal closure at higher temperatures. The water vapor conductance decreases an average of about 20% from nighttime temperatures of 5°C to 15°C, 50% from 15°C to 25°C, and 70% from 25°C to 35°C. Such stomatal closure restricts water loss at higher

nighttime temperatures, but it also substantially decreases net CO_2 uptake for these species at the higher temperatures. When *Agave deserti* is operating in the C_3 mode with daytime stomatal opening, stomatal closure occurs as daytime temperature increases (Nobel and Hartsock, 1979), but the closure is somewhat less than that for increasing nocturnal temperature when this species is in the CAM mode (Fig. 4.1).

Transpiration depends on the product of the water vapor conductance and the drop in concentration or mole fraction of water vapor from the chlorenchyma to the ambient air (Eq. 2.6). The water vapor content of the intercellular air spaces of the chlorenchyma increases rapidly with temperature, because water vapor saturation values increase nearly exponentially with temperature (Fig. 2.16). For *Ferocactus acanthodes*, the increasing water vapor level in the chlorenchyma caused by higher temperatures is overriden by the decrease in water vapor conductance with increasing temperature (Fig. 4.1), resulting in transpiration values that are relatively independent of temperature over essentially the entire range of temperature expected in the field (Lewis and Nobel, 1977; we are assuming that the mole fraction of water vapor, N_{wv}, is constant in the ambient air). This constancy of transpiration as temperature changes is an extremely interesting aspect of stomatal control for this species. For *Opuntia stricta*, most of the stomatal closure that occurs at higher temperatures in the field is due to changes in ΔN_{wv} (Osmond et al., 1979a; also see Table 3.3), but for the five species represented in Figure 4.1 the stomatal response to temperature generally dominates the response to ΔN_{wv}.

OPTIMAL NOCTURNAL TEMPERATURES FOR CO_2 UPTAKE – ACCLIMATION

Although the water vapor conductance increases as the nocturnal temperature is reduced to 5°C for four of the five species considered in Figure 4.1, optimal temperatures for net CO_2 uptake are considerably higher than 5°C because of the biochemical reactions involved in nighttime CO_2 fixation. Actually, net CO_2 uptake for *Opuntia polyacantha* in midwestern United States can be optimal near 5°C, an extremely low temperature (Gerwick

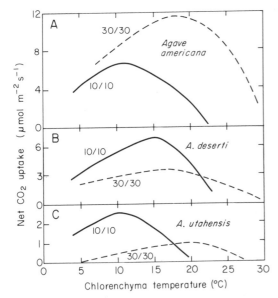

Figure 4.2. Influence of nighttime chlorenchyma temperature on net CO_2 uptake by three species of agave. The day/night temperatures for growth are indicated in °C/°C next to the curves. Modified from Nobel and Hartsock (1981).

maulipas (Gentry, 1982), where mean annual air temperatures are about 25°C; consistent with this, it has higher net CO_2 uptake rates at higher temperatures when grown at 30°C/30°C than when grown at 10°C/10°C, unlike the other two species (Fig. 4.2), which come from colder habitats. (It is ironic that the native habitat of this most widely distributed agave, which was carried to the Old World by Columbus, is uncertain, and indeed *A. americana* is not now found in native stands.) *Agave deserti* is common in the northwestern Sonoran Desert where mean annual temperatures are about 21°C (Shreve and Wiggins, 1964), and *A. utahensis* occurs in the Mojave Desert where mean annual temperatures are about 15°C (Beatley, 1976). The latter two species have much lower relative nocturnal CO_2 uptake at 30°C/30°C than at 10°C/10°C, especially *A. utahensis*. Thus, at least part of the gas-exchange responses to temperature for these species is apparently genetically fixed, although seasonal acclimation does occur.

Because the water vapor conductance of *A. deserti* and many cacti tends to decrease progressively with temperatures above 5°C (Fig. 4.1), whereas the optimal temperature for net CO_2 uptake for agaves and cacti is generally 12°C to 19°C, we expect that the CO_2 liquid-phase conductance (Eqs. 2.9 and 2.10) would also be maximal at low temperatures, but at temperatures that are above those optimal for net CO_2 uptake (see Eq. 2.12). Consistent with this, the maximal liquid-phase conductance occurs 1–4°C above the temperatures optimal for net CO_2 uptake by two species of cacti (Fig. 4.3). Also, $g_{CO_2}^{liq}$ is maximal about 2°C above the nighttime temperature optimal for net CO_2 uptake by *A. deserti* (Nobel and Hartsock, 1978). Therefore, the progressive stomatal closure with increasing temperature above 5°C (Fig. 4.1), coupled with the thermal properties of the biochemical reactions with optimal temperatures closer to 20°C, leads to maximal rates of net CO_2 uptake at intermediate temperatures. Also, acclimation of the optimal temperature for net CO_2 uptake occurs for *Carnegiea gigantea* (Fig. 4.3A) and *Opuntia ficus-indica* (Fig. 4.3B), in large measure reflecting the acclimation of the CO_2 liquid-phase conductance to higher am-

and Williams, 1978). As ambient temperatures increase in the field from January through August for *O. phaeacantha*, the upper nighttime temperature at which net CO_2 uptake just becomes zero progressively increases from 11°C to 27°C (Nisbet and Patten, 1974), indicating the capacity to acclimate. The mean optimal temperatures for net CO_2 uptake by three species of agaves grown at day/night air temperatures of 10°C/10°C is 12°C (Fig. 4.2). When the day/night air temperatures are increased to 30°C/30°C for the agaves, the mean optimal temperature shifts upward to 19°C. Thus, the optimal temperatures for net CO_2 uptake at night are low for these agaves and cacti, and the plants can acclimate as the ambient temperatures change.

We should consider what happens not only to the optimal temperature for net CO_2 uptake as the ambient temperature changes but also to the magnitude of net CO_2 uptake. Actually, the highest rates of net CO_2 uptake for the three species of agave considered (Fig. 4.2) reflect the ambient temperatures in their typical habitats. In particular, *Agave americana* is a subtropical species and probably originates in eastern Mexico, perhaps in Ta-

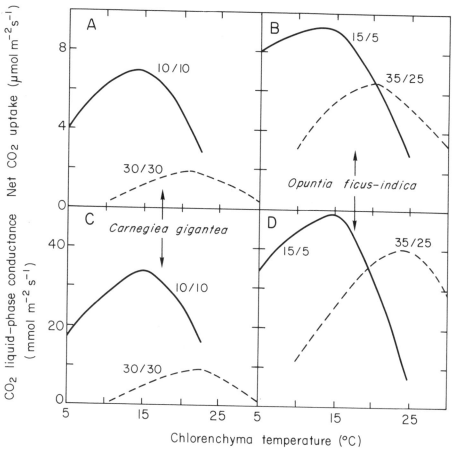

Figure 4.3. Influence of nighttime chlorenchyma temperature on net CO_2 uptake (**A, B**) and liquid-phase conductance (**C, D**) for *Carnegiea gigantea* (**A, C**) and *Opuntia ficus-indica* (**B, D**). The day/night temperatures for growth are indicated in °C/°C next to the curves. Data are adapted from Nobel and Hartsock (1981) for *C. gigantea* and from Nobel and Hartsock (1984) for *O. ficus-indica*.

bient temperatures for these two species (Fig. 4.3C,D).

Shifts in the nocturnal temperature optimal for net CO_2 uptake are triggered by changes in nighttime temperature, at least for *Coryphantha vivipara* (Fig. 4.4). As might be expected for a process occurring at night, daytime temperature does not influence the optimal temperature for net CO_2 uptake at night. As the nighttime temperature is increased from 10°C to 30°C, the optimal temperature increases by 13°C with a halftime of about 8 days; when the nighttime temperature is shifted back to 10°C, the optimal temperature for net CO_2 uptake at night returns to its original value, with a halftime of nearly 4 days (Fig. 4.4).

Many agaves and cacti commonly have low nighttime temperatures for optimal net CO_2 uptake, and the optimal temperatures can shift as growth temperatures change (Table 4.1). When the nighttime temperature is 10°C, the optimal temperature for nocturnal CO_2 uptake averages 12.3°C for the nine species considered. When the nighttime temperature is increased by 20°C, the optimal temperature increases by 7.7°C (Table 4.1). The largest shift occurs for *C. vivipara* (Fig. 4.4), which ranges from northern Mexico to southern Canada and thus occupies habitats that differ greatly in temperature. Smaller but faster shifts occur for the other species, with halftimes averaging 1 day for *A. americana* and *A. utahensis* and just under 2 days for *C. gi-*

Table 4.1. *Optimal temperatures for nocturnal CO_2 uptake by agaves and cacti, emphasizing those studies using multiple, controlled temperatures*

Species	Day/night air temperatures (°C/°C)	Optimal temperature for CO_2 uptake (°C)	Reference
Agave americana	10/10	11.6	Nobel and Hartsock (1981) (**A**)
	30/30	18.6	A
A. deserti	10/10	15.2	A
	30/30	17.8	A
A. utahensis	10/10	10.4	A
	30/30	19.8	A
Carnegiea gigantea	10/10	14.2	A
	30/30	21.2	A
Coryphantha vivipara	10/10	10.2	A
	30/30	23.2	A
Echinocereus fendleri	30/15	10.0	Dinger and Patten (1972) (**B**)
E. ledingii	30/15	12.8	B
E. triglochidiatus	30/15	8.4	B
Ferocactus acanthodes	10/10	12.0	A
	30/30	21.8	A
F. viridescens	10/10	13.0	A
	30/30	19.2	A
Mammillaria dioica	10/10	12.6	A
	30/30	16.0	A
M. woodsii	~25/15	15	Winter et al. (1986) (**C**)
Opuntia bigelovii	10/10	11.4	A
	30/30	22.0	A
O. ficus-indica	10/0	11.4	Nobel and Hartsock (1984) (**D**)
	15/5	14.0	D
	25/15	17.8	D
	35/25	20.2	D
	45/35	23.4	D
O. vulgaris	~25/15	15	C

Note: Studies cited more than once are assigned a boldface letter the first time they appear, this letter being used for subsequent citations.

gantea, F. acanthodes, and *O. bigelovii* (Nobel and Hartsock, 1981). Such rapid temperature responses allow for acclimation to individual weather fronts, increasing the capacity for CO_2 uptake in the field.

Three species of *Echinocereus* occur with nonoverlapping ranges in the Pinaleño Mountains of southeastern Arizona, where the mean air temperature decreases 0.71°C for each 100-m increase in elevation (Dinger and Patten, 1972). The elevational ranges are 1,100 to 1,400 m for *E. fendleri*, 1,500 to 2,000 m for *E. ledingii*, and 2,200 to 2,900 m for *E. triglochidiatus*. The optimal temperatures for net

CO_2 uptake at 30°C/15°C (Table 4.1) are consistent with distribution for *E. ledingii* (optimum at 12.8°C) and *E. triglochidiatus* (8.4°C) but not for *E. fendleri* (10.0°C). However, we must consider acclimation and total CO_2 uptake over the night as well as low-temperature tolerances for plant survival before trying to explain the observed ranges of these cacti in terms of temperature responses.

DAILY PATTERNS FOR CO_2 EXCHANGE

Now that we have discussed the instantaneous responses of transpiration and CO_2 exchange to temperature, let us consider CO_2

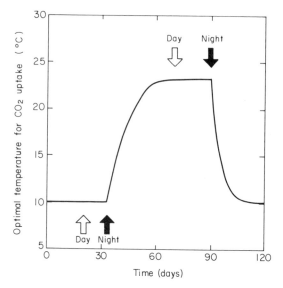

Figure 4.4. Influence of day and night temperatures on the optimal temperature for net CO_2 uptake by *Coryphantha vivipara*. Initially the day/night air temperatures were 10°C/10°C. The upward open arrow ("Day") indicates when the day temperature was increased to 30°C, and the downward open arrow indicates when it was returned to 10°C. The upward filled arrow ("Night") indicates when the night temperature was increased to 30°C, and the downward filled arrow indicates when it was returned to 10°C. Modified from Nobel and Hartsock (1981).

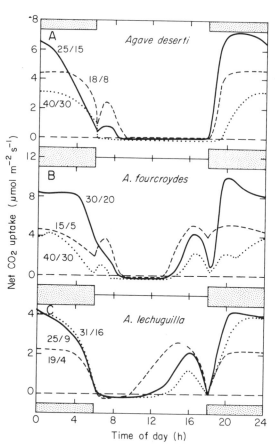

Figure 4.5. CO_2 exchange over a 24-h period by leaves of three species of agaves at the day/night growth temperatures indicated in °C/°C next to the curves. Mature plants were maintained under wet conditions, the PAR level was at least 90% of the saturation value, and the CO_2 exchange was measured at the same temperatures as used for growth. Solid lines indicate data for the mean annual nighttime temperatures at the field sites. Data for *Agave deserti* and *A. fourcroydes* are from unpublished observations under the conditions of Nobel (1984a) and Nobel (1985a), respectively, and those for *A. lechuguilla* are from Nobel and Quero (1986).

exchange over 24-h periods for agaves and cacti maintained under various growth temperatures.

For *Agave deserti* (Fig. 4.5A), *A. fourcroydes* (Fig. 4.5B), and *A. lechuguilla* (Fig. 4.5C), the highest net CO_2 uptake rate occurs when the nighttime temperatures are maintained at about the mean annual values observed in the native habitats of these species. Raising the day/night air temperatures above the mean annual values substantially decreases the net CO_2 uptake during the daytime for all three species, and lowering the temperatures increases the daytime uptake. The maximum rates at night for the lowest nighttime temperatures (4 or 5°C) and for the highest nighttime temperature (30°C) are substantially reduced compared with those measured at or near the mean annual temperatures (Fig. 4.5). For *A. lechuguilla*, the highest nighttime temperature used is only 16°C, which is the mean minimum air temperature for the hottest

part of the year at the study site (Nobel and Quero, 1986); such a moderate nighttime temperature does not severely limit net CO_2 uptake at night (Fig. 4.5C).

As we mentioned in Chapter 2, *Agave vilmoriniana* has much of its net CO_2 uptake during the daytime. When grown at day/night air temperatures of 20°C/10°C, fully 92% of the net CO_2 uptake over a 24-h period occurs during the daytime, and daytime uptake still represents 36% of the total when the air temper-

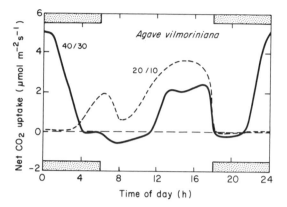

Figure 4.6. CO_2 exchange for leaves of *Agave vilmoriniana* at the day/night growth temperatures indicated in °C/°C next to the curves. Mature plants were growing at the indicated temperatures under wet conditions at the same mean total daily PAR as in the field. Modified from Nobel and McDaniel (1988).

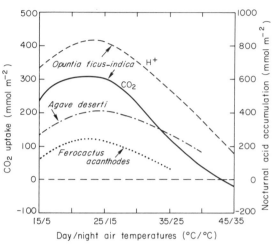

Figure 4.7. Influence of day/night growth temperatures on total CO_2 uptake over 24-h periods for *Agave deserti* and *Ferocactus acanthodes*. Also shown are the total nocturnal CO_2 uptake and nocturnal acid accumulation for *Opuntia ficus-indica*. Gas exchange, which was measured at the same day/night temperatures as for growth, are from Nobel (1984a, 1986a) and Nobel and Hartsock (1984).

atures are raised to 40°C/30°C (Fig. 4.6). Moreover, this species exhibits only 8% less total CO_2 uptake at the higher air temperatures compared with the lower ones. We note that less reliance on the daytime C_3 mode at the higher temperatures (Fig. 4.6) is consistent with maintaining a higher water-use efficiency.

Similar effects of temperature on the daily pattern of CO_2 exchange have been observed for cacti. For instance, the highest rates of net CO_2 uptake for *Ferocactus acanthodes* occur for day/night growth temperatures of 23°C/14°C, which represent the mean daily maximum temperature/mean daily minimum temperature at the Agave Hill field site (Nobel, 1986a). Even though stomatal opening is greater at typical mean wintertime day/night temperatures of 11°C/5°C (Fig. 4.1), the CO_2 liquid-phase conductance of *F. acanthodes* is 70% lower and the maximal rate of nocturnal net CO_2 uptake is 40% lower than when measured for plants at 23°C/14°C. The maximal rate of net CO_2 uptake for *Opuntia ficus-indica* occurs when day/night temperatures are 25°C/15°C; at night temperatures of 0°C, the rate of net CO_2 uptake is still 30% of the maximum value, but at 35°C no CO_2 uptake is observed at night (Nobel and Hartsock, 1984).

For *O. ficus-indica* as well as other species, some stomatal opening and net CO_2 uptake can occur at the beginning of the day,

especially at lower growth temperatures. For *O. erinacea*, a high temperature of 35°C at dawn can apparently increase the rate of malate decarboxylation, which eventually raises the internal CO_2 level; higher internal CO_2 levels reduce the driving force for inward CO_2 diffusion as well as cause some stomatal closure, thus limiting net CO_2 uptake (Littlejohn and Ku, 1985). Increasing the leaf temperature for *A. deserti* in the field can even eliminate the net CO_2 uptake occurring at dawn (Nobel, 1976).

TOTAL DAILY CO_2 UPTAKE AND
ACIDITY CHANGES

Once daily patterns for net CO_2 exchange have been determined at various growth temperatures, the fluxes over a 24-h period can readily be integrated and thus the temperature dependence of total daily CO_2 uptake can be determined (Fig. 4.7). For *Agave deserti*, daily net CO_2 uptake is approximately maximal for day/night growth temperatures of 25°C/15°C, decreasing 35% at 15°C/5°C and 60% at 40°C/30°C (Fig. 4.5A). The optimal temperatures for CO_2 uptake are slightly lower for *Ferocactus acanthodes*, which has approximately 50%

less maximal net CO_2 uptake over 24 h than does *A. deserti*, and for *Opuntia ficus-indica*, which has about 50% more CO_2 uptake than does *A. deserti* (Fig. 4.7; *O. ficus-indica* has very little daytime CO_2 uptake, so its nighttime CO_2 uptake is essentially the same as the CO_2 uptake over 24 h).

Many of the effects of temperature on gas exchange over 24-h periods warrant further discussion, not the least of which is the accompanying change in tissue acidity. For *O. ficus-indica*, the changes in nocturnal acid accumulation parallel the changes in net nocturnal CO_2 uptake (Fig. 4.7), but there are important differences. For instance, net CO_2 uptake for day/night growth temperatures of 45°C/35°C is negative, but the nocturnal acid accumulation is positive and fully 20% of the maximal value (Fig. 4.7). Thus, the acidity changes give not only the wrong quantitative assessment of net CO_2 uptake, as we discussed in Chapter 2, but also the wrong qualitative assessment of the effect of high temperatures on net CO_2 uptake. On the other hand, the relative changes in the two parameters at moderate temperatures can be consistent with the production of two H^+ per atmospheric CO_2 fixed; as growth temperatures are increased from 15°C/5°C to 20°C/10°C, net CO_2 uptake increases nearly 65 mmol m^{-2} and net H^+ increases 130 mmol m^{-2} for *O. ficus-indica*. From 25°C/15°C to 35°C/25°C, CO_2 decreases 170 mmol m^{-2} and H^+ decreases 290 mmol m^{-2} (Fig. 4.7); on a 2 H^+ : 1 CO_2 basis, the reduction in CO_2 uptake would be accompanied by a 340 mmol m^{-2} reduction in H^+ accumulation. The smaller observed decrease in H^+ indicates slightly more H^+ production per atmospheric CO_2 fixed at the higher temperatures, which is attributable to fixation of CO_2 released internally by respiration.

Nocturnal acid accumulation by *Opuntia erinacea* progressively increases as the minimum nighttime temperature increases seasonally from 1°C to 17°C, and the magnitude of the acidity change is related to the seasonal growth rate (Littlejohn and Williams, 1983). Also, when the nighttime temperature is increased in 5°C increments from 1°C to 16°C, the changes in nocturnal acid accumulation for *Agave salmiana* are consistent with changes in net CO_2 uptake; acidity increases are 100 mmol H^+ m^{-2} at 1°C and 540 mmol H^+ m^{-2} at 16°C (Nobel and Meyer, 1985). Nocturnal acid accumulation can also be substantial for cacti under the continuously high minimum nighttime temperatures that occur in the tropics of northern Venezuela. For example, for nighttime temperatures above 25°C, nocturnal acid accumulation is just over 300 mmol H^+ m^{-2} for *Acanthocereus tetragonus* and *Opuntia wentiana* and nearly 200 mmol H^+ m^{-2} for *Stenocereus griseus* (Díaz and Medina, 1984), which are similar to the nocturnal acid accumulation for *O. stricta* under lower temperatures but threefold less than for *O. basilaris* (Fig. 2.14). Actually, many cacti growing at higher latitudes have a much reduced nocturnal acid accumulation and no net nocturnal CO_2 uptake above 25°C.

As for other plants, the respiration rates of agaves and cacti increase as temperatures increase. Respiration in the dark doubles per 10°C increase in temperature for *Opuntia versicolor* from 30°C to 45°C (Richards, 1915). For *O. basilaris*, respiration at night approximately doubles from 10°C to 20°C and again from 20°C to 30°C, reaching a maximum at 40°C (Szarek and Ting, 1974a). For *O. bigelovii*, respiration can increase with increasing temperatures up to 53°C (Didden-Zopfy and Nobel, 1982). The higher respiration at higher temperatures leads to more CO_2 released internally at night, which can be incorporated into malate. For *Mammillaria woodsii* and *Opuntia vulgaris* at nighttime temperatures above 30°C, substantial nocturnal acid accumulation can occur even though net CO_2 uptake is negative (Winter et al., 1986), just as for *O. ficus-indica* at day/night temperatures of 45°C/35°C (Fig. 4.7). Nocturnal CO_2 uptake also becomes negative at nighttime temperatures of about 35°C for *Agave americana* (Neales, 1973), *A. lechuguilla* (Eickmeier and Adams, 1978), *O. basilaris* (Hanscom and Ting, 1978a), and *O. humifusa* (Koch and Kennedy, 1980). Indeed, the acclimation to higher temperatures for net CO_2 uptake by cacti (Table 4.1) may reflect changes in the temperature dependency of respiration (Gulmon and Bloom, 1979).

For *O. ficus-indica*, not only does the total daily CO_2 uptake decrease during drought (Fig. 3.16), but also the optimal temperature becomes progressively lower. For instance, the optimal temperature for net CO_2 uptake shifts from 17°C under wet conditions to 14°C after four weeks of drought to 11°C after seven weeks of drought (Nobel and Hartsock, 1984). Apparently CO_2 fixation is more sensitive to drought than is respiration. Also, respiration has a greater increase with temperature and a higher optimal temperature than does CO_2 fixation. Thus, respiration represents a greater absolute proportion of CO_2 fixation during water stress, which, coupled with the greater increase with temperature for respiration (CO_2 releasing) than for carboxylation (CO_2 consuming), causes the optimal temperature for net CO_2 uptake to decrease progressively during drought.

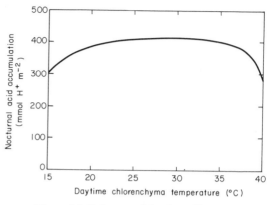

Figure 4.8. Influence of daytime chlorenchyma temperature on nocturnal acid accumulation by *Agave deserti*. Mature plants were maintained under wet conditions, about 90% PAR saturation, and a nighttime temperature of 15°C in all cases. Modified from Nobel and Hartsock (1978).

DAYTIME TEMPERATURE EFFECTS

We have already indicated that stomatal closure occurs as daytime temperature increases for an overwatered *Agave deserti* in the C_3 mode, just as for the nocturnal stomatal response to temperature in the CAM mode (Fig. 4.1). However, the daytime closure is only about half as great as that at night for a particular temperature increase (Nobel and Hartsock, 1979). For *A. deserti* as well as *A. fourcroydes* and *A. lechuguilla*, daytime CO_2 uptake under normal watering conditions also decreases as temperature is increased from the mean annual temperature observed in their native habitats (Fig. 4.5). Indeed, increasing daytime temperatures above about 20°C decreases daytime stomatal opening and net CO_2 uptake for CAM plants in general (Kluge and Ting, 1978). For leaves of *Pereskia aculeata* and *P. grandifolia*, which use the C_3 pathway and have only daytime stomatal opening (Figs. 2.7A,B and 3.16A), the water vapor conductance decreases about 40% from 15°C to 25°C and again 40% from 25°C to 35°C (Gibson and Nobel, 1986). Net CO_2 uptake during the day for these two pereskias is maximal at about 24°C, and the CO_2 liquid-phase conductance is maximal at 27°C, which are much higher optimal temperatures than for nocturnal CO_2 uptake by cacti in the CAM mode, except for certain tropical species.

Of course, when agaves and cacti are in the CAM mode, CO_2 released internally can be fixed into photosynthetic products in the chloroplasts during the daytime (Fig. 2.13). For chloroplasts isolated from *Opuntia polyacantha*, the electron transport reactions and ATP (adenosine triphosphate) synthesis, both of which are necessary for CO_2 fixation, have maximal rates near 40°C (Gerwick, Williams, and Uribe, 1977). The CO_2 fixation rate by chlorenchyma cells isolated enzymatically from *O. polyacantha* and *O. vulgaris* approximately doubles for every 10°C increase in temperature from 10°C up to the optimal temperature, 40°C; some CO_2 fixation occurs up to about 50°C (Gerwick et al., 1978).

Although daytime temperatures markedly affect C_3 photosynthesis, many other processes are involved in CO_2 uptake in the CAM mode (Fig. 2.13). Also, CO_2 is evolved by respiration involved in plant maintenance and growth. As a consequence, daytime temperatures generally have a rather small effect on nocturnal CO_2 uptake and acid accumulation for agaves and cacti that are predominantly in the CAM mode. For *A. deserti* at nighttime temperatures of 15°C, nocturnal acid accumulation increases about 30% as the daytime temperature is raised from 15°C to 20°C, but then remains fairly constant as daytime temperature is raised all the way up to 35°C; it decreases about 30% from the maximum for daytime temperatures of 40°C (Fig. 4.8). For many species in the CAM mode, nocturnal CO_2 uptake is greater when the daytime temperature is at least 10°C higher than the nighttime temperature. This is the case for *Agave americana* (Neales, 1973), *Coryphantha vivipara* and *Echinocereus viridiflorus* (Green and Williams, 1982), three other species of *Echinocereus* (Dinger and Patten, 1972, 1974), and *Opuntia basilaris* (Hanscom and Ting, 1978a). When the daytime temperature is about 10°C higher than the nighttime temperature, which would normally be the situation in the field, further increases of 10–20°C usually have little effect on nocturnal CO_2 uptake, similar to the case for *A. deserti* (Fig. 4.8). Thus, when discussing the productivity of agaves and cacti in various regions (Chapter 7), we will emphasize nocturnal temperatures.

Low-temperature tolerances

We have thus far discussed the gas-exchange responses of agaves and cacti to intermediate temperatures. We now turn to the effects of extreme temperatures, which can have as much or even more biological importance. As our emphasis shifts from physiological parameters, such as net CO_2 uptake, to survival, we need criteria by which to assess plant damage and death. This may seem like a straightforward matter until one realizes that cacti can live for a few years after being exposed to an ultimately fatal, low-temperature episode. For instance, death of *Carnegiea gigantea* in response to a severe freeze in January 1937 was not evident for some plants until three years later (Gill, 1942; Steenbergh and Lowe, 1976, 1977), and some *C. gigantea* "killed" by a freeze in 1971 did not die until nine years later (Steenbergh and Lowe, 1983).

Even though waiting around for a few years may provide unambiguous evidence for the effect of a particular low-temperature episode, it is not practical for laboratory experimentation on temperature tolerances. Therefore, assays are needed that indicate whether a particular treatment will lead to cell death and local *necrosis* (death of cells – the rest of the plant may remain living). The uptake of a "vital" stain such as neutral red [3-amino-7-dimethyl-amino-2-methylphenazine (HCl)] has proved quite useful in this regard (Gurr, 1965). Neutral red accumulates in the central vacuoles of living cells only and provides a convenient, quantitative index for the viability of cells. In particular, the location of any red

coloring can be readily observed for tissue slices examined with a light microscope, and the percentage of the cells taking up the stain can be determined (Stadelmann and Kinzel, 1972; Onwueme, 1979; Didden-Zopfy and Nobel, 1982). After an agave or cactus (or even a piece thereof) is exposed to a specific temperature treatment, tissue slices are placed in a neutral red solution. Subsequent microscopic examination of the tissue slices will reveal whether the cells have taken up the vital stain and thus are alive, which helps indicate whether the plant can survive the imposed conditions.

COOLING CURVES AND CELL VIABILITY

As the air temperature is steadily lowered over a matter of hours at a rate that can occur in the field, the temperatures of the stems of cacti also become lower (Fig. 4.9). However, when the stems of *Carnegiea gigantea* and *Coryphantha vivipara* reach about −5°C, the chlorenchyma temperature increases by 3 or 4°C, even though the air temperature is still being lowered. This increase in temperature signifies that heat is being produced within the stems, a process that we will discuss in detail in the next subsection. After the rise in temperature, the stems progressively cool as the air temperature is further lowered during the rest of the night (Fig. 4.9).

As the stems are being cooled, small pieces of the chlorenchyma can be removed and tested for uptake of a vital stain by the cells (Fig. 4.10). In the absence of an imposed temperature stress, uptake of neutral red into the vacuoles can be demonstrated for 90% or so of the chlorenchyma cells (sometimes the tissue slices are incubated with the stain for 24 h to maximize stain uptake; Nobel, 1981a, 1982a). Such a high percentage of cells taking up stain is also observed for stem pieces removed during the initial cooling phase, the subsequent phase involving a rise in temperature, and part of the next cooling phase (Figs. 4.9 and 4.10). However, once the stem temperature is lowered below about −6°C for *Carnegiea gigantea* and −9°C for *Coryphantha vivipara*, the percentage of chlorenchyma cells taking up stain begins to decrease. For *C. gigantea*, stain uptake is halved at −9°C and

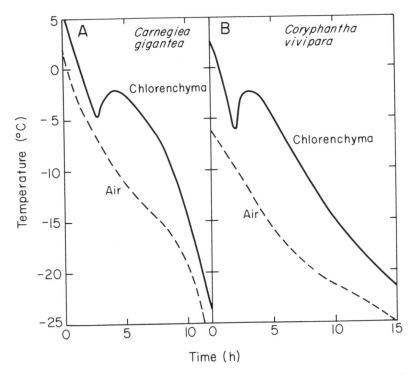

Figure 4.9. Cooling curves, showing the changes in chlorenchyma temperature as the air temperature is progressively lowered through the night for *Carnegiea gigantea* (**A**) and *Coryphantha vivipara* (**B**). Data are adapted from Nobel (1981a, 1982a).

eliminated at $-11°C$, and for *C. vivipara* the analogous temperatures are $-14°C$ and $-20°C$, respectively (Fig. 4.10).

How do changes in stain uptake with temperature compare with other measures of cell viability? Chlorenchyma cells and the functioning of their plasmalemmas were carefully examined, because disruption of membranes is often an early sign of freezing damage. For *Coryphantha vivipara*, all the protoplasts appear normal when the stem temperature is initially $-5°C$, as the plasmalemma is then adjacent to the cell wall; the protoplasts appear slightly shrunken at $-5°C$ after the rise in chlorenchyma temperature (Fig. 4.9), considerably shrunken at $-10°C$, and all protoplasts of chlorenchyma cells are severely shrunken at $-15°C$ (Nobel, 1981a). Also tested was the ability of the cells to *plasmolyze* (shrinking of the protoplast and pulling away of the plasmalemma from the cell wall in response to an increase in osmotic pressure of an external solution bathing the tissue slices, which causes loss of turgor and then the draw-

ing of water out of the cells by *osmosis*). Again for *C. vivipara*, the ability of the chlorenchyma cells to plasmolyze decreases concomitant with the decrease in their ability to take up stain (Fig. 4.10) as the subzero temperature is lowered, suggesting that membrane damage has occurred at the lower temperatures (Nobel, 1981a).

For stems of *C. vivipara* whose temperatures were reduced to those temperatures causing stain uptake to become zero ($-20°C$), no net CO_2 uptake was observed for the next two months after returning the plants to $10°C$, and all five of the treated plants eventually died (Nobel, 1981a). When stem temperature was reduced to $-15°C$, net CO_2 uptake two months later had almost recovered to its initial value for three plants (some tissue necrosis was apparent), whereas two other plants of *C. vivipara* had no net CO_2 uptake and eventually died. Thus, the ability of the chlorenchyma cells to take up a vital stain immediately after the low-temperature treatment can be closely related to the ability of the plants to have a net

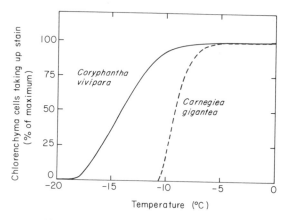

Figure 4.10. Influence of subzero temperatures on stain uptake by *Carnegiea gigantea* and *Coryphantha vivipara*. Uptake of neutral red was determined for pieces of clorenchyma isolated at various temperatures along the cooling curves portrayed in Figure 4.9.

CO_2 uptake over the next few months and, more importantly, to survive. We can conclude that for agaves and cacti in general, plants usually die at the low temperature that reduces stain uptake to zero, which averages about 4°C below the subzero temperature that halves the percentage of cells taking up neutral red for a 1-h low-temperature treatment.

MECHANISM OF DEATH BY FREEZING

As we indicated in Chapter 1, *Opuntia ficus-indica* and its hybrids can be injured by temperatures of −5°C to −15°C, and we have just indicated that *Coryphantha vivipara* can be killed by −15°C. Apparently, it is not the average cellular constituents that are freezing. For instance, the cellular liquid that can be squeezed out of *O. ficus-indica* and its hybrids freezes at −0.5°C (Uphof, 1916). Because the *freezing-point depression* (the temperature to which the freezing point of a solution is lowered by the presence of solutes) is 1.86°C for a 1 *molal* solution (1 mole of solute dissolved in 1 kg of water; molality is approximately equal to molarity, moles/liter, for dilute solutions), freezing at −0.5°C occurs for a (0.5°C)/(1.86°C per 1 molal) or 0.3 molal solution. By the van't Hoff relation ($\pi = cRT$, Eq. 3.2), such a solution would have an osmotic pressure of (0.3 molal)(2.44 MPa molal^{-1}) or 0.7 MPa at 20°C. Such an osmotic pressure is common for cacti and agaves under wet conditions

(Table 3.1). Because freezing of cellular constituents is generally accompanied by death for vascular plants, the cells apparently do not freeze when the temperature is reduced below the freezing point of their mean initial contents.

Let us now provide an explanation for the initial dip and the subsequent rise in temperature observed for the cooling curves of the stems of cacti as the air temperature is progressively lowered (Fig. 4.9). As the stems are initially cooled to about −5°C, no freezing of cellular contents occurs, even though the temperature is below that at which freezing is expected, a nonequilibrium condition referred to as *supercooling*. Although supercooling is not the mechanism by which freezing damage is avoided for the whole night by agaves and cacti, it can help us interpret some of the ensuing cellular events. The melting of ice requires an input of energy, known as the *heat of fusion*; exactly the same amount of energy is released when ice is formed. The increases in temperature observed for the stems of *Carnegiea gigantea* and *Coryphantha vivipara* 3 to 4 h after cooling begins (Fig. 4.9) represent the release of heat within the stems as freezing of water occurs. However, as noted long ago by Uphof (1916), ice initially forms in the intercellular air spaces of the chlorenchyma as well as between the cell wall and the plasmalemma, not within the cells.

Based on the heat of fusion of water, the stem water content, and the extent of the rise in temperature observed for *C. vivipara*, the rise in temperature represents the heat released upon freezing of about 10% of its stem water, which is microscopically identified as extracellular water. This freezing leads to the formation of ice crystals outside of the protoplasts, a condition that is not lethal. After the initial formation of extracellular ice, water diffuses out of the protoplasts and becomes incorporated into the ice crystals, which grow with time. As the subzero air temperature continues to be lowered (Fig. 4.9), more water leaves the chlorenchyma cells (the cooling rate employed was slow enough to avoid intracellular freezing). Such gradual dehydration leads to the shrunken appearance of the protoplasts of *C. vivipara* just noted and is known as *frost plasmolysis* (Levitt, 1980). Thus, the eventual death primarily reflects an intracellular dehydration, although some membrane disruption from the ice crystals is possible.

Cellular death occurs near $-15°C$ for *Coryphantha vivipara* under the conditions employed. Based on the freezing-point depression, the remaining intracellular solution has an osmotic pressure of about 17 MPa at this temperature. To obtain a π of 17 MPa based on the van't Hoff relation (Eq. 3.2), 94% of the cellular water would have to be lost (although conditions of ideality assumed by Eq. 3.2 are not met here, the van't Hoff relation nevertheless gives us a useful estimate of the solution osmotic pressure; see Nobel, 1983a). In close agreement with the predicted water loss at death, a drought dehydration of 91% of the cellular water leads to death for *C. vivipara* (Nobel, 1981a). Thus, survival of drought and of low temperature events are apparently related for cacti, just as they are for other plants (Levitt, 1978, 1980).

For certain winter-hardy platyopuntias, the cladodes noticeably shrink at low temperatures and become more cold tolerant when desiccated (Anonymous, 1883). Despite rainfall events, the water content of *Opuntia erinacea* decreases in the field as winter approaches (Littlejohn and Williams, 1983). This could represent decreased water uptake from the soil at lower temperatures because of slower water movement in the soil or lower root hydraulic conductivity (Eq. 3.3), as occurs for other plants (Berry and Raison, 1981; Tranquillini, 1982), or perhaps excess transpiration just before winter. For *O. humifusa*, the lowest mortality due to low temperatures occurs at the driest sites, possibly because of the high tissue solute contents under the more xeric conditions (Koch and Kennedy, 1980). The lower tissue water content may lead to less water that can freeze extracellularly and therefore less ice that could damage the cells as the ice crystals eventually become large enough to disrupt the plasmalemmas. For *Agave deserti*, the subzero temperature that halves the percentage of chlorenchyma cells taking up neutral red decreases from $-6°C$ for a leaf osmotic pressure of 0.5 MPa to $-8°C$ for a π of 1.3 MPa. A similar decrease of 2°C in the low-temperature tolerance as the stem osmotic pressure increases over this range occurs for *Ferocactus acanthodes* (Nobel, 1984c), again indicating greater cold tolerance at higher π. In addition, *Agave macroacantha* can tolerate nighttime air temperatures of $-6°C$ when droughted but not if watered (MacEwan, 1973).

Another environmental factor that changes as winter approaches is photoperiod. However, shortening of the photoperiod may have only small effects on cold tolerance, although changes in light quality (especially for wavelengths in the red and near infrared) accompanying changes in photoperiod have not been examined. For *Coryphantha vivipara*, a large decrease in photoperiod from 21 h to 3 h per day lowers the subzero temperature for halving of stain uptake by only 0.5°C (Nobel, 1983c). Similarly, reducing the photoperiod from 20 h to 4 h lowers the low-temperature tolerance by only about 0.6°C for *A. deserti* and *F. acanthodes* (Nobel, 1984c). Future studies on the effects of photoperiod on agaves and cacti are clearly called for.

TOLERABLE TEMPERATURES AND
COLD HARDENING

The effects of subzero temperatures on stain uptake have also been measured for the three species considered throughout this book: *Agave deserti* (Fig. 4.11A), *Ferocactus acanthodes* (Fig. 4.11B), and *Opuntia ficus-indica* (Fig. 4.11C). To standardize the assay for the low-temperature treatments, as well as for the high-temperature treatments to be discussed in the next section, stem or leaf pieces are routinely maintained at a particular temperature for 1 h, a time representative of the duration of an extreme temperature in the field for agaves and cacti. The percentage of chlorenchyma cells taking up stain decreases for all three species as the treatment temperature is lowered. Moreover, the temperature for 50% inhibition of cells taking up stain is lower for plants grown at day/night air temperatures of 10°C/0°C than at 30°C/20°C, especially for *A. deserti* (Fig. 4.11). In other words, all three species acclimate to lower temperatures, or *cold harden*, by being exposed to lower growth temperatures.

Cold hardening is widespread among agaves and cacti, occurring for essentially all species that have been studied (Table 4.2). For

Table 4.2. *Low-temperature tolerances and cold-hardening ability of agaves and cacti*

Species	Temperature leading to 50% inhibition of the fraction of chlorenchyma cells taking up stain for plants at day/night temperatures of 10°C/0°C (°C)	Cold hardening per 10°C decrease in day/night temperatures (°C)
Agaves		
Agave americana	−7.4	1.8
A. angustifolia	−9.4	2.5
A. bovicornuta	−8.3	0.4
A. deserti	−16.3	3.9
A. lechuguilla	−13.8	1.8
A. multifilifera	−12.7	2.9
A. murpheyi	−11.2	1.6
A. palmeri	−10.4	0.4
A. parryi	−19.6	4.8
A. parryi var. *huachucensis*	−10.9	1.2
A. patonii	−9.2	0.8
A. pedunculifera	−8.4	0.9
A. rhodacantha	−9.4	2.2
A. schottii	−11.3	2.1
A. shawii	−7.9	0.4
A. sisalana	−6.4	0.0
A. utahensis	−17.5	3.9
A. vilmoriniana	−8.0	1.2
A. weberi	−9.8	1.7
Cacti		
Carnegiea gigantea	−8.6	0.5
Coryphantha vivipara	−20.3	1.7
C. vivipara var. *deserti*	−18.6	
C. vivipara var. *rosea*	−22.1	
Denmoza rhodacantha	−10.4	1.0
Eriosyce ceratistes	−10.1	0.8
Ferocactus acanthodes	−8.7	0.3
F. covillei	−7.2	0.0
F. viridescens	−6.1	0.3
F. wislizenii	−8.4	0.3
Lophocereus schottii	−6.8	0.5
Opuntia bigelovii	−7.3	0.8
O. ficus-indica	−8.8	1.1
O. polyacantha	−17.1	
O. ramosissima	−4.4	
Pediocactus simpsonii	−18.3	
Stenocereus thurberi	−9.0	0.3
Trichocereus candicans	−7.4	1.0
T. chilensis	−7.8	0.9

Note: Data were obtained graphically (see Fig. 4.11). Cold hardening indicates the decrease in the 1-h treatment temperature that leads to a 50% inhibition in the fraction of chlorenchyma cells taking up neutral red as the day/night air temperatures are reduced from 20°C/10°C to 10°C/0°C. Data are adapted from Nobel (1982a, 1984b) and Nobel and Smith (1983).

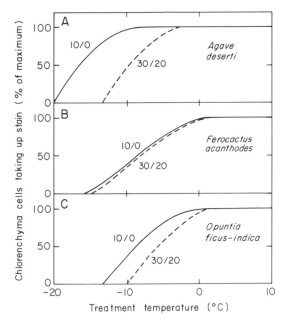

Figure 4.11. Influence of day/night growth temperatures, indicated in °C/°C next to the curves, on the low-temperature tolerance of *Agave deserti* (**A**), *Ferocactus acanthodes* (**B**), and *Opuntia ficus-indica* (**C**). Stem pieces were incubated for 1 h at a particular treatment temperature, and the percentage of chlorenchyma cells taking up neutral red was subsequently determined. Data are adapted from Nobel (1984b) and unpublished observations.

Figure 4.12. *Agave patonii* at 2,240 m near Durango, Durango, Mexico. Freezing damage is evident near the tips of the leaves, which are lighter in color (bleached) and wrinkled.

the fourteen species of agave examined, the decrease in the low temperature that is tolerated averages 2.0°C per 10°C decrease in growth temperature. The greatest cold hardening of 4–5°C per 10°C decrease in growth temperature occurs for *A. deserti, A. parryi*, and *A. utahensis*, all of which can occupy cold habitats, such as in northern Arizona. These three species can tolerate a leaf temperature maintained for 1 h at an average of about −18°C when grown at 10°C/0°C (Table 4.2). Also, *A. parryi* and *A. utahensis* are among the most cold tolerant of agaves grown as ornamental plants in Europe (Köhlein, 1976). Both cold hardening and the fact that stain uptake does not drop to zero until about 4°C below the temperature for 50% inhibition of cells taking up stain suggests that at least some agaves can tolerate temperatures as low as −25°C, or even slightly lower. On the other hand, *A. americana* is not very cold tolerant (Table 4.2), stain uptake being reduced 50% at

about −10°C, indicating survival to only −14°C or so, a temperature where 1 h reduces stain uptake to zero. Indeed, horticultural evidence indicates that *A. americana* is best restricted to regions where the average minimal annual temperatures are above −20°C to −15°C (Mekhtiev, 1972; Klingman, 1979). *Agave patonii* is also not very cold tolerant, damage to its leaves occurring when plants in the field are exposed to temperatures of about −15°C for a few hours (Fig. 4.12).

The cold hardening of fourteen species of cacti amounts to an average of an 0.7°C decrease in the subzero temperature leading to 50% inhibition of the cells taking up stain as the growth temperature is reduced by 10°C (Table 4.2). Although this is about threefold less acclimation per unit temperature decrease than the average for the agaves considered, the wide-ranging *Coryphantha vivipara* has a cold hardening of 1.7°C per 10°C decrease in growth temperature. Indeed, *C. vivipara* along with

Opuntia polyacantha and *Pediocactus simp-sonii*, when grown at day/night air temperatures of 10°C/0°C, can tolerate 1 h at an average of −19°C with only a halving of the chlorenchyma cells that accumulate stain (Table 4.2). These three species occur in habitats with considerable annual snowfall (see Fig. 1.19); for example, all are native in southeastern Wyoming (at 41°N) at elevations of up to 3,000 m (Nobel, 1982a). Snow cover can provide a relatively moist microhabitat in which water loss from the stems is greatly reduced, thus preventing possible injury due to desiccation. In addition, snow cover can provide thermal insulation from cold nighttime air temperatures (Geiger, 1965). Other studies also indicate that *O. polyacantha* can survive −24°C (Rajashekar, Gusta, and Burke, 1979). Field tests show that the low temperature limit can be quite abrupt. For instance, *O. erinacea, O. humifusa, O. phaeacantha, O. spinosior,* and *O. stricta* can tolerate −17°C or −18°C in Karakum (USSR) but die at −20°C (Nardina and Mukhammedov, 1973).

Acclimation to cold can occur in a matter of days when growth temperatures change. For two species of cacti from northcentral Argentina, *Denmoza rhodacantha* and *Trichocereus candicans*, the halftime for cold hardening in response to a sudden decrease in growth temperatures is about 3 days (Fig. 4.13). Similarly, the halftime for the low-temperature acclimation of *A. deserti* is 4 days (Nobel and Smith, 1983). Thus, cold hardening would be expected to occur in the field in response to the seasonal decrease in temperatures from fall to winter.

The cold temperatures that can be tolerated vary considerably among cacti. For instance, a severe freeze in December 1983, when the temperature in the southern part of Texas was 0°C or below for two and a half days, killed *Cereus pentagonus* and *Opuntia ficus-indica* but left *Echinocereus enneacanthus, O. leptocaulis,* and *O. lindheimeri* unharmed (Lonard and Judd, 1985). As would be expected, cacti from regions with prolonged periods of subzero temperatures, such as *Coryphantha vivipara, Opuntia polyacantha,* and *Pediocactus simpsonii,* tolerate very low temperatures (Table 4.2), such as −30°C when cold acclimation is taken into consideration.

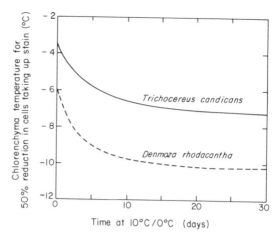

Figure 4.13. Time course for changes in the low temperature leading to halving of the percentage of chlorenchyma cells taking up a vital stain. Plants that had been maintained at day/night air temperatures of 50°C/40°C were switched to 10°C/0°C on day 0, and the low temperature leading to a 50% inhibition in the percentage of cells taking up neutral red for a 1-h treatment was determined for subsequent days. Modified from Nobel (1982a).

On the other hand, cacti from warm regions, such as *Ferocactus viridescens* and *O. ramosissima*, are not very tolerant of low temperatures and are killed by −10°C. The genetic variability in low-temperature tolerance may be quite important in future horticultural endeavors and breeding experiments with platyopuntias of potential agronomic importance, such as when trying to extend their cultivation to higher latitudes or to higher elevations.

High-temperature tolerances

High-temperature tolerances of cacti have received much greater attention than the low-temperature tolerances of agaves and cacti combined. Indeed, high-temperature tolerances of cacti have been reported since the early part of the twentieth century (Table 4.3). Early research indicates that the lowest high temperature tolerated occurs for *Pereskia columbiana* (52°C), most cacti tolerate 60°C, and a few can tolerate temperatures of 65°C or even higher (Table 4.3). Yet, despite these early studies and others that will be described here, we still do not understand the cellular mechanism for high-temperature damage to agaves and cacti. Just as for other plants, mul-

Table 4.3. *Early observations on high temperatures tolerated by various cacti*

Species	High temperature tolerated (°C)	Reference
Carnegiea gigantea	60	Biebl (1962)
Gymnocalycium denudatum	56	Nagano, Kuraishi, and Nito (1980) (**A**)
G. mihanovichii	57	**A**
Opuntia ficus-indica	60	Uphof (1916) (**B**)
	63	Konis (1950)
Hybrids of *O. ficus-indica*	60	**B**
O. phaeacantha	55	McGee (1916)
	65	Huber (1932)
Various platyopuntias	55–62	MacDougal (1921); MacDougal and Working (1921, 1922)
Pereskia columbiana	52	Schnetter (1971)
Tephrocactus articulatus	66	Taylor (1972)

Note: Studies cited more than once are assigned a boldface letter the first time they appear, this letter being used for subsequent citations.

tiple factors may be involved. Protein denaturation was initially thought to cause the high-temperature damage, but more recently attention has been focused on the breakdown of membrane function (Steponkus, 1981).

We will begin by summarizing the high-temperature tolerance and its acclimation for agaves and cacti based on inhibition of stain uptake, which will complement our previous discussion of low-temperature tolerance and cold hardening. We will indicate cellular and physiological changes that accompany alterations in treatment time or growth temperature. In the last section of this chapter, we will further consider ecological implications of the responses of agaves and cacti to extreme temperatures.

TOLERABLE TEMPERATURES AND HEAT HARDENING

Just as for lowering the treatment temperatures imposed for 1 h (Fig. 4.11), raising the treatment temperatures will eventually lead to a decrease in the percentage of the chlorenchyma cells that take up a vital stain, neutral red (Fig. 4.14). For *Agave deserti*, *Ferocactus acanthodes*, and *Opuntia ficus-indica* growing at moderately warm day/night air temperatures of 30°C/20°C, half-inhibition of stain uptake occurs at 57°C to 59°C, which are

quite high temperatures. Indeed, most vascular plants succumb to high-temperature damage at 50°C to 55°C (Larcher, 1980; Kappen, 1981). Moreover, the high temperatures for 50% inhibition of cells taking up stain when these plants are grown at hot day/night temperatures of 50°C/40°C range from 63°C to 67°C (Fig. 4.14), which are remarkably high temperatures for vascular plants to tolerate.

All three species acclimate to increasing growth temperatures. For *A. deserti*, the number of degrees Celsius of heat hardening for a 20°C increase in temperature (Fig. 4.14A) are similar to the number of degrees Celsius of cold hardening for a 20°C decrease in temperature (Fig. 4.11A), but for the two cacti the heat hardening is considerably greater than the cold hardening (Fig. 4.14B,C compared with Fig. 4.11B,C). For all three species, the heat hardening upon shifting from 30°C/20°C to 50°C/40°C is more than half complete in 3 days (Fig. 4.14). For *Carnegiea gigantea*, heat hardening upon shifting to higher growth temperatures is half complete in 3 days, and for *Opuntia bigelovii* it is half complete in only 1 day (Smith, Didden-Zopfy, and Nobel, 1984).

The ability to tolerate high temperatures and the ability to acclimate to increasing growth temperatures are widespread among agaves and cacti (Table 4.4). For fourteen spe-

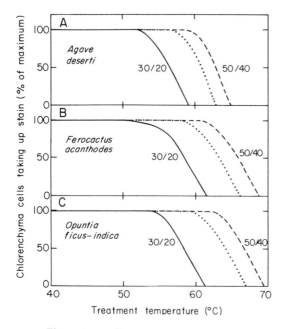

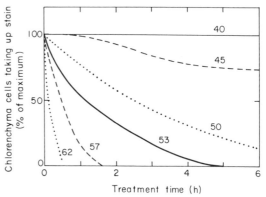

Figure 4.15. Influence of treatment time at various high temperatures on stain uptake by chlorenchyma cells of *Opuntia bigelovii*. Treatment temperatures in °C are indicated next to the curves for plants maintained at day/night air temperatures of 30°C/20°C. Data are adapted from Didden-Zopfy and Nobel (1982).

Figure 4.14. Influence of day/night growth temperatures (indicated in °C/°C next to the curves) on the high-temperature tolerance of *Agave deserti* (**A**), *Ferocactus acanthodes* (**B**), and *Opuntia ficus-indica* (**C**). The dotted line is the high-temperature response three days after shifting from 30°C/20°C to 50°C/40°C, and the dashed line is three weeks after the temperature shift. Stem pieces were incubated for 1 h at a particular treatment temperature, and the percentage of chlorenchyma cells taking up neutral red was subsequently determined. Modified fron Nobel (1984b), Kee and Nobel (1986), and unpublished observations.

cies of agaves, the temperatures leading to a halving of cells taking up stain for plants at day/night temperatures of 50°C/40°C average 62.1°C and the heat hardening for a 10°C increase in growth temperatures averages 4.0°C. For eighteen species of cacti, the analogous high temperature tolerated is 63.9°C, and the heat hardening is fully 5.3°C per 10°C increase in growth temperatures (Table 4.4). Both the high temperatures tolerated by agaves and cacti and the extent of heat hardening are apparently unsurpassed among the higher vascular plants that have been studied so far.

HIGHEST TEMPERATURES TOLERATED

The ability of chlorenchyma cells to take up a vital stain decreases with time at a particular treatment temperature in the stressful range (Fig. 4.15). For instance, for *Opuntia bigelovii* incubated at 57°C, the percentage of the cells taking up stain decreases from 50% at 0.5 h to 17% at 1 h to 0% at 2 h. Thus, incubation for a period longer than the 1-h treatment routinely used (Fig. 4.14, Table 4.4) would lower the high temperature at which the percentage of chlorenchyma cells taking up stain is halved (Fig. 4.15). *Ariocarpus fissuratus* (Fig. 4.16A), *Epithelantha bokei* (Fig. 4.16B), and *Mammillaria lasiacantha* (Fig. 4.16C) can survive three days at the incredibly high day/night air temperatures of 58°C/48°C; *A. fissuratus* and *M. lasiacantha* can even survive three days at 60°C/50°C, but all succumb within three days at temperatures of 63°C/53°C. On the other hand, these plants are uninjured by a 1-h treatment at temperatures of 65°C to 70°C (Nobel et al., 1986).

For plants growing at 50°C/40°C, stain uptake is reduced to zero at about 3°C above the temperature leading to 50% inhibition of the percentage of cells taking up stain (Fig. 4.14). Thus, we expect agaves and cacti to succumb to high temperatures about 3°C above the temperature leading to the readily measured halving of the fraction of cells taking up stain, in general agreement with observation (Nobel and Smith, 1983; Nobel et al., 1986). For instance, for *O. bigelovii* growing at 30°C/20°C, 1 h at 58°C causes the stain uptake to just become zero and killed all five of the

Table 4.4. *Summary of high-temperature tolerance and high-temperature hardening ability of agaves and cacti*

Species	Temperature leading to 50% inhibition of the fraction of chlorenchyma cells taking up stain for plants at day/night temperatures of 50°C/40°C (°C)	High-temperature hardening per 10°C increase in day/night temperatures (°C)
Agave americana	63.8	3.3
A. angustifolia	62.6	5.6
A. bovicornuta	64.6	5.2
A. deserti	62.8	3.1
A. lechuguilla	64.7	5.1
A. multifilifera	61.5	4.3
A. palmeri	61.9	2.8
A. parryi	60.4	2.4
A. parryi var. *huachucensis*	58.8	0.8
A. patonii	63.7	4.4
A. pedunculifera	64.1	6.0
A. rhodacantha	61.1	3.3
A. schottii	61.1	5.9
A. shawii	61.2	4.0
A. utahensis	57.2	1.4
Ariocarpus fissuratus	66.3	6.2
Carnegiea gigantea	62.7	2.8
Coryphantha vivipara	64.1	5.4
Epithelantha bokei	60.5	3.1
Ferocactus acanthodes	66.0	4.0
F. covillei	67.2	8.2
F. viridescens	65.4	6.1
F. wislizenii	67.0	6.6
Lophocereus schottii	64.5	5.0
Mammillaria dioica	64.4	4.9
M. lasiacantha	62.8	4.0
Opuntia acanthocarpa	60.0	6.1
O. basilaris	62.6	5.3
O. bigelovii	60.6	2.8
O. chlorotica	63.9	7.2
O. ficus-indica	66.6	4.2
O. ramosissima	62.5	7.9
Stenocereus thurberi	64.2	5.4

Note: Data were obtained graphically (see Fig. 4.14). High-temperature hardening indicates the increase in the 1-h treatment temperature leading to a 50% inhibition in the fraction of the chlorenchyma cells taking up neutral red as the day/night air temperatures were increased from 40°C/30°C to 50°C/40°C. Data are adapted from Didden-Zopfy and Nobel (1982), Nobel and Smith (1983), Nobel (1984b), Smith et al. (1984), and Nobel et al. (1986).

plants tested, but some stain uptake occurs after 1 h at 57°C (Fig. 4.15) and most plants survive 1 h at 56°C (Didden-Zopfy and Nobel, 1982 and unpublished observations). For this species, the 1-h treatment at 58°C has only moderate effects on leakage of solutes out of the cells, which is an indicator of plasma-lemma damage, but abolishes day/night differences in stomatal opening and nocturnal acid accumulation; some nocturnal stomatal

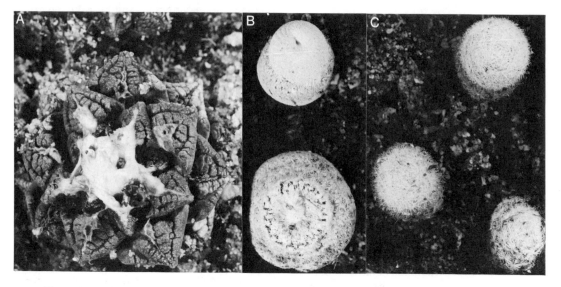

Figure 4.16. Three sympatric species of dwarf cacti from the Chihuahuan Desert: (**A**) *Ariocarpus fissuratus*, (**B**) *Epithelantha bokei*, and (**C**) *Mammillaria lasiacantha*. Plants were collected near Lajitas, Texas (29°16'N, 103°34'W, 730 m) and then maintained in the laboratory (Nobel et al., 1986). The mean stem diameter was 5 cm for the *A. fissuratus*, 3 cm for *E. bokei*, and 2 cm for *M. lasiacantha*.

opening and acid accumulation occur after 1 h at 56°C (Didden-Zopfy and Nobel, 1982), further indicating that high temperatures affect different processes to different extents.

For many C_3 and C_4 species, plants tolerate higher temperatures as their osmotic pressure increases (Levitt, 1980). However, as π^{stem} increases from 0.5 MPa to 1.7 MPa for *O. bigelovii* under drought conditions, the high temperature leading to 50% inhibition of the cells taking up stain decreases by 2.0°C (Nobel, 1983c). Similarly, as π increases from 0.5 MPa to 1.3 MPa for *A. deserti* and *F. acanthodes*, the high temperature for 50% inhibition decreases by 2.3°C (Nobel, 1984c). Thus, the greatest high-temperature tolerances of these desert succulents apparently occur for individuals in the fully hydrated state. Photoperiod variation from 4 h to 20 h has essentially no effect on the high-temperature tolerance of *A. deserti* and *F. acanthodes* (Nobel, 1984c).

It is human nature to be interested in records, including the highest temperature that can be tolerated by agaves and cacti. We will limit our attention to heat treatments lasting 1 h, because extreme temperatures are often experienced for such periods in the field. When maintained at day/night temperatures of 55°C/ 45°C, halving of cells taking up stain occurs at 69°C for *Ferocactus covillei* and *F. wislizenii* (Smith et al., 1984), so presumably they could survive temperatures up to 72°C. For a basal cladode of *Opuntia ficus-indica* in the field, halving of stain uptake at ground level, which is the hottest location on the cladode, occurs at 70°C (Nobel et al., 1986). The other cactus species whose stain uptake is not halved until treatment temperatures exceed 66°C for plants maintained at 50°C/40°C is *Ariocarpus fissuratus* (Table 4.4), and it apparently can survive exposure to 74°C (Nobel et al., 1986). Taking both high-temperature tolerance and heat acclimation into consideration (Table 4.4), we see that other agaves and cacti that should be able to survive temperatures of 70°C for 1 h when properly acclimated to high temperatures include *Agave bovicornuta, A. lechuguilla, A. pedunculifera, Coryphantha vivipara, Ferocactus acanthodes, Lophocereus schottii, Opuntia chlorotica, O. ramosissima,* and *Stenocereus thurberi*. It is remarkable that nearly half of the thirty-two species of agaves and cacti whose high-temperature tolerances have been determined (Table 4.4) may be capable of surviving 70°C, a temperature that is lethal to all other higher vascular plants reported so far!

CHANGES ACCOMPANYING HIGH-TEMPERATURE STRESS

The underlying causes of the ability of agaves and cacti to tolerate such extremely high temperatures have proved elusive to identify. Nevertheless, studies of high-temperature responses have provided insights into various aspects of their physiology.

Let us begin by examining the fatty acid content of the chlorenchyma cells of agaves and cacti (Table 4.5), which has a major influence on membrane structure and function because essentially all of these fatty acids are incorporated into membrane lipids. Compared with other plants, mainly C_3 agronomic species, these desert succulents have more palmitic acid (16:0, meaning sixteen carbon atoms and no double bonds), stearic acid (18:0), and oleic acid (18:1) but much less linolenic acid (18:3); similar conclusions also hold for their chloroplast and mitochondrial lipids (Kee and Nobel, 1985). Adding up all the double bonds between carbon atoms indicates the overall degree of *unsaturation* of a fatty acid. The membrane lipids of the C_3 plants are more unsaturated, meaning that they have more double bonds per carbon atom than do the membrane lipids of agaves and cacti. Because more saturation leads to less membrane fluidity, which is considered to be advantageous in coping with high temperatures (Raison et al., 1980), the membranes of the three desert succulents (Table 4.5) should be better adapted to high temperatures than are representative C_3 plants.

Shifts in gross fatty acid saturation with day/night temperatures cannot account for the observed acclimation of net CO_2 uptake or survival for all three species considered (Kee and Nobel, 1985). In particular, for *A. deserti* the degree of unsaturation (fraction of double bonds) increases at the higher temperatures for chlorenchyma but not for chloroplasts, is unchanged for *C. gigantea*, and decreases as 18:3 acid increases for *F. acanthodes* (Table 4.5). We must conclude either that only certain membranes or parts thereof are involved, and so gross fatty-acid changes mask critical changes in specific regions (hard-to-detect changes in a small fraction of the lipids could affect membrane fluidity), or that fatty-acid changes alone cannot account for the observed high-temperature acclimations for all three species. The increases in fatty-acid saturation for chlorenchyma lipids as the growth temperature increases are consistent with greater high-temperature tolerance only for *F. acanthodes* (Tables 4.4 and 4.5).

Various plants when exposed for a few hours to high temperatures (as well as other stresses) synthesize *heat-shock* proteins, which are relatively small proteins that have been proposed to afford protection against otherwise lethal temperatures (Key et al., 1985; Ananthan, Goldberg, and Voellmy, 1986). Also, new proteins can be synthesized over a period of days when plants are shifted to higher growth temperatures. For *A. deserti*, *C. gigantea*, and *F. acanthodes*, a net synthesis of relatively small proteins with molecular weights from 25,000 to 27,000 (typical of proteins in membranes), as well as other proteins, occurs as the growth temperatures are increased from 30°C/20°C to 50°C/40°C (Kee and Nobel, 1986). Some of these proteins may be involved with the high-temperature acclimation, because they are more prevalent during the acclimation process, such as three days after shifting to a higher day/night temperature (Kee and Nobel, 1986). Others of the proteins may help maintain the high-temperature tolerance, because they remain after the development of the increased high-temperature tolerance is complete. In any case, much remains to be learned about the nature and the function of the heat-shock and other proteins induced by changes in temperature.

PHOTOSYNTHETIC ASPECTS

Photosynthesis involves the absorption of photons by chlorophyll and certain other pigments, eventually leading to the incorporation of CO_2 into specific carbohydrates (Eq. 2.1). Underlying this process is the excitation of an electron in the chlorophyll–protein complex referred to as Photosystem II. This electron, which is replaced by one coming from water, moves along a chain of molecules to another chlorophyll–protein complex, Photosystem I. Accompanying the electron movement is the synthesis of ATP from ADP and phosphate. Electrons excited away from Pho-

Table 4.5. *Abundance of six major fatty acids for chlorenchyma and chloroplast lipids*

Fatty acid		C₃ plants (moderate temperatures)	Agave deserti		Carnegiea gigantea		Ferocactus acanthodes	
Name	Carbon atoms: double bonds		(30°C/20°C)	(50°C/40°C)	(30°C/20°C)	(50°C/40°C)	(30°C/20°C)	(50°C/40°C)
Chlorenchyma lipids (%)								
Myristic	14:0	2	7	5	0	2	2	3
Palmitic	16:0	14	32	25	25	22	28	29
Stearic	18:0	2	9	5	6	7	4	10
Oleic	18:1	7	14	18	29	31	19	27
Linoleic	18:2	16	19	28	11	9	19	18
Linolenic	18:3	58	19	18	29	30	28	14
Chloroplast lipids (%)								
	14:0	1	4	3	1	1	3	2
	16:0	11	28	26	21	19	27	27
	18:0	2	5	4	7	6	4	5
	18:1	4	12	17	25	29	19	25
	18:2	7	20	29	8	9	18	21
	18:3	75	31	22	39	36	29	20

Note: Plants are indicated together with day/night air temperatures during growth. Adapted from Kee and Nobel (1985).

Table 4.6. *High temperatures leading to abrupt increases in chlorophyll fluorescence*

Plant group	Temperature of fluorescence increase (°C)
Ten species of cool climate plants (mean)	42.2
Twelve species of warm climate plants (mean)	47.4
Thirteen perennial species from Death Valley (mean)	49.8
Agave	
Agave americana	55.9
Cacti	
Carnegiea gigantea	56.0
Echinocactus polycephalus	49.8
Ferocactus covillei	57.2
Lophocereus schottii	57.9
Opuntia basilaris	51.5
O. echinocarpa	56.2
O. fulgida	55.9
O. phaeacantha	55.0
Stenocereus thurberi	54.5

Note: Data generally refer to summertime field conditions and are adapted from Downton, Berry, and Seemann (1984).

tosystem I by photon absorption lead to the production of NADPH (a form of nicotinamide adenine dinucleotide phosphate), which, together with ATP, is required for CO_2 fixation (Cogdell, 1983; Nobel, 1983a; Salisbury and Ross, 1985). For many plants, such reactions occurring in chloroplast membranes are the ones that are most sensitive to high temperature (Berry and Raison, 1981; Steponkus, 1981).

When the photosynthetic apparatus is damaged by high temperatures, *fluorescence* by chlorophyll increases abruptly, indicating that some of the energy from the absorbed light can no longer lead to electron flow and so is reradiated as light (Seeman, Berry, and Downton, 1984). Although temperatures for increases in fluorescence only indirectly indicate the high temperatures that photosynthesis can tolerate, the relative temperatures for fluorescence increases can help rank various species in terms of their high-temperature tolerance. Using this criterion, *Agave americana* and nine species of cacti are more tolerant of high temperatures than are the other plant groups tested (Table 4.6). In particular, the fluorescence increase occurs at an average

temperature of 55°C for the ten species of desert succulents and at 47°C for the thirty-five other species.

For plants in general, electron transport involved with Photosystem II is more sensitive to high-temperature disruption than is that associated with Photosystem I, and the intermediate electron transport chain is even more temperature sensitive. In agreement with this, for *A. deserti* and *O. ficus-indica* maintained at day/night temperatures of 30°C/20°C, intermediate electron transport is 50% inhibited when the chlorenchyma is incubated for 1 h at an average temperature of 50°C, Photosystem II activity at 52°C, and Photosystem I activity is 50% inhibited at 55°C (Fig. 4.17).

The temperature responses of all three electron transport activities can acclimate to higher growth temperatures. Specifically, the 50%-inhibition temperature increases an average of 5°C as the growth temperature is raised 15°C (Chetti and Nobel, 1987). At the higher growth temperature, about 19% of the Photosystem I activity remains after 1 h at 70°C (Fig. 4.17) and 7% of the Photosystem II activity remains, but intermediate electron transport is essentially abolished. Even at

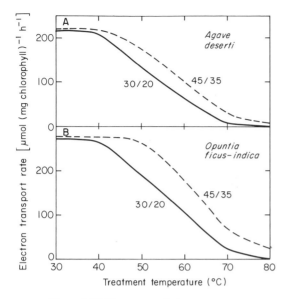

Figure 4.17. Responses of electron transport associated with Photosystem I to high temperatures for *Agave deserti* (**A**) and *Opuntia ficus-indica* (**B**) growing at the day/night air temperatures indicated in °C/°C next to the curves. Sections of chlorenchyma were incubated at a particular treatment temperature for 1 h and then chloroplasts were isolated. Electron transport refers to electrons moving from ascorbate-dichlorophenol indophenol to the electron transport chain and then to methyl viologen. Modified from Chetti and Nobel (1987).

Lowe, 1984). Smaller plants of *Carnegiea gigantea* are more susceptible to fire than are larger plants, especially in regions with considerable combustible material near the ground, such as dried grasses (Steenbergh and Lowe, 1977). Although mortality of *C. gigantea* can be clearly evident one year after a fire (McLaughlin and Bowers, 1982), some deaths may not be evident until five years after the fire (Rogers, 1985).

The frequent fires in the coastal chaparral of California greatly reduce the native cacti there. However, as mentioned in Chapter 1, a hybrid has arisen between native species such as *Opuntia littoralis* var. *vaseyi* and the introduced species *O. ficus-indica*, which was cultivated around the missions founded by Franciscan fathers in the late 1700s (Benson and Walkington, 1965; Gibson and Nobel, 1986). The hybrids and backcrosses with one or the other parent have formed dense thickets that reproduce vegetatively. When fire occurs, only the outermost plants of these hybrid swarms succumb, the inner plants surviving unharmed. Yet neither of the parent species would have survived the fire, because they do not form the requisite dense thickets that lead to protection of the inner plants, and so the hybrid is more successful against fire than either of its parents.

80°C, some Photosystem I activity remains for both *A. deserti* and *F. acanthodes* (Fig. 4.17). Although the damage to the photosystems can be severe at temperatures above 60°C, for these two species Photosystem I apparently can tolerate higher temperatures than can plasmalemma integrity or the plants themselves (Chetti and Nobel, 1987).

FIRE

Another type of high-temperature damage is caused by fire. Fire causes cellular death, usually for cells of the epidermis or outer part of the chlorenchyma, because of exposure to extremely high temperatures for generally very short periods.

Fires in desert grasslands tend to kill agaves and cacti, such as platyopuntias in Texas and *Agave schottii* in Arizona, while favoring grasses, which have their meristems at ground level where it is cooler during passage of the fire (Humphrey, 1958; Niering and

Predicting plant temperatures

The temperatures at particular locations on individual plants can be both predicted and measured. Although accurate prediction entails knowledge of the various interacting energy exchanges that lead to a particular temperature and thus is a worthwhile goal, biology is predominantly an empirical science and so measurements are more convincing to most people. Perhaps this is as it should be – an inaccurate model is of little use. Yet models allow one to do "experiments" that are virtually impossible in nature, such as examining the thermal consequences of changing the spine pattern on a particular cactus or the orientation of leaves on a particular agave. The knowledge gained by empirical testing, when properly combined with the insights of the modeling process, should lead to greater understanding of the environmental biology of agaves and cacti, as well as of other species.

Table 4.7. *Cases where shoot temperatures of agaves and cacti are at least 15°C above air temperature*

Species	Amount above air temperature (°C)	Reference
Agave americana	19	Kuraishi and Nito (1980)
A. salmiana	18	Herzog (1938) (**A**)
Copiapoa cinerea	20	Ehleringer et al. (1980)
C. haseltoniana	20	Mooney et al. (1977)
Ferocactus acanthodes	15	Smith (1978) (**B**)
Mammillaria dioica	17	**B**
Opuntia basilaris	20	**B**
O. bergeriana	26	Huber (1932) (**C**)
O. ficus-indica	16	Konis (1950) (**D**)
	22[a]	**D**
	25	Wallace and Clum (1938)
O. leucotricha	26	**C**
O. "occidentalis"	23	Schroeder (1975)
Unidentified opuntia	16	Gates, Alderfer, and Taylor (1968)
Eleven species of cacti	≥15	**A**

Note: Studies cited more than once are assigned a boldface letter the first time they occur, this letter being used for subsequent citations.
[a] Cladodes horizontal.

Just as there have been many more reports on the high temperatures tolerated by cacti than on the low temperatures tolerated by agaves and cacti combined, there are many more reports on how much above air temperature the stems of cacti can be during the daytime than on how much below air temperature the shoots of agaves or cacti can be at night. Indeed, shoots of over twenty species of agaves and cacti can be at least 15°C above air temperature sometime during the day, and a few can even be 25°C above air temperature (Table 4.7). Such large elevations above air temperature relate to the massiveness of these desert succulents. On the other hand, the thin leaves of *Pereskia columbiana* rarely are even 5°C above daytime air temperatures (Schnetter, 1971). Time lags occur between the highest air temperatures and the highest internal temperatures for massive stems, amounting to 6 h for the center of a *Ferocactus acanthodes* 36 cm in diameter (Mozingo and Comanor, 1975), and 8–12 h for large specimens of *Carnegiea gigantea* (Gates, 1980). The cladodes of *O. acanthocarpa* and *O. bigelovii* can be a few degrees Celsius below air temperature at night

(Gibbs and Patten, 1970), as can the stems of *C. gigantea* (Steenbergh and Lowe, 1976, 1977). These data summarize many of the empirical reports on agave and cactus temperatures published before energy budget predictions facilitated consideration of temperatures of desert succulents in the field.

We should also comment on soil surface temperatures, which can be much greater than maximum air temperatures reported by a weather bureau; the soil temperatures can also have much greater daily variations than reported air temperatures. Maximum surface temperatures of desert soils can exceed 70°C (Sinclair, 1922; Buxton, 1925; Konis, 1950; Hadley, 1970; Despain, 1974; Oke, 1978; Nobel, 1984c) and can even reach 82°C (Körner and Cochrane, 1983). As we will see, such high temperatures play a crucial role for the temperatures experienced by agaves and cacti, especially during the seedling stage.

ENERGY BUDGET TERMS
Although the plant and soil temperatures that we have just described were measured, they can also be predicted using energy bud-

gets, which can also lead to predictions for new environmental conditions or different plant shapes. As we have already indicated, an energy budget summarizes the various ways that energy can enter, leave, or be distributed within an organism. Underlying the use of energy budgets is the *first law of thermodynamics* – known as the conservation of energy – which states that energy cannot be created or destroyed but only changed from one form to another. We will recognize and describe six specific terms: (1) shortwave radiation, (2) longwave radiation, (3) heat conduction, (4) heat convection, (5) latent heat, and (6) heat storage (Derby and Gates, 1966; Lewis and Nobel, 1977; Nobel, 1983a). Energy budgets for agaves and cacti are more complicated than the typical energy budgets for leaves of most plants, because we can usually ignore heat conduction and heat storage for thin leaves, whereas these terms are quite important for the massive shoots of succulents.

Shortwave radiation

Shortwave radiation, characterized by wavelengths that are generally less than 4,000 nm, is emitted by the sun (sunlight) as well as by tungsten, fluorescent, and other types of lamps in environmental growth chambers. The shortwave radiation impinging on a surface is termed shortwave *irradiation* or *irradiance*, which has units of energy area^{-1} time^{-1}, such as J m^{-2} s^{-1} or W m^{-2} (a watt is a joule per second). We recognize two forms of irradiation, *direct* and *diffuse*, the latter being scattered by molecules and particles in the air or reflected from surrounding surfaces before striking the surface of interest. We can represent the shortwave irradiation, S, absorbed by a particular surface by

Shortwave irradiation absorbed

$$= a(S^{\text{direct}} + S^{\text{diffuse}}) \quad (4.1)$$

where the *shortwave absorptance a* indicates the fraction of the impinging shortwave energy that is absorbed by the surface.

Longwave radiation

Longwave radiation, also known as *infrared* or *thermal* radiation, is emitted by all surfaces above a temperature of "absolute zero" ($-273.2°C$, or 0 K), including the surroundings, the sky, leaves of agaves, and stems of cacti. Nearly all of such radiation has wavelengths greater than 4,000 nm (Nobel, 1983a). The amount emitted can be predicted by the Stefan–Boltzmann law for a surface at temperature T^{surf}, a temperature that must be expressed on the absolute, or Kelvin, scale:

Longwave radiation emitted

$$= e\sigma(T^{\text{surf}})^4 \quad (4.2)$$

where e is the *emissivity* (or *emittance*) of the surface, a dimensionless parameter having a maximum value of unity for a perfect or "blackbody" radiator, and σ is the Stefan–Boltzmann constant (5.67×10^{-8} W m^{-2}K^{-4}). For *Agave lechuguilla* and eight species of cacti, e is 0.96–1.00 (Idso et al., 1969; Arp and Phinney, 1980), meaning that their emitted longwave radiation is near the maximum possible. A surface has a net longwave exchange, representing the difference between the longwave irradiance from the environment times the longwave absorptance (a_{IR}), which gives the longwave radiation absorbed, and the longwave radiation emitted (Eq. 4.2).

Heat conduction

Whenever a temperature exists between two regions, heat can be conducted:

$$\text{Heat conduction} = K\frac{\Delta T}{\Delta x} \quad (4.3)$$

where K is the *thermal conductivity coefficient*, a measure of the effectiveness of heat conduction across a specific material, and $\Delta T/\Delta x$ is the temperature gradient, here represented by the drop in temperature ΔT over some distance Δx. We will apply Equation 4.3 when considering heat conducted within the leaves of agaves or stems of cacti; it also applies to heat conducted across an *air boundary layer* (layer of relatively still air adjacent to any surface of a shoot) and between adjacent soil layers.

Heat convection

The heat conducted across an air boundary layer is convected away by the moving, generally turbulent air outside the boundary layer, a process referred to as heat convection.

Using Equation 4.3 and representing the boundary layer thickness by δ^{bl}, we thus obtain

$$\text{Heat convection} = K^{air}\frac{T^{surf} - T^{air}}{\delta^{bl}} \quad (4.4a)$$

$$= h_c(T^{surf} - T^{air}) \quad (4.4b)$$

where K^{air} is the thermal conductivity coefficient of air at the mean temperature of the boundary layer, T^{surf} is the surface temperature of the part of the shoot being considered, T^{air} is the temperature of the air just outside the boundary layer, and h_c is the *heat convection coefficient*, which is generally empirically derived and is quite useful for those surfaces of complex geometry whose air boundary layer thicknesses are not known or not easily calculated.

The average thicknesses of air boundary layers usually depend primarily on the dimensions of the object and the local wind speed. In particular, boundary layers are thinner, and hence heat convection is greater (Eq. 4.4a), for smaller objects and at higher wind speeds. Relatively simple formulae have been developed for certain geometric shapes (Nobel, 1974, 1975, 1983a):

$$\text{Flat plate: } \delta^{bl}_{(mm)} = 4.0 \sqrt{\frac{l_{(m)}}{v_{(m\,s^{-1})}}} \quad (4.5a)$$

$$\text{Cylinder: } \delta^{bl}_{(mm)} = 5.8 \sqrt{\frac{d_{(m)}}{v_{(m\,s^{-1})}}} \quad (4.5b)$$

$$\text{Sphere: } \delta^{bl}_{(mm)} = 2.8 \sqrt{\frac{d_{(m)}}{v_{(m\,s^{-1})}}}$$

$$+ \frac{0.25}{v_{(m\,s^{-1})}} \quad (4.5c)$$

where $\delta^{bl}_{(mm)}$ indicates the average boundary layer thickness in millmeters, $l_{(m)}$ is the mean distance across the flat plate in the wind direction in meters, $d_{(m)}$ represents the diameter of the cylinder or sphere in meters, and $v_{(m\,s^{-1})}$ is the ambient wind speed in meters per second measured just outside the boundary layer. The air boundary layers next to the shoots of agaves and cacti are often about 1 mm thick at modest wind speeds but can become much thicker for large plants under calm air conditions.

Latent heat

During transpiration, water evaporates at air–liquid interfaces in the cell walls within the shoot and then diffuses out across the epidermis, usually by going through the stomatal pores. Because energy is required to evaporate water, generally termed *latent* heat for this change of state from a liquid to a gas, transpiration represents a mode for heat loss. Using Equation 2.6 to define the flux density of water vapor constituting transpiration (J_{wv}), we can represent this term in the energy budget as follows:

$$\text{Latent heat} = H_{vap}J_{wv}$$

$$= H_{vap}g_{wv}\,\Delta N_{wv} \quad (4.6)$$

where the *heat of vaporization* H_{vap} is the energy required to vaporize a unit amount of water at the local shoot temperature.

Heat storage

Massive plant parts, such as the leaves of agaves or the stems of cacti, can store or release considerable amounts of energy as their temperatures change. We can represent the rate of heat storage for a region of volume V as follows:

$$\text{Heat storage} = C_P V \frac{\Delta T}{\Delta t} \quad (4.7)$$

where C_P is the *volumetric heat capacity* (amount of energy required to raise the temperature of a unit volume of some substance by 1°C) at constant pressure, and $\Delta T/\Delta t$ is the temperature change (ΔT) occurring in the time interval Δt.

Five of the energy budget terms (Eqs. 4.1–4.4 and 4.6) are expressed in units of energy area^{-1} time^{-1}, such as watts per square meter, and so represent energy flux densities for a particular surface. On the other hand, heat storage (Eq. 4.7) is a rate for a particular volume and is in the units of energy time^{-1}, such as watts.

MODELS

Once the appropriate equations are identified, the next step in predicting plant temperatures involves the development of *models* – sets of equations and rules governing oper-

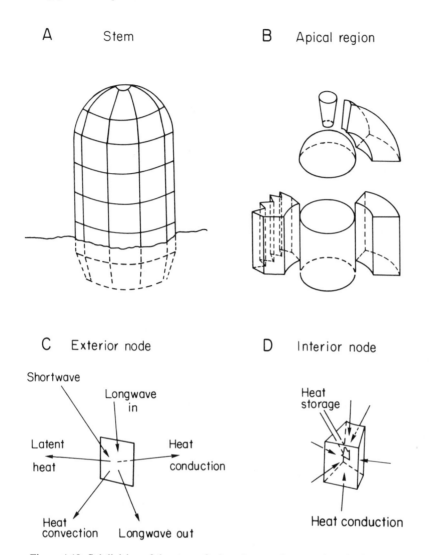

A   Stem

B   Apical region

C   Exterior node

Shortwave

Longwave in

Latent heat

Heat conduction

Heat convection   Longwave out

D   Interior node

Heat storage

Heat conduction

Figure 4.18. Subdivision of the stem of a barrel cactus into a series of subvolumes or nodes, each of which has a balanced energy budget (energy in − energy out = energy stored). (**A**) Overall divisions of stem. (**B**) Nodal geometries near stem apex. (**C**) Energy exchange for an exterior node. (**D**) Energy exchange for an interior node. Modified from Lewis and Nobel (1977) and Gibson and Nobel (1986).

ations between specific choices. Assumptions are inevitable in modeling, because a model represents a simplified abstraction of the interactions involved, and the predictions resulting from solving the equations need to be carefully tested. With the recent advent of high-speed computing, models are playing an increasingly important role in interpreting biological and physical phenomena, such as the influences of plant morphology on surface temperature, which can even affect plant survival. The models for predicting the surface temperatures of agaves and cacti considered here and those for predicting the interception of photosynthetically active radiation for complex geometries (Chapter 5) currently require an hour or so of computing time on a minicomputer for a particular 24-h period of field data. The same computations would require over 200 years of nonstop calculations if done by a hand-held calculator, and thus were impossible a few decades ago. Clearly, we have entered a new era in which plant morphology, plant physiology, and environmental parameters can be considered together in a quantitative fashion.

Let us see how we can approach the problem of an energy budget for a cactus stem. We begin by subdividing the stem into a series of about 200 *nodes* (Fig. 4.18A; Lewis and Nobel, 1977; Nobel, 1978). We can have more nodes in regions of special interest, such as near the stem apex (Fig. 4.18B). Also, we distinguish between surface or exterior nodes, with no volume and hence no heat storage, and interior nodes (Lewis and Nobel, 1977). Each exterior node (Fig. 4.18C) has an interception of shortwave radiation (Eq. 4.1), net longwave exchange (see Eq. 4.2), heat convection (Eq. 4.4), and latent heat loss (Eq. 4.6); any energy imbalance from these terms represents energy conducted to interior nodes (Eq. 4.3). The energy budget of an interior node (Fig. 4.18D) involves heat conduction from all the surrounding nodes plus heat storage (Eq. 4.7).

To appreciate the computing difficulty involved in an energy budget, we note the different temperature dependencies of the various processes (Eqs. 4.1–4.4, 4.6, and 4.7). Specifically, absorption of shortwave radiation is independent of temperature, emission of longwave radiation depends on T^4, heat conduction depends on $\Delta T/\Delta x$, heat convec-

tion depends on $T^{\text{surf}} - T^{\text{air}}$, the latent heat term depends approximately exponentially on temperature because the water vapor content in the chlorenchyma depends approximately exponentially on temperature (Fig. 2.16), and heat storage depends on $\Delta T/\Delta t$. Only with modern computers can such nonlinearity with temperature be readily handled when simultaneously solving the sets of equations involved in the energy budgets of all the nodes.

The development of a particular model involves myriad choices; it also helps identify areas of relative ignorance that require future research. First, a *flow diagram* or logical connection must be established between the different variables and parameters involved (Fig. 4.19). Next the flow diagram must be translated into computer language. At this stage, the programmer must be aware of recent software developments and programming logic, otherwise an inefficient model will be developed. The programmer must also allow for changes in the values of those parameters and variables of particular biological interest. Determining the importance of specific parameters and variables for affecting plant temperatures is a crucial feature of computer

Figure 4.19. (**A**) Flow diagram indicating relationship between programs (CACTUS, GEOM, SUN), primary input files (ENVDAT, TAIR), intermediate-data files (SUNDAT, SPECS), and output files (PLOT, TOUT, TSHORT) for a model used to calculate the temperatures of small cacti and the upper part of the soil (Nobel et al., 1986; Nobel and Geller, 1987). ENVDAT contains hourly microclimate data; TAIR calculates hourly air temperature profiles near the ground; SUN calculates the hourly solar irradiance for each cactus surface node; GEOM calculates nodal volumes, internodal thermal conductances, and other geometry-dependent parameters; CACTUS is the main program that calculates plant temperatures; PLOT contains output data for use by a plotting program; TOUT is a long-format output file containing hourly temperatures for all nodes; and TSHORT is a brief-format output file containing temperatures of only selected nodes. (**B**) Logic flow of program CACTUS. After program initialization, the microclimatic data for each hour are read, and subroutines EXTEMP and INTEMP are called to calculate the temperatures of surface (exterior) nodes and interior nodes, respectively. To pass the convergence test (diamond at bottom left), temperatures of all interior and exterior nodes must be within 0.1°C of the value obtained during the previous iteration. If any of the nodes have not converged, EXTEMP and INTEMP are called again and all node temperatures are recalculated. After all nodes have converged, the temperatures are written out, the next hourly record in ENVDAT is read, and the node temperatures for that hour are calculated. When the end-of-file (EOF) is detected in ENVDAT, indicating that the day has ended, a summary page is written and the program stops. (**C**) Logic flow of subroutine EXTEMP. It is called from CACTUS and calculates the temperature of exterior nodes based on energy exchange with the underlying interior nodes and the external environment. During each pass through the indicated loop, the temperature of one node is determined by an iterative procedure whereby the node temperature is adjusted until the sum, or balance, of the energy budget terms (BAL) for that node is essentially zero, after which control is returned to the main program. (**D**) Logic flow of subprogram INTEMP. It is called from CACTUS and calculates the temperature of each interior node based on the temperatures of the surrounding nodes, the thermal conductances to the surrounding nodes, and heat-storage characteristics. Convergence occurs when all nodes are within 0.1°C of their temperature calculated in the previous iteration. (Diagrams and caption prepared by Gary N. Geller.)

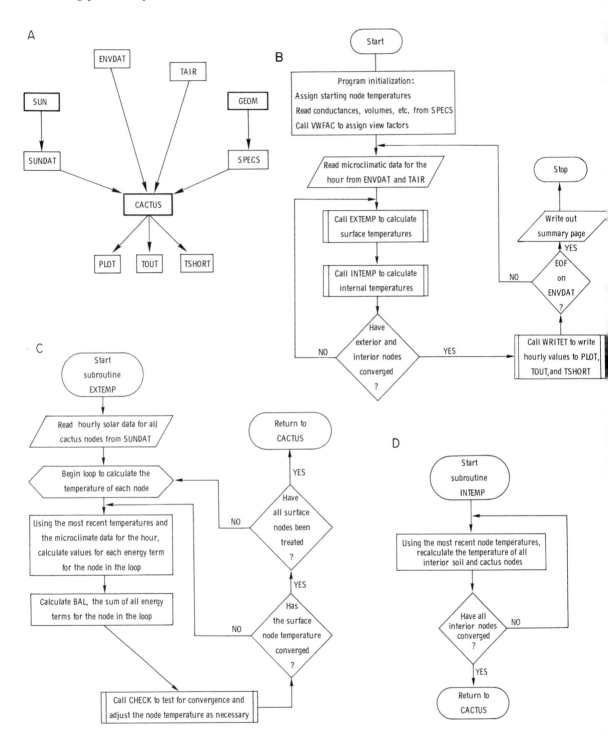

Table 4.8. *Sensitivity analysis by thermal model for daily mean temperatures in response to changes in environmental variables*

Condition (variable)	Change in chlorenchyma temperature averaged over 24-h period (°C)	
	Agave deserti	*Ferocactus acanthodes*
Halve shortwave irradiation ($S^{direct} + S^{diffuse}$)	−1.4	−1.2
Halve longwave irradiation (effectively lowers T^{sky})	−1.1	−1.6
Decrease air temperature 5.0°C (T^{air})	−4.7	−4.3
Decrease wind speed fivefold (v)	+0.1	+0.5
Halve air water vapor content (N_{wv})	−0.7	−0.4

Note: Simulations are for relatively clear days during which air temperature in standard weather bureau enclosures ranged from a minimum of about 10°C to a maximum of about 30°C. The symbols are from Equations 4.1 through 4.6. Data are adapted from Lewis and Nobel (1977), Nobel (1978), Woodhouse et al. (1983), and unpublished observations.

simulations, known as a *sensitivity analysis*, which we will consider in the next subsection. Also, predictions of the model must be compared with field measurements to help "validate" the model (Fig. 4.19). Indeed, the models to be discussed here for predicting temperatures of agave leaves (Woodhouse, Williams, and Nobel, 1983; Nobel, 1984c), cactus stems (Lewis and Nobel, 1977; Nobel, 1978, 1983b; Nobel et al., 1986), and the soil (Nobel and Geller, 1987) have all been appropriately validated in the field, where mean surface temperatures predicted for hourly intervals are generally within 1°C of those measured.

SENSITIVITY ANALYSES

Once a model has been developed and validated, it can be used to examine the influence of various parameters on the physiological processes of agaves and cacti. The thermal model (Figs. 4.18 and 4.19) has permitted an understanding of how changes in size and other variations in morphological features influence the temperatures of the chlorenchyma, the apical meristem, and other regions. We can also investigate the influence of changes in specific environmental variables on plant temperature, as we will discuss first. Then we will

focus on morphological aspects, most of which we will consider again in the next section in a more ecological context.

Environmental factors

To be specific, we will use our thermal model to predict the temperature of a mature *Agave deserti* (Fig. 4.20) and a medium-sized *Ferocactus acanthodes* whose stem is 0.4 m tall. The simulations show the major effect that ambient air temperature has on plant temperature; a 5.0°C decrease in air temperature, with all other variables remaining the same, lowers the average chlorenchyma temperature by 4.3–4.7°C (Table 4.8). If the air temperature were to decrease 5.0°C in nature, the soil temperature and the temperature of the surroundings would also decrease, leading to an even greater decrease in plant temperature than 4.3–4.7°C.

Halving the shortwave irradiation reduces the chlorenchyma temperatures averaged over 24 h by 1.2–1.4°C (Table 4.8), but nearly all the effect occurs during the daytime, when the average temperature decrease is about twice as great. Lowering the longwave irradiation, such as occurs in going from a cloudy to a clear-sky condition, lowers average plant temperatures, but changes in wind-

Table 4.9. *Influence of shortwave absorptance on predicted maximum temperature*

Species	Shortwave absorptance	Maximum temperature (°C)
Agave deserti	0.30	38.9
	0.45	41.9
	0.60	45.1
	0.75	48.4
Ariocarpus fissuratus	0.67	57.8
Epithelantha bokei	0.34	52.6
Mammillaria lasiacantha	0.49	54.2

Note: Simulations for the leaves of *A. deserti,* for which the measured *a* is 0.45, are for a winter day (Woodhouse et al., 1983); simulations for the stems of dwarf cacti used the measured *a*'s with spines intact and are for a summer day (Nobel et al., 1986).

speed have little effect (Table 4.8). Halving the air water vapor content lowers the chlorenchyma temperatures only 0.4–0.7°C (Table 4.8), indicating the relatively small effect of transpirational cooling on average temperatures of agaves and cacti. Actually, most of the stomatal opening and hence transpiration occurs at night, when the lowering of the temperature by halving N_{wv} would be about twice as great as the average for the full 24-h period. In any case, transpirational cooling has a relatively small influence on the temperature of these desert succulents.

Shortwave absorptance

Shortwave absorptance (Eq. 4.1) varies considerably among agaves and cacti, as we can judge, albeit rather inaccurately, based on our own visual perceptions of plant color (although human vision is acute for wavelengths from about 420 to 720 nm, shortwave radiation for sunlight at the earth's surface involves a wavelength span from about 280 to 4,000 nm; Nobel, 1983a). For instance, *Ariocarpus fissuratus* (Fig. 4.16A) appears fairly dark, and indeed its stem has a relatively high shortwave absorptance of 0.67 (Table 4.9). On the other hand, much more shortwave irradiation is reflected by *Epithelantha bokei* (Fig. 4.16B), which has an *a* of only 0.34, and by *Mammillaria lasiacantha* (Fig. 4.16C), which has an *a* of 0.49 (Table 4.9). The lower shortwave absorptance of the latter two species is mainly due to their highly reflective spines.

Figure 4.20. *Agave deserti* at Agave Hill (Figs. 1.1, 1.2, 3.20) with thermocouple wires for temperature determination in place. This plant was employed for validation of a thermal model (Woodhouse et al., 1983). Used by permission from Nobel (1986c). © 1986 by Cambridge University Press.

Of course, a higher *a* means a greater absorption of shortwave irradiation, which will lead to higher plant temperatures (Table 4.9). For *Agave deserti*, simulations indicate that as *a* increases by about 0.1, maximum plant temperatures increase by about 2°C. Similarly, for the three species of dwarf cacti considered, the maximum stem surface temperature increases nearly 2°C for each 0.1 increase in shortwave absorptance (Table 4.9). In a study of fourteen species of agaves, the two with the lowest shortwave absorptance (*A. deserti* with an *a* of 0.49 and *A. lechuguilla* with an *a* of 0.50) occur in open desert scrub

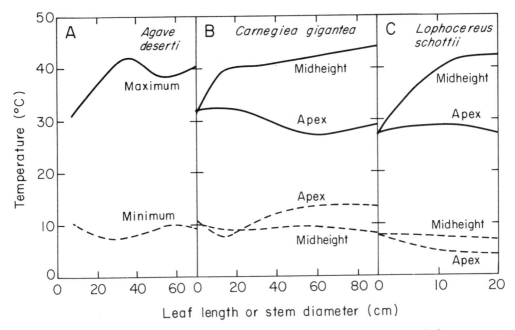

Figure 4.21. Influence of plant size on predicted maximum (——) and minimum (– – –) shoot temperatures over a 24-h period. **(A)** Chlorenchyma temperature of *Agave deserti* when all leaf dimensions are varied proportionally (Woodhouse et al., 1983). **(B)** Surface temperature of stem apex and at midheight of *Carnegiea gigantea* (Nobel, 1978). **(C)** Surface temperature of stem apex and at midheight of *Lophocereus schottii* (Nobel, 1980c).

habitats where they are exposed to high levels of shortwave irradiation; the two with the highest absorptance (*A. multifilifera* and *A. rhodacantha* with *a*'s of 0.65) tend to occur in cooler regions where other vegetation leads to considerable shading (Nobel and Smith, 1983).

Plant size

We recognize that mature agaves and cacti vary greatly in size within a species (the particular case of seedlings will be considered in the next section). Yet, without an energy budget model (Figs. 4.18 and 4.19), it would be difficult to predict the effect of agave leaf length or cactus stem diameter on chlorenchyma temperature. Indeed, even with a detailed model it is hard to interpret all the subtle effects of size on plant temperature (Fig. 4.21).

For a small plant, the air boundary layers are thin (Eq. 4.5), so the convective heat loss is great (Eq. 4.4). High convective heat losses tend to cause the temperature of the shoot surface to approach that of the air, so small plants tend to be near ambient air temperature (Fig. 4.21). As leaf lengths of agaves or stem di-

ameters of cacti increase, the maximum temperatures at the shoot surface or in the chlorenchyma can increase over 10°C under the conditions employed (Fig. 4.21; chlorenchyma and surface temperatures usually differ by less than 0.6°C for these desert succulents under field conditions). However, less change occurs in the maximum temperature at the apex (Fig. 4.21B,C), where the meristem responsible for the growth of the stems of cacti is located. The minimum temperature of the stem apex initially decreases slightly as stem diameter of cacti increases up to about 20 cm, and then it tends to increase, at least for *Carnegiea gigantea* (Fig. 4.21B). The minimum temperature at midheight on the sides of the stem is affected little by stem diameter (Fig. 4.21B,C). Although predicted temperatures depend on the interplay of all the terms in an energy budget, certain processes can be singled out to help interpret the rather complicated changes in temperature with plant size (Fig. 4.21). For example, as size increases, maximal temperatures first increase as the heat convection coefficient decreases, then they decrease as heat conduction into the shoot becomes relatively more important, and then they increase again because the shortwave irradiation cannot be as readily dissipated and thus more of it becomes stored in the larger plants (Nobel, 1978).

Spines and pubescence

Spines and pubescence affect air flow, heat conduction, heat convection, and shortwave absorptance (Hadley, 1972), with important consequences for temperatures near the apical meristem, which affect cactus development and survival. The presence of spines decreases the overall heat convection coefficient h_c (Eq. 4.4b) about 20% for *Ferocactus acanthodes* (Lewis and Nobel, 1977). Spines also decrease the shortwave radiation incident on the stem. Simulated removal of spines of *F. acanthodes* increases the average daytime surface temperature by up to 4°C and reduces the nighttime temperature by up to 2°C, indicating that spines moderate the daily variation of surface temperatures. Simulations increasing the shading by spines from 0% (no spines to 100% (complete shading) approximately halve the day/night variations in mean stem surface temperature for both winter and summer conditions (Nobel, 1983b).

Simulations for *Carnegiea gigantea* show the moderating influence of both spines and pubescence on temperatures in the crucial apical region (Fig. 4.22). Specifically, the minimum temperature in their absence is about 5°C, whereas under the same environmental conditions but with 10 mm of apical pubescence and complete spine shading of the apex, the minimum temperature is just over 12°C. Spines and pubescence exert even greater influences on maximum temperatures (Fig. 4.22); however, minimum temperatures generally have greater influences on the distributional boundaries of this species and other columnar as well as barrel cacti, as we shall see in the next section.

Other factors affecting temperature

As we have already indicated, halving of N_{wv} in the ambient air has only minor effects on the chlorenchyma temperatures of agaves and cacti. Simulations similarly indicate that a large water vapor conductance (100 mmol

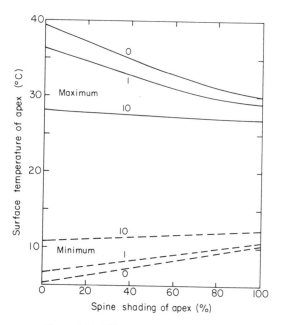

Figure 4.22. Effect of apical spine shading and apical pubescence depth (indicated in mm next to the curves) on predicted maximum (———) and minimum (– – –) apical surface temperatures for *Carnegiea gigantea*. Modified from Nobel (1980c).

$m^{-2}\ s^{-1}$) during the hottest part of the year reduces the nighttime surface temperature of *Ferocactus acanthodes* up to only 4°C below the temperature predicted for no stomatal opening (Lewis and Nobel, 1977). Of course, because of high nighttime temperatures, full stomatal opening by *F. acanthodes* would not be expected in the summertime (Fig. 4.1), even under wet conditions, so the maximal cooling in the field would be less than 4°C. Similarly, full stomatal opening by *Agave deserti* reduces its mean leaf temperature at night by only 3°C compared with no stomatal opening at all (Woodhouse et al., 1983). We again conclude that under field conditions transpirational cooling does not greatly reduce shoot temperatures at night and generally has very little effect during the daytime for these CAM plants.

Simulations have shown that changes in K^{tissue} (Eq. 4.3) and C_P^{tissue} (Eq. 4.7) have only small effects on surface temperatures of agaves and cacti. Specifically, decreasing the thermal conductivity coefficients or the volumetric heat capacities by 10% changes the hourly surface temperature by only up to 1°C and the mean surface temperatures over a 24-h period by only up to 0.1°C for *A. deserti* (Woodhouse et al., 1983), *Carnegiea gigantea*, and *Mammillaria dioica* (Nobel, 1978). Indeed, little error is incurred in substituting the values for water for tissue K and C_P (Lewis and Nobel, 1977), which is consistent with the high water content of these desert succulents.

Simulations also reveal asymmetries in the temperatures of the shoots of these massive plants. For instance, the peak temperature for the west side of the stem of *C. gigantea* is predicted to occur at 3 PM, a time when the east side has been cooling for 6 h, in agreement with field measurements (Nobel, 1978). Also, daytime mean temperatures are higher on the west-facing than on the east-facing sides of *F. acanthodes* and *Opuntia basilaris*, which is consistent with the higher treatment temperatures required to halve the fraction of chlorenchyma cells taking up a vital stain for chlorenchyma from the west-facing side (Smith et al., 1984). As would be expected because of their massiveness, temperatures near the center of large stems of cacti lag considerably behind surface temperatures and also have much smaller daily variations. For a *C. gigantea* 40 cm in diameter, the daily variation in simulated surface temperature at midheight is 33°C; the daily variation at the center is only 6°C, and the peak temperature at the center lags 10 h behind that at the surface (Nobel, 1978). For an individual of *F. acanthodes* 36 cm in diameter, the highest core temperature of 36°C occurs 6 h after the highest surface temperature of 49°C (Mozingo and Comanor, 1975).

Further ecological aspects

In this section we bring together predictions of the thermal model and measured tolerances of extreme temperatures for agaves and cacti. We will consider seed germination and seedling establishment, especially from a thermal point of view. Temperature aspects of nurse plants and of flowering will also be discussed.

SEED GERMINATION

Seed germination requires wet conditions. For instance, soaking or leaching the seeds of various cacti for 0.5 to 3 days can enhance germination (Alcorn and Kurtz, 1959; Wiggins and Focht, 1967; Potter, Petersen, and Ueckert, 1984). Yet temperature is often the variable of primary interest in research on seed germination. We will review temperature responses for germination of seeds of agaves and cacti together with some brief comments on the light requirements (pH responses of germination are discussed in Chapter 6).

For three species of agaves, the optimal temperature for germination averages about 25°C (Table 4.10). The fraction germinating is on average half as great at 12°C below the optimal temperature and at 8°C above it. Light apparently is not required for germination of agave seeds. Germination for *Agave deserti* under wet conditions at 25°C has a halftime of 3 days, and the final germination percentage, achieved in 2 weeks, is usually about 95% (D. A. Baxter and P. S. Nobel, unpublished observations). For nineteen species of cacti, the optimal temperature for seed germination ranges from 17°C to 34°C, again with a mean of 25°C (Table 4.10). Seed germination for cacti is reduced 50% an average of 9°C above or below the optimal temperature. In general, at least a low level of light is required for the greatest fractional germination of cactus seeds (Table 4.10).

Optimal germination of seeds can be influenced by factors other than the temperature, usually held constant, for seeds on moist filter paper. For instance, seeds of *Maihuenia poeppigii* may require a chilling period, as higher fractional germination occurs after two to three months at 5°C to 10°C (Zimmer, 1972). The fractional germination of seeds of *Ferocactus latispinus* and *F. wislizenii* steadily increases for the first three years after harvest (Zimmer, 1980a). On the other hand, seeds of *A. deserti*, *A. lechuguilla*, and *A. parryi* have no dormancy period, as they can exhibit maximum germination immediately upon harvesting (Freeman, 1973, 1975; Nobel, 1977a). Fractional germination of seeds of various species of *Opuntia* increases about 50% upon passage through the digestive tracts of rabbits or cattle (Timmons, 1942; Potter et al., 1984). We also note that a temperature of 60°C, as can occur near the soil surface, does not change the fractional germination for dehydrated

Table 4.10. *Temperature and light responses for the germination of seeds of agaves and cacti*

Species	Temperature (°C)			Light requirements	Reference
	Low for 50% inhibition	Optimal	High for 50% inhibition		
Agave deserti	13	23	33	None (same in dark as in continuous light)	Nobel (1977a)
	14	21	31	None	Jordan and Nobel (1979)
A. lechuguilla	~12	~27	~33	None	Freeman (1973)
A. parryi	~13	~25	~32	None	Freeman (1975)
Carnegiea gigantea	19	25	32	Red light stimulated	Alcorn and Kurtz (1959)
	18	26	36	Required	McDonough (1964) (**A**)
Cereus peruvianus	13	~20	35	—	Zimmer (1969a) (**B**)
Cleistocactus jujuyensis	13	22	32	Low level	Zimmer (1968b) (**C**)
Coryphantha gladiispina	12	26	—	Low level	Zimmer (1968a) (**D**)
Echinocactus grusonii	18	30	—	Low level	**D**
E. platyacanthus	—	24	—	Low level	Trujillo Argueta (1982)
Ferocactus acanthodes	21	29	34	—	Jordan and Nobel (1981)
F. histrix	—	24	—	Low level	del Castillo Sanchez (1982)
F. wislizenii	17	25	36	—	Zimmer (1980a)
Gymnocalycium saglione	17	23	36	Low level	**B**
Mammillaria zeilmanniana	13	~23	32	Low level	**B**
Melocactus caesius	24	34	43	Required, long period best	Arias and Lemus (1984)
Opuntia lindheimeri	20	30	35	—	Potter et al. (1984) (**E**)
O. phaeacantha	18	28	35	—	**E**
Parodia maassii	11	17	26	Low level	**C**
Pereskia aculeata	12	32	38	None	Dau and Labouriau (1974)
Rebutia minuscula	11	17	26	Increases with level	**B**
Rhipsalis fasciculata	17	23	29	—	Zimmer (1980b)
Stenocereus thurberi	19	26	38	Required	**A**

Note: Seeds were usually germinated on moist filter paper at constant temperatures for periods of about two weeks. Emphasis is on the more recent studies examining effects of low, optimal, and high temperatures (only representative examples are taken from the reports of Zimmer, whose light studies are summarized in Zimmer, 1969b). Studies cited more than once are assigned a boldface letter the first time they occur, this letter being used for subsequent citations.

seeds of *A. lechuguilla*, even when such a high temperature is maintained for seven days (Freeman, 1973).

The various environmental and biological factors affecting seed germination are generally viewed as promoting seed germination at times of the year most favorable for seedling establishment. For the three species of agave considered (Table 4.10), soil temperatures close to those optimal for seed germination occur in protected microhabitats toward the end of the summer rainy period. This is when seedling establishment has been proposed to occur (Freeman, 1975; Jordan and Nobel, 1979), especially if there are cloudy periods favoring retention of soil moisture (Freeman, Tiffany, and Reid, 1977). Also, seed germination for *Carnegiea gigantea* occurs mainly in the summer, but establishment of seedlings is less than 1% successful because of ensuing droughts (discussed in Chapter 3) and an even greater threat from insect and rodent damage (Steenbergh and Lowe, 1977). In any case, the generally observed optimal temperatures near 25°C for germination of seeds of agaves and cacti are close to the optimal temperatures for seedling growth.

SEEDLING ESTABLISHMENT AND GROWTH

In the few studies that have been reported, the optimal temperatures for the growth of roots and shoots of seedlings of agaves and cacti are 24°C to 30°C. For 10-day-old seedlings of *Agave deserti*, growth is maximal at 27°C for root length and at 30°C for leaf area (Jordan and Nobel, 1979). For seedlings of *Ferocactus acanthodes*, total plant dry weight gain is maximal at 24°C (Jordan and Nobel, 1981; P. S. Nobel, unpublished observations). In both cases, growth is halved approximately 10°C above or below the optimal temperatures. Initiation of new stem segments is optimal at 30°C for *Schlumbergera truncata*, and their subsequent growth is optimal near 25°C (Rünger, 1979). Such thermal responses are similar to the thermal dependency of seed germination, but observations on other species are needed.

Although seedlings of *A. deserti* tend to occur in microhabitats protected by nurse plants (Figs. 1.1 and 3.20), seedlings of *F.*

Figure 4.23. Seedling of *Ferocactus acanthodes* occurring at an exposed, south-facing, rocky location in the Kingston Range of eastern California (35°47′N, 115°55′W, 1,630 m altitude). Stem diameter is 5 cm.

acanthodes can occur in exposed regions (Fig. 4.23). When the tolerance to extreme temperatures of these two species is compared and thermal acclimation is taken into account, we find that seedlings of *A. deserti* are about 3°C less sensitive to low temperatures but 6°C more sensitive to high temperatures than are seedlings of *F. acanthodes* (Nobel, 1984c). Thus, the association of seedlings of *A. deserti* with nurse plants may be related to protection from high temperatures. Also, high-temperature damage is more evident on *A. deserti* than on sympatric *F. acanthodes* at similar low-elevation, high-temperature sites (Nobel, 1984c). Compared with plants four or more years old, one- and four-month-old seedlings of both species have similar sensitivity to low temperatures but are about 2°C more sensitive to high temperatures (Nobel, 1984c).

Calculations based on models have indicated that temperatures can be most extreme for the smallest seedlings, primarily reflecting the high temperatures near the soil surface, but also relating to the increase in wind speed with distance above the ground and the increase in heat storage as stem volume increases. As seedlings of *F. acanthodes* grow to 1 cm in height, average minimum stem temperatures increase 1°C and average maximum temperatures decrease 3°C (Nobel, 1984c; Nobel et al., 1986). By a stem height of 5 cm, the mean minimum stem temperature has increased about 2°C and the maximum has decreased 5–9°C. Thus, the smallest seedling is the most vulnerable, both in terms of temperature tolerance and the temperatures experienced.

Cacti tnat propagate vegetatively by detached stem segments (Holthe and Szarek, 1985) have an interesting way of avoiding the high temperatures experienced by small seedlings. For *Opuntia bigelovii*, the detachable stem segments are about 4 cm in diameter and 8 cm long, and thus they are exposed to a thermal environment similar to that just discussed for a 5-cm-tall *F. acanthodes* (see also Didden-Zopfy and Nobel, 1982). Moreover, because of high-temperature acclimation (Table 4.4), detached stem segments after being on the ground for a week can tolerate about 4°C higher temperatures than can comparable segments that are still attached (Fig. 4.24). Simulations suggest that the spines also play a thermal role by raising the stem segments of *O. bigelovii* off the ground; this reduces the maximum mean stem temperature by about 1°C if 1 cm off the ground (Nobel et al., 1986) and by 5°C if 2 cm off the ground. Spine shielding of the stem from shortwave irradiation further lowers maximum stem temperatures by up to 3°C. Detachable stem segments are apparently crucial for the reproduction of *O. bigelovii*, because all fifty plants monitored in the northwestern Sonoran Desert during a typical rainfall year produced no viable seed (Nobel et al., 1986).

ROOTS

Exposure of the shoots of agaves and cacti to extreme air temperatures near the soil surface implies that roots could be exposed to similar extreme temperatures. We need to consider both the temperatures that occur in the upper part of the soil and the thermal tolerances of roots. As we have already indicated, many studies have shown that surface temperatures of desert soils can exceed 70°C, and so very high temperatures can be involved.

To evaluate the thermal environments of roots, we need to know how soil temperature T is attenuated as we move a distance z down through the soil (Nobel, 1983a):

$$T = \overline{T}^{\mathrm{surf}} + \Delta T^{\mathrm{surf}} e^{-z/d} \cos\left(\frac{2\pi t}{p} - \frac{z}{d} - \alpha\right)$$

$$(4.8)$$

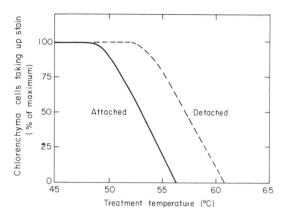

Figure 4.24. High-temperature tolerance for stem segments of *Opuntia bigelovii* either attached 0.5 m above the ground ("Attached") or stem segments of similar size that had become detached and were rooted near the bases of the same plants ("Detached"). Data are adapted from Nobel et al. (1986).

where $\overline{T}^{\mathrm{surf}}$ is the average surface temperature for the period p of interest, which is generally either a day or a year (p is 24 h or 8.64×10^4 s for a day and 365 times longer for a year); ΔT^{surf} is the daily or annual amplitude of the variation in T^{surf} about the mean; d is the *damping depth*, which is the depth in the soil where the variations in temperature have been damped down to $1/e$ or 0.37 of the value at the soil surface; t is time (in the same units as p); and α is the phase shift for the soil surface temperature, whose maximum occurs at a t of $\alpha p/2\pi$. The damping depth for daily temperature changes for both wet and dry desert soils is about 10 cm (Nobel and Geller, 1987). Under representative summer conditions for which the soil surface temperature varies 50°C daily (a ΔT^{surf} of 25°C), maximum temperatures relative to the soil surface would thus be 5°C less at a soil depth of 2 cm and 8°C less at 4 cm.

Let us next consider the high-temperature tolerances and high-temperature acclimation for roots of desert succulents. As for the shoots (Table 4.4), the roots of *Agave deserti* and *Ariocarpus fissuratus* can tolerate quite high temperatures and are also capable of acclimating to higher growth temperatures (Fig. 4.25). However, for day/night air temperatures of 50°C/40°C, the leaves of *A. deserti* can tolerate 7°C higher temperatures than its

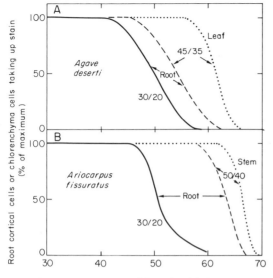

Let us next consider the high-temperature tolerances and high-temperature acclimation for roots of desert succulents. As for the shoots (Table 4.4), the roots of *Agave deserti* and *Ariocarpus fissuratus* can tolerate quite high temperatures and are also capable of acclimating to higher growth temperatures (Fig. 4.25). However, for day/night air temperatures of 50°C/40°C, the leaves of *A. deserti* can tolerate 7°C higher temperatures than its roots and the stem of *A. fissuratus* can tolerate 3°C higher temperatures than its roots. Similarly greater high-temperature tolerances for shoots compared with roots have also been observed for *Ferocactus acanthodes* (Nobel, 1984c), *Epithelantha bokei*, and *Mammillaria lasiacantha* (Nobel et al., 1986).

Because of the greater sensitivity of roots to high temperatures, we would not expect to find roots at the surface of bare soil, because small shoots are generally just at the limit of their high-temperature tolerances there. Indeed, the attenuation of soil surface temperatures by about 7°C, as occurs in the upper 3 cm of soil, is necessary before roots of *A. deserti* and *F. acanthodes* become as vulnerable as shoots to a particular high-temperature episode (Jordan and Nobel, 1984). In agreement with this, roots of these two species are essentially absent in the upper 3 cm of the soil (Figs. 3.8 and 4.26; Jordan and Nobel, 1984), except directly under the shoots or where other vegetation provides considerable shade.

DISTRIBUTIONS

The distributional boundaries of various desert succulents can be affected by environmental factors, especially low temperatures. For instance, *Opuntia imbricata* occurs northward in the southcentral United States through regions where the mean January temperatures are above −1°C (Kinraide, 1978). The northern and the high elevational limits of *Carnegiea gigantea* and other columnar cacti of the Sonoran Desert are primarily dictated by episodic, low, wintertime temperatures (Shreve, 1911; Turnage and Hinckley, 1938). Indeed, the main environmental factor leading to the death of mature plants of *C. gigantea* in Arizona is freezing damage (Steenbergh and Lowe, 1983). Yet cacti do occur in cold regions. *Coryphantha vivipara* ranges at least to 50°N in Alberta and Manitoba, Canada (Nobel, 1981a); *O. polyacantha* to 53°N in Saskatchewan, Canada; and *O. fragilis* all the way up to 58°N in northern Alberta and British Columbia, Canada (Benson, 1982). The last species, which is the most northerly occurring of any native cactus or agave, tends to occur on south-facing rock outcrops in the northern part of its range (Frego and Staniforth, 1986).

We have already indicated that species of agave occurring in high-radiation environments tend to have low shortwave absorptances. Leaf morphology can also vary for a single species in different habitats. For instance, leaves of *Agave deserti* tend to be smaller at higher elevations (Burgess, 1985), resulting in higher minimum leaf temperatures for a particular air temperature (Fig. 4.21A), at least for the smaller leaves with their thinner air boundary layers. Species such as *Agave schottii* and *A. utahensis* (Fig. 2.3) that occupy seasonally very cold habitats also tend to have small leaves, which would again result in higher minimum leaf temperatures and thus would help to mitigate freezing damage. Another species that occupies very cold habitats is *A. parryi*, which has the greatest low-temperature tolerance and the greatest cold hardening of the agaves tested, followed by *A. utahensis* in these categories (Table 4.2). In contrast with the freezing damage that can be observed for agaves in the field (for instance, Fig. 4.12), unambiguous high-temperature damage for mature plants is hard to detect, in

Figure 4.26. Excavated root system of *A. deserti* at Agave Hill. Excavation was at a dry time of the year, so only established roots are visible. Note the shallowness of the roots and yet their absence from the very upper part of the soil, except directly under the shoot (photographed by E. Raymond Hunt, Jr.).

part because of the great tolerance to high temperature for agaves (Table 4.4). Thus, low temperatures apparently restrict the distributions of agaves more than do high temperatures, except possibly during seedling establishment.

High-temperature damage is also hard to observe for mature cacti in the field. Nevertheless, certain relationships exist between morphology, high-temperature sensitivity, and distribution. For instance, the shortwave absorptance increases from *Epithelantha bokei* to *Mammillaria lasiacantha* to *Ariocarpus fissuratus* (Table 4.9), which is the same ranking as the increasing tolerance to high temperatures for these sympatric dwarf cacti (Table 4.4). Of the four species of *Ferocactus* occurring in the southwestern United States, only *F. viridescens* could not survive one week at day/night air temperatures of 55°C/45°C, and it is restricted to coastal habitats, whereas the other three species can inhabit warmer desert regions (Smith et al., 1984). Cylindropuntias like *Opuntia acanthocarpa*, *O. bigelovii*, and *O. ramosissima* are slightly more sensitive to high temperatures than are the more massive barrel and columnar cacti (Table 4.4), and the cylindropuntias also do not attain as high chlorenchyma temperatures in the field.

The diameters of the three columnar cacti considered here – *Carnegiea gigantea*, *Lophocereus schottii*, and *Stenocereus thurberi* – all decrease with decreasing latitude (southward) in the Sonoran Desert (Table 4.11; Niering, Whittaker, and Lowe, 1963; Felger and Lowe, 1967). Because the thickness of the air boundary layer increases with the square root of diameter for cylinders (Eq. 4.5b), the effectiveness of cooling by heat convection increases as the stem diameter becomes smaller (Eq. 4.4a). Simulations indicate that the observed decreases in diameter of the columnar cacti cause the maximum temperature at midheight for a particular air temperature to decrease by 2–6°C from about 32°N to 28°N (see Fig. 4.21B,C), which could be important in preventing high-temperature damage for the mature plants. Indeed, the smallest diameters for these three columnar cacti occur near the hottest part of the Sonoran Desert (Smith et al., 1984).

The changes in morphology with latitude also influence the susceptibility of columnar

Table 4.11. *Summary of morphological parameters affecting temperatures of the apical meristem of various barrel and columnar cacti*

Species	Site (latitude, longitude, altitude above sea level)	Diameter at midheight (cm)	Spine shading of apex (%)	Apical pubescence depth (mm)
Columnar cacti of the northern Sonoran Desert				
Carnegiea	34°50'N, 113°39'W, 650 m	44.1	41	9.8
gigantea	33°31'N, 112°33'W, 400 m	41.0	35	9.2
	31°58'N, 112°48'W, 530 m	36.2	21	9.5
	30°25'N, 112°27'W, 290 m	33.2	9	9.6
Lophocereus	31°47'N, 113°2' W, 430 m	14.3	17	0.0
schottii[a]	30°42'N, 111°56'W, 430 m	14.0	20	0.1
	28°52'N, 110°45'W, 310 m	11.8	19	0.1
	27°38'N, 109°54'W, 80 m	7.3	17	0.1
Stenocereus	32°14'N, 112°48'W, 540 m	15.1	71	0.1
thurberi	30°38'N, 111°36'W, 580 m	13.2	60	0.1
	28°31'N, 111°3' W, 170 m	11.8	44	0.1
Barrel cacti of the southwestern United States				
Ferocactus	36°52'N, 115°16'W, 1590 m	25.0	93	8.5
acanthodes	34°10'N, 112°9' W, 980 m	27.2	65	6.8
F. covillei	31°59'N, 111°39'W, 1160 m	32.6	14	8.2
F. viridescens	33°13'N, 117°22'W, 40 m	13.7	22	1.9
F. wislizenii	32°37'N, 110°48'W, 1340 m	28.5	49	7.4
Cacti from central Chile				
Eriosyce	28°53'S, 70°5' W, 2300 m	14.1	67	6.7
ceratistes	32°54'S, 70°15'W, 1900 m	30.7	23	7.0
Trichocereus	30°11'S, 70°48'W, 2050 m	10.3	40	1.0
chilensis	34°53'S, 70°33'W, 1630 m	14.2	41	0.4

Note: Generally fifteen to twenty plants were averaged for each site. Data are from Nobel (1980c, 1980e).
[a] This species has two morphologically distinct stem forms. Data are for the more freezing-vulnerable immature stems from which grow the thinner, flower-bearing stems whose upper parts are usually covered by long, thin spines.

cacti to freezing damage, which is quite common at the northernmost sites listed for each species in Table 4.11. *Carnegiea gigantea* and *S. thurberi* have approximately the same low-temperature tolerance and the same cold hardening (Table 4.2), and yet *C. gigantea* ranges much further north. The critical tissue with regard to freezing damage is the apical meristem, where necrosis can arrest the growth of the entire stem. This region is protected by a fairly dense covering of spines and a deep pubescent layer for *C. gigantea*, both of which will substantially raise its minimum apical tempera-tures (Fig. 4.22). Also, the increase in stem diameter with increasing latitude tends to raise minimum apical temperatures for *C. gigantea* (Fig. 4.21B); the diameter changes with latitude for *S. thurberi* occur in a diameter range that has little effect on minimum apical temperatures. Thus, the more northerly distributional boundary for *C. gigantea* than for *S. thurberi* is attributed to the greater depth of apical pubescence and the larger stem diameters of the former species. *Stenocereus thurberi* can survive in colder habitats than can *Lophocereus schottii* (Felger and Lowe, 1967), most likely because of reduced apical shading by spines (Table 4.11) and the somewhat greater tissue sensitivity to low temperatures (Table 4.2) for the latter species.

Two mound-forming species, *Echinocereus engelmannii* and *E. triglochidiatus*, have similar stem diameters of 5 cm but form different types of mounds. Mounds formed by *E. engelmannii*, which ranges up to 1,500 m in southeastern California, are relatively open and contain about twenty stems. On the other hand, *E. triglochidiatus* ranges to a higher elevation, 2,500 m, and can have about eighty closely packed stems per mound. In terms of an energy budget, the latter mounds tend to act as one large stem, leading to higher nocturnal temperatures that may account for its distribution extending to higher elevations (Yeaton, 1982). The more open canopy of *E. engelmannii* leads to more convective heat exchange and hence lower daytime maximum temperatures, consistent with it occurring at lower elevations than does *E. triglochidiatus*. However, the tolerances of low and high temperatures have so far not been reported for these species.

Let us next consider the effects of low temperature on the four species of *Ferocactus* that occur in the southwestern United States, all of which had freezing damage on at least 10% of the stems for the sites listed in Table 4.11, except for *F. viridescens*, which had no damage. The greater spine shading of the apex presumably helps *F. acanthodes* range further north than *F. wislizenii*, because tissues of these two species are similar with regard to low-temperature sensitivity (Table 4.2). Less spine shading and slightly less low-temperature tolerance occur for *F. covillei. Ferocactus viridescens* has much less apical pubescence than the other three species, has little spine shading of the apex (Fig. 1.33B), and its tissues are even more sensitive to low temperature (Table 4.2). Consistent with these features, *F. viridescens* occurs in regions that do not experience the low temperatures experienced by the other three species of *Ferocactus* considered.

Two species of cactus from Chile, *Eriosyce ceratistes* and *Trichocereus chilensis*, were studied at their upper elevational limits for a series of sites with similar slopes and exposures at about 1° latitude intervals in the Andes, the most northerly and the most southerly sites being indicated in Table 4.11. The sites range over 4–5° latitude, for which the upper elevational limit of each species decreases about 400 m poleward. A thermal model was used to predict the apical temperatures, adjusting for the morphological differences that occur between sites (Table 4.11). Assuming that each species has the same low-temperature tolerance and acclimation potential throughout its distribution, the predicted upper elevational limits are within 30 m of the measured elevational limits for each species over the entire range (Nobel, 1980e). Hence, the upper elevations for each of these species can be accurately interpreted based on limitations posed by low temperatures at the stem apex.

NURSE PLANTS

As already indicated, nurse plants are important for the seedling survival of *Agave deserti* (Fig. 1.1) and other agaves, presumably because they protect against high-temperature damage (Gentry, 1972; Jordan and Nobel, 1979; Nobel, 1984c; Franco and Nobel, 1988). Nurse plants are also important for seedling establishment by *Carnegiea gigantea*. For example, *Cercidium microphyllum* (palo verde) provides shade that reduces local soil and air temperatures, which is needed for the establishment of *C. gigantea* (Shreve, 1931b). Indeed, nearly all unshaded seedlings of *C. gigantea* in the field in southern Arizona succumbed to high-temperature damage within three months (April–June; Despain, 1974) and none survived one year, even when provided with supplemental watering (Turner et al., 1966). Nurse plants or other shaded microhabitats, such as those provided by rocks, are also important for seedling establishment of *Echinocactus platyacanthus* (Trujillo Argueta, 1982) and *Ferocactus histrix* (del Castillo Sanchez, 1982).

Nurse plants can lower maximum seedling temperatures much more than we might expect based on the influence of shortwave irradiation on mean chlorenchyma temperatures for mature plants (Table 4.8). This disparity reflects the major influence of soil temperature on seedling temperature. Also, wind

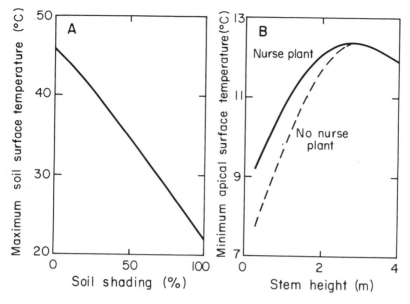

Figure 4.27. Simulated influence of shading by nurse plants on (**A**) maximum soil surface temperature and (**B**) minimum apical temperature for *Carnegiea gigantea* of various heights. Soil surface temperatures for various amounts of shade provided by *Hilaria rigida* were calculated using a soil thermal model for dry conditions in the early spring (Nobel and Geller, 1987). Stem temperatures of *C. gigantea* were calculated using the thermal model (Figs. 4.18 and 4.19) for a field site in northern Arizona near Wikieup (34°56'N, 113°28'W, 845 m altitude) using either the longwave irradiation actually measured at the apex of cacti of various heights that were associated with nurse plants ("Nurse plant") or the longwave irradiation in adjacent exposed regions ("No nurse plant") for early spring environmental conditions (Nobel, 1980d and unpublished observations).

speeds are lower near the ground, so convective heat exchange is not as effective for seedlings as their size might at first indicate. Simulations with a thermal model show that modest shading (40%) by the bunchgrass *Hilaria rigida* (Fig. 1.1) can reduce maximum soil surface temperature by about 10°C, and appreciable shading (80%) can reduce it by nearly 20°C (Fig. 4.27A). It is therefore not surprising that *H. rigida* often acts as a nurse plant for *A. deserti* (Jordan and Nobel, 1979; Franco and Nobel, 1988). Both the thermal model and field measurements show that a 70% reduction in shortwave irradiation reduces the maximum stem temperature by 17°C for a 2-cm-tall seedling of *Ferocactus acanthodes* (Nobel et al., 1986) and by 9°C for a 5-cm-tall seedling (Nobel, 1984c), indicating that shading is more effective in reducing the maximal temperatures of smaller seedlings. Such 70% shading reduces the maximal shoot surface temperature by 7°C for an 11-cm-tall *A. deserti* (Nobel, 1984c). In addition to the protection of seedlings from high-temperature damage, the lowering of soil temperatures by shading can decrease the rate of evaporation of water from the soil surface.

Nurse plants can also provide protection from low temperatures. For instance, survival of a freezing episode in southern Arizona during the winter of 1967/68 by 1-cm-tall seedlings of *Carnegiea gigantea* increased from 58% for exposed plants to 70% when protected by *Cercidium microphyllum* (Steenbergh and Lowe, 1976). Survival of a freeze in January 1971 by similar-sized seedlings increased from 35% when exposed to 78% when protected by screens. For sites where freezing could occur, seven out of twenty *C. gigantea* less than 50 cm tall in exposed locations had freezing damage, but only two out of thirty-three protected by nurse plants were damaged (Nobel, 1980d).

Simulations indicate that smaller plants of *C. gigantea* have lower apical temperatures and thus are more vulnerable to freezing damage (Fig. 4.27B). By increasing the longwave irradiation incident on the stem at night, nurse plants can increase the apical temperatures of the smallest plants of *C. gigantea* by up to 2°C; the finer branches of the nurse plant are close to air temperature and thus replace part of the exposure to the "cold" sky by a radiator at a higher effective temperature (see Eq. 4.2). As a stem increases in height to over 2 m so that its apex projects above the nurse plant, minimum apical temperatures increase (also observed as plant diameter and height increase in exposed locations; Fig. 4.21B), and therefore the protection of the nurse plant is lost at a stage when it is not as useful anyway (Fig. 4.27B). For *Trichocereus chilensis* in central Chile, nearly all freezing damage occurs for stems less than 50 cm tall; simulations again indicate that these stems have lower temperatures at the stem apex than do taller stems (Nobel, 1980e). Thus, nurse plants can extend the distributions of cacti into colder regions (Steenbergh and Lowe, 1977; Nobel, 1980d).

FLOWERING

Temperature can also influence flowering and other aspects of the reproductive process. For instance, the freeze of January 1971 in Arizona reduced seed production by *Carnegiea gigantea* to about 20% of that for the previous year (Steenbergh and Lowe, 1977). Flowers for this species tend to develop more prominently on the eastern and southeastern side of the stem apex, which may lead to more hours with temperatures suitable for flower development than on the west side (Johnson, 1924). For *Trichocereus chilensis, T. litoralis,* and *Eulychnia castanea* in Chile, most or all of the flowers occur on the north (equatorial facing) side of the top of the stems (Rundel, 1974). Ambient temperature also influences the dates for flowering, such periods occurring later for *C. gigantea* at higher latitudes and higher elevations (Steenbergh and Lowe, 1977).

The equatorial tilting of species of *Copiapoa* and *Ferocactus* (Fig. 1.10) has been proposed to be of adaptive value by increasing apical temperatures in the springtime and thus favoring the development of reproductive structures. In particular, at noon in the springtime, *Copiapoa haseltoniana* points toward the sun, which will lead to maximal apical temperatures (Mooney et al., 1977). Similar conclusions were reached for another species from the Atacama Desert of Chile, *C. cinerea* (Ehleringer et al., 1980). Equatorial tilting of *Ferocactus acanthodes* (Ehleringer and House, 1984) and the other three *Ferocactus* species in the southwestern United States (Nobel, 1981b) would have similar consequences. Clearly, the interesting effects of temperature on the reproductive processes of agaves and cacti require further research.

5 Photosynthetically active radiation

Photosynthesis requires the absorption of light energy by pigment molecules, including those in the chloroplasts of agaves and cacti. In particular, light leads to the excitation of chlorophyll or other pigments, such as carotenoids, that pass the excitation induced by light on to chlorophyll. Such excitation of chlorophyll leads to the formation of ATP and NADPH, which are necessary for CO_2 fixation. The range of wavelengths used for CO_2 fixation in photosynthesis is termed *photosynthetically active radiation* (PAR) or *photosynthetic photon flux density* (PPFD) and is conventionally considered as the wavelengths from 400 nm (violet) to 700 nm (red; Nobel, 1983a). Because photons represent packets of energy that can be counted, their number can be expressed in moles; thus instantaneous PAR (or PPFD) can have units of μmol m^{-2} s^{-1}, and *total daily PAR*, the summation or integral of instantaneous PAR levels over the entire daytime, can have units of mol m^{-2} day^{-1}. In this chapter our emphasis shifts from the wavelengths embodied in sunlight, which affect plant temperature (Chapter 4), to those that can induce a photosynthetic response.

CAM plants are sharply distinct from C$_3$ and C$_4$ plants in that most of the net CO_2 uptake by CAM plants occurs at night when there is no light and hence no photosynthesis. For C$_3$ and C$_4$ plants, an instantaneous PAR value in the field during the daytime can be related to the rate of net CO_2 uptake at that time. This contrasts with CAM plants, for which instantaneous daytime PAR levels cannot be related to net CO_2 uptake at a particular time of the night. For CAM plants, the total daily PAR is more relevant to the total net CO_2 uptake at night. Also as we shall discuss in this chapter, PAR and morphology of agaves and cacti are interrelated. The arrangement of the photosynthetic surfaces in space sets limits on PAR interception and hence on CO_2 uptake and productivity. On the other hand, the available PAR levels influence shoot orientation and the production of new shoots.

Gas exchange

We will begin by describing the daytime net CO_2 uptake by agaves and cacti in relation to the instantaneous value of daytime PAR. We will then consider how nocturnal CO_2 uptake and nocturnal acid accumulation depend on the total daily PAR. Net CO_2 uptake in response to various levels of total daily PAR for most other CAM plants is similar to that for agaves and cacti (Kluge and Ting, 1978; Winter, 1985), whose PAR responses have been studied more extensively than those of other groups of CAM plants.

RESPONSE OF DAYTIME CO$_2$ UPTAKE TO INSTANTANEOUS PAR

Although phosphoenolpyruvate (PEP) carboxylase is present in the chlorenchyma cells, most of the daytime CO_2 fixation accompanying net CO_2 uptake for agaves and cacti is presumably catalyzed by ribulose-1,5-bisphosphate carboxylase (Fig. 2.13) using the C$_3$

photosynthetic pathway. This daytime CO_2 uptake would be expected to increase as the instantaneous level of photosynthetically active radiation increases, in a manner similar to that observed for leaves of C_3 plants. In particular, CO_2 uptake often increases up to about 30% of the instantaneous PAR level representing maximal direct sunlight, which is approximately 2,000 μmol m^{-2} s^{-1}, meaning that net CO_2 uptake generally increases up to a PAR level of about 600 μmol m^{-2} s^{-1}. We say that CO_2 uptake is then PAR *saturated*, because further increases in the instantaneous PAR level do not increase net CO_2 uptake for leaves of most C_3 plants (exceptions exist for plants from shady or very sunny habitats; see Larcher, 1980; Nobel, 1983a; Salisbury and Ross, 1985). In general agreement with this, chloroplasts isolated from the cactus *Cereus sylvestrii* operating in the CAM mode exhibit a PAR response consistent with C_3 photosynthesis, reaching 90% saturation of CO_2 uptake at a PAR level of about 300 μmol m^{-2} s^{-1} (Seeni and Gnanam, 1980).

The daytime CO_2 uptake for adult agaves generally occurs during a short period at the beginning of the day and a longer period at the end of the day (Figs. 2.2, 2.4, 2.5, 3.17, and 4.5). Daytime CO_2 uptake can even exceed nocturnal net CO_2 uptake for young agave seedlings (Fig. 2.6) and for *Agave vilmoriniana* at low day/night temperatures (Fig. 4.6). Thus, the daytime stomatal opening and net CO_2 uptake tend to occur just after dawn, when the lower temperature reduces water loss, or in the late afternoon, when the internal CO_2 concentration drops if all of the nocturnally stored malate has been decarboxylated. Moreover, the daytime net CO_2 uptake reaches 90% of PAR saturation at 420 μmol m^{-2} s^{-1} for young seedlings of *Agave deserti*, 400 μmol m^{-2} s^{-1} for overwatered adult plants, and 410 μmol m^{-2} s^{-1} for adult plants of *Agave lechuguilla* (Fig. 5.1A), a PAR response that is consistent with using the C_3 pathway.

As we discussed in Chapter 2, leaves of cacti in subfamilies Pereskioideae and Opuntioideae take up CO_2 primarily during the daytime (Figs. 2.7 and 2.10, respectively). The rate of such net CO_2 uptake also responds to the instantaneous value of PAR (Fig. 5.1B).

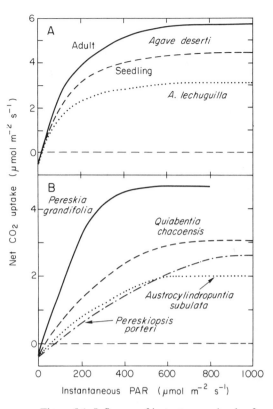

Figure 5.1. Influence of instantaneous levels of photosynthetically active radiation on the daytime rate of CO_2 uptake for agaves (**A**) and cacti (**B**). Data on agaves were obtained in the laboratory for 50-day-old seedlings of *Agave deserti*; overwatered, adult plants of *A. deserti*; and *Agave lechuguilla* (unpublished observations of P. S. Nobel and T. L. Hartsock). Data for cacti are from Nobel and Hartsock (1986a). PAR was measured in the planes of the photosynthetic surfaces, all plants were maintained under wet conditions and at approximately optimal temperatures, and measurements were made during the part of the daytime with the highest sustained rates of net CO_2 uptake.

Net CO_2 uptake at zero PAR (darkness) is negative for the leaves of all four cactus species considered, as is expected for C_3 plants. Net CO_2 uptake reaches 90% of the maximum at a PAR of 310 μmol m^{-2} s^{-1} for *Pereskia grandifolia*, 520 μmol m^{-2} s^{-1} for *Austrocylindropuntia subulata* and *Quiabentia chacoensis*, and 710 μmol m^{-2} s^{-1} for *Pereskiopsis porteri* (Fig. 5.1B). These values for 90% PAR saturation of net CO_2 uptake are within the range usually found for C_3 plants. Moreover, values for 90% PAR saturation near 300 μmol m^{-2} s^{-1} for *Pereskia grandifolia* (Fig.

5.1B) and *P. aculeata* (Gibson and Nobel, 1986) are typical of C_3 species adapted to shady habitats. Consistent with this, these two pereskias are native to tropical woodland or tropical scrub, where they can be considerably shaded by other vegetation.

The net CO_2 uptake by agaves and cacti that occurs just after dawn can generally be reduced by shading, as expected for C_3 photosynthesis. Shading that reduces PAR by 95% nearly eliminates net CO_2 uptake at the beginning of the daytime for *Agave deserti* in the field (Nobel, 1976). However, for *Opuntia erinacea*, a PAR of 700 μmol m^{-2} s^{-1} after dawn apparently initially increases malate decarboxylation and thus increases the internal CO_2 level, thereby decreasing CO_2 diffusion from the atmosphere into the cladodes (Littlejohn and Ku, 1985). Shading late in the afternoon reduces C_3 photosynthesis, which will decrease daytime net CO_2 uptake if the stomata are open; but such shading can increase net CO_2 uptake at the early part of the night by inducing the CAM mode. The situation is complicated not only because more than one enzyme (both PEP carboxylase and Rubisco; Fig. 2.13) can bind CO_2, but also CO_2 can come from multiple sources. Thus, various factors must be kept in mind when interpreting the PAR responses of daytime CO_2 uptake of agaves and cacti.

RESPONSE OF NOCTURNAL CO_2 UPTAKE TO TOTAL DAILY PAR

Even when the stomata are closed and virtually no CO_2 exchange occurs with the environment, daytime photosynthesis is still important for agaves and cacti in the CAM mode. In particular, the CO_2 released from the stored malate must be incorporated into starch or another glucan (Fig. 2.13), a process that requires the conversion of light energy into chemical energy by photosynthesis. The glucan can then lead to PEP, which serves as the acceptor for CO_2 taken up at night. It should thus come as no surprise that nocturnal CO_2 uptake by agaves and cacti is related to total daily PAR (Fig. 5.2), because more PAR will lead to a greater synthesis of carbohydrates during the daytime and hence more acceptor will be available for CO_2 binding during the subsequent night. In fact, for the four CAM

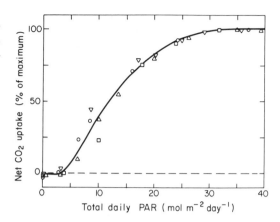

Figure 5.2. Net CO_2 uptake at night as a function of total daily photosynthetically active radiation for two species of agaves and two species of cacti. Data were obtained for *Agave deserti* in the laboratory (O; Nobel, 1984a), for *Agave fourcroydes* in the laboratory ($\triangledown$; Nobel, 1985a), for *Ferocactus acanthodes* in the field ($\triangle$; Nobel, 1977b), and for *Opuntia ficus-indica* in the laboratory ($\square$; Nobel and Hartsock, 1983). Total daily PAR in the planes of the photosynthetic surfaces was obtained by integrating the instantaneous PAR over the daytime. All plants were maintained under wet conditions and at approximately optimal temperatures.

species studied in detail – *Agave deserti, A. fourcroydes, Ferocactus acanthodes*, and *Opuntia ficus-indica* – the responses of nocturnal CO_2 uptake to total daily PAR are remarkably similar (Fig. 5.2); over 95% of the net CO_2 uptake by these species occurs at night under the conditions employed, except for *A. fourcroydes*, where 85% is at night.

Net nocturnal CO_2 uptake is negative up to a total daily PAR of about 3 mol m^{-2} day^{-1} for the four species considered, which is the *PAR compensation* level on a total daily PAR basis (Fig. 5.2). Such a total daily PAR corresponds to an instantaneous PAR level of 70 μmol m^{-2} s^{-1} for a 12-h photoperiod (an illumination of about 3,500 lux or 350 foot-candles). Although such an illumination level is considered adequate for reading, as well as for certain house plants and understory species in forests, it will lead to no net CO_2 uptake and hence no growth for agaves and cacti. We thus surmise that agaves and cacti require rather high PAR environments, as was long ago suspected by Shreve (1931a). For the four species considered (Fig. 5.2), 90% PAR saturation of nocturnal net CO_2 uptake occurs at 23 mol m^{-2} day^{-1}, a number that should be kept in mind when we predict available PAR levels at various latitudes and seasons later in this chapter. Such a PAR level for 90% saturation of nocturnal net CO_2 uptake is consistent with studies on other CAM plants such as *Ananas comosus* (pineapple; Nose et al., 1977; Sale and Neales, 1980), *Bryophyllum daigremontianum* (*Kalanchoe daigremontiana*; Kaplan, Gale, and Poljakoff-Mayber, 1976), and *Tillandsia usneoides* (Spanish moss; Kluge et al., 1973; but see Martin, Eades, and Pitner, 1986).

NOCTURNAL ACID ACCUMULATION

Nocturnal CO_2 uptake is accompanied by nocturnal acid accumulation for both agaves and cacti (Chapter 2). Because acidity increases are far easier to measure than are CO_2 exchanges, especially in the field, they have been measured for more species. Specifically, for five species of agaves (Fig. 5.3A) and eight species of cacti (Fig. 5.3B), nocturnal acid accumulation increases up to a total daily PAR of about 30 mol m^{-2} day^{-1}. The increase is fairly linear with total daily PAR up to about 5 mol m^{-2} day^{-1} (Fig. 5.3), a range over which CO_2 uptake is mainly negative (Fig. 5.2). At low PAR, respiration, which releases CO_2 internally, dominates CO_2 assimilation, causing net CO_2 uptake to be negative; the CO_2 released internally by respiration can be incorporated into malate (Fig. 2.13) and thus lead to a positive nocturnal acid accumulation. Acid accumulation reaches 90% of maximal at a total daily PAR of about 20 mol m^{-2} day^{-1} for the agaves (Fig. 5.3A) and the cacti (Fig. 5.3B) considered. Indeed, the PAR responses are remarkably similar for many species in the two taxa.

The relatively easy measurement of nocturnal acid accumulation has provided a number of insights into the PAR responses of agaves and cacti. The youngest three-quarters of the leaves of a rosette of *Agave deserti* are basically indistinguishable in their PAR responses; the oldest one-quarter of living leaves (including those for which only 20% of the surface area is still green) have an average of 40% less nocturnal acid accumulation at a particular total daily PAR than do the youngest

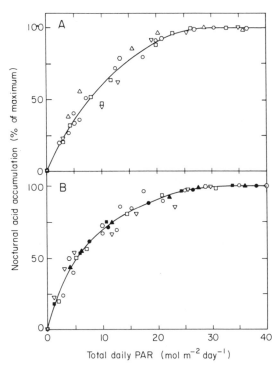

Figure 5.3. Nocturnal acid accumulation as a function of total daily photosynthetically active radiation for agaves (**A**) and cacti (**B**). Data were obtained in the field for *Agave angustifolia* (○; unpublished observations of P. S. Nobel on six-year-old plants examined 30 km eastsoutheast of Oaxaca, Oaxaca at 16°56′N, 96°25′W, 1,550 m), *A. deserti* (△; Woodhouse, Williams, and Nobel, 1980), *A. fourcroydes* (○; Nobel, 1985a), *A. salmiana* (□; Nobel and Meyer, 1985), *A. tequilana* (▽; Nobel and Valenzuela, 1987), *Ferocactus acanthodes* (○; Jordan and Nobel, 1981), *Opuntia basilaris* (●; Nobel, 1980a), *O. bigelovii* (■; Nobel, 1983b), *O. chlorotica* (▲; Nobel, 1980a), *O. echios* (○; Nobel, 1981b), *O. ficus-indica* (△; Acevedo et al., 1983), *Stenocereus gummosus* (□; Nobel, 1980a), and *Trichocereus chilensis* (▽; Nobel, 1981b).

leaves (Woodhouse et al., 1980). When the photoperiod is kept constant at 12 h by a low instantaneous background PAR and the total daily PAR is kept at 10 mol m^{-2} day^{-1} by providing various instantaneous levels for specific periods, nocturnal acid accumulation by *Opuntia ficus-indica* is 90% of maximal when the instantaneous daytime PAR is 700 μmol m^{-2} s^{-1} (Fig. 5.4). Again, this is consistent with the PAR response of C_3 photosynthesis taking place during the daytime for C_3 and CAM plants. We also note that at a

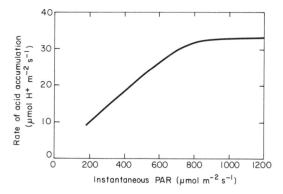

Figure 5.4. Influence of instantaneous PAR level on nocturnal acid accumulation by *Opuntia ficus-indica*. A background level of 40 μmol m^{-2} s^{-1} was provided throughout the 12-h photoperiod; for one week, the indicated instantaneous PAR was provided near the center of the day so that the total daily PAR was 10 mol m^{-2} day^{-1} in all cases; and the rate of nocturnal acid accumulation indicates the total acid accumulated at night divided by the time at the elevated PAR (acid accumulation in response to the background PAR was corrected for). Modified from Nobel and Hartsock (1983).

constant instantaneous PAR, which is generally the case for plants in environmental growth chambers, nocturnal acid accumulation is about 8% greater than for the same total daily PAR in the field, because diurnal variations in the field cause the PAR for some of the time to be close to or above saturation, in which case the excess photons are not effective for photosynthesis (Nobel and Hartsock, 1983).

Before further discussing the PAR relations of agaves and cacti, we will comment on the consequences of a total daily PAR of 20 mol m^{-2} day^{-1}, which leads to about 90% of maximal nocturnal CO_2 uptake (Fig. 5.2) and of maximal nocturnal acid accumulation (Fig. 5.3). Such a total daily PAR corresponds to an average instantaneous PAR during a 12-h day of approximately 500 μmol m^{-2} s^{-1}, which is close to the saturation value for C_3 photosynthesis. Because CAM plants utilize the C_3 pathway during the daytime, a total daily PAR near 20 mol m^{-2} day^{-1} can thus lead to maximal fixation of CO_2 released internally from malate during the daytime, which then sets the stage for maximal CO_2 uptake at night by synthesizing the maximal

amount of the necessary glucan (Fig. 2.13). Thus, orientation of the photosynthetic surfaces of agaves and cacti leading to the interception of about 20 mol m^{-2} day^{-1}, if the instantaneous PAR is appropriately distributed over the daytime, would be optimal for nocturnal CO_2 uptake; higher total daily PAR levels would lead to PAR saturation of CO_2 fixation and hence inefficient use of the additional photons. At the same time, a total daily PAR of 20 mol m^{-2} day^{-1} is on the verge of PAR limitation, because any reduction in PAR reduces net CO_2 uptake.

Agaves

We next examine the influence of specific levels of total daily PAR on the daily pattern of CO_2 exchange for three species of agave. Then we will consider the PAR levels in the planes of the leaves, using both field observations and models, to investigate the influence of leaf arrangement or the slope of the ground on PAR interception for agaves. We will also use PAR models to investigate spacing and, hence, productivity of agaves under various environmental conditions in Chapter 7.

DAILY PATTERNS FOR CO₂ EXCHANGE

As the total daily PAR in the planes of the leaves is raised from a low level (3–4 mol m^{-2} day^{-1}), which could occur on an overcast day, to an intermediate level (15–16 mol m^{-2} day^{-1}) to a high level (23–28 mol m^{-2} day^{-1}) representative of a sunny day in the summer, net CO_2 uptake during the day and at night increases (Fig. 5.5). At the low PAR level, only net CO_2 release occurs during the daytime for *Agave deserti* (Fig. 5.5A) and *A. fourcroydes* (Fig. 5.5B), and net CO_2 uptake over the 24-h period is close to zero, whereas *A. lechuguilla* (Fig. 5.5C) exhibits a small amount of daytime CO_2 uptake and a positive net CO_2 uptake over a 24-h period. Increasing the total daily PAR from the intermediate to the high level increases net CO_2 uptake over the 24-h period by an average of 30% (Fig. 5.5). Such responses of net CO_2 uptake to PAR level underlie the importance of the interactions among the morphology, PAR interception, and net CO_2 uptake of agaves.

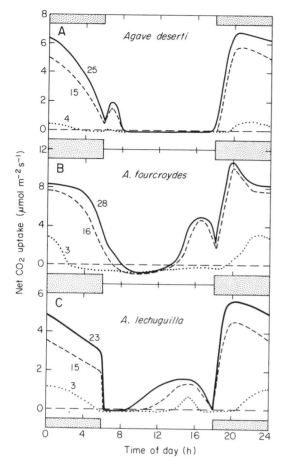

Figure 5.6. Geometric model for *Agave deserti* in which each leaf is represented by nine planar surfaces. For simplicity, only thirty of the sixty leaves on the *A. deserti* in Figure 4.20 are shown. Stippled areas represent the stem. Used by permission from Woodhouse et al. (1980). © 1980 by the Botanical Society of America.

Figure 5.5. Net CO_2 uptake over a 24-h period by leaves of agaves at the total daily PAR indicated in mol m^{-2} day^{-1} next to the curves. The plants were maintained under wet conditions and at approximately optimal temperatures, and the PAR was measured in the planes of the leaves. Data for *Agave deserti* (A) are from unpublished observations under the conditions of Nobel (1984a), for *A. fourcroydes* (B) from Nobel (1985a), and for *A. lechuguilla* (C) from Nobel and Quero (1986).

MODEL

The geometry of a basal rosette of leaves radiating into various directions, characteristic of agaves and many other taxa, is difficult to represent analytically. Nevertheless, a computer model for an individual of *Agave deserti* with sixty living leaves (Fig. 4.20) has been developed (Fig. 5.6; Woodhouse et al., 1980). Each leaf is represented by three planar surfaces for its upper side and six for its lower side, giving a total of 540 surfaces for the whole plant. The main geometric problem in the development of this particular model is determining which leaf shades what part of another leaf at various times of the day. Specifically, when the entire plant with its 540 surfaces delimited by 1,080 lines is presented on a computer monitor, deciding which line belongs to which leaf surface is difficult. And although the "plant" can be easily rotated on the monitor and the leaves exposed to direct PAR can be readily indicated, determining the location of shadows on individual leaves is challenging (another modeling approach will be described for agaves and platyopuntias in Chapter 7).

The predictions by the model for PAR interception at four canopy positions, in the four cardinal directions, and at various locations on the leaves are generally within 10% of measured values (Woodhouse et al., 1980). Indeed, one reason for choosing a rosette with so many leaves (Fig. 4.20) is to allow validation for many different surface orientations. Once the PAR model has been validated, various *simulations* can be performed in which the values of certain parameters or variables are changed – changes that are often not feasible for actual plants. In one simulation, the effect of removing every other leaf is tested. Although this increases the average total daily PAR per unit leaf surface area for the remain-

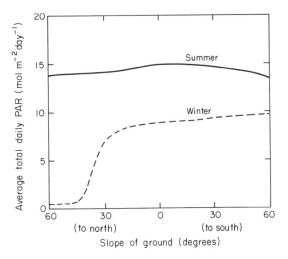

Figure 5.7. Influence of a north or south slope of the ground on the total daily PAR averaged over the leaf surfaces of an *A. deserti* with sixty leaves (Fig. 4.20), as simulated by a PAR interception model (Fig. 5.6). Data are for clear days near the indicated solstices for the Agave Hill site at 33°38′N and are adapted from Woodhouse et al. (1980).

ing leaves by about 50%, it lowers nocturnal acid accumulation for the whole plant by 30% because of the halving of leaf surface area. In another simulation, the direct PAR is halved throughout the day and the diffuse PAR is maintained unchanged, which might occur on a partially cloudy day. This reduces the nocturnal acid accumulation for the whole plant by only 20% for a winter day at 34°N, indicating the importance of diffuse radiation for nocturnal acid accumulation. Indeed, the lower surfaces of leaves in the lower part of the canopy and on the northern side of the plant as well as the upper surfaces of the leaves on the southern side of the plant receive essentially only diffuse PAR throughout a day near the winter solstice.

The model can also be used to study the influence of the slope of the ground on PAR interception by *A. deserti* (Fig. 5.7). Near the summer solstice, the total daily PAR averaged over the leaf surfaces remains at 14 ± 1 mol m^{-2} day^{-1} for simulated slopes of the ground from 60° to the north all the way to 60° to the south. However, because of the lower trajectory of the sun across the sky near the winter solstice, the angle of slope to the north

has a marked effect on PAR interception at that time of the year. Compared with the total daily PAR near the winter solstice for plants on level ground, which averages about 9 mol m^{-2} day^{-1} in the planes of the leaves, northward slopes reduce the intercepted PAR at that time of the year by 21% at a slope of 30° and by 91% at 45° (Fig. 5.7).

As we indicated in Chapter 3, rainfall tends to occur both in the summer and in the winter for the Agave Hill site, although occurrence of the winter rainfall is somewhat more reliable from year to year (Fig. 3.3). Thus, the lower PAR during the winter for the steeply north-facing slopes should reduce the productivity and perhaps the number of *A. deserti* per unit ground area compared with south-facing slopes. In addition, slopes influence soil water relations as well as soil and plant temperatures. High temperature can limit the distribution of *A. deserti* at low elevations and low temperature at high elevations (Nobel and Smith, 1983). At intermediate elevations where limitations by temperature would be minimized, compared with slopes of 16–30° the number of *A. deserti* per square meter of ground area decreases 4% for south-facing slopes of 31–45° and 30% for 46–60°; presumably, this mainly reflects influences of slope on soil water (Woodhouse et al., 1980). Although the number of *A. deserti* per unit area is the same for north-facing slopes of 16–30° as for south-facing slopes of the same magnitude at the site considered, it decreases 26% more for north-facing slopes of 31–45° and 62% more for slopes of 46–60° than for the same slopes to the south. This reduction in number of plants per unit ground area for north-facing compared with south-facing slopes at intermediate elevations can be attributed to decreases in intercepted PAR during the wintertime for steeply north-facing slopes (Fig. 5.7). Specifically, on north-facing slopes of more than 40°, the average PAR during the wintertime decreases to levels insufficient for CO_2 uptake over a 24-h period by *A. deserti* (Fig. 5.2), so no growth would be expected. Thus, the model allows a quantitative estimate of the influence of plant geometry and ground slope on PAR interception at various times of the year, which can be directly related to CO_2 uptake by the leaves.

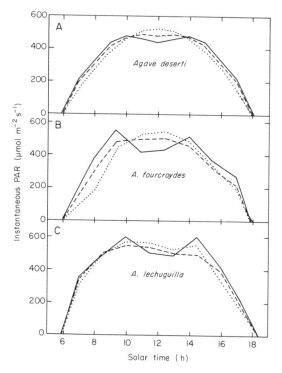

Figure 5.8. Average values of instantaneous photosynthetically active radiation for three species of agave at various canopy positions during the course of a day near an equinox. **(A)** *Agave deserti*, determined by simulation for 34°N (Woodhouse et al., 1980; Nobel, 1985c). **(B)** *A. fourcroydes*, measured at 21°N (Nobel, 1985a). **(C)** *A. lechuguilla*, measured at 25°N (Nobel and Quero, 1986; Nobel, 1986c). Data are averaged over both sides of the upper (youngest) one-third of the leaves (———), middle one-third (– – –), or lowest (oldest) one-third (· · · ·) on predominantly clear days.

PAR DISTRIBUTION OVER THE ROSETTE

The PAR model can also be used to predict the PAR intercepted by leaves in various parts of the canopy of an agave (Fig. 5.8A), leading to a very interesting conclusion. In particular, the orientation of the leaves of *Agave deserti* is such that the average PAR intercepted is essentially the same throughout the day at the top (youngest leaves), middle, and lower (oldest leaves) parts of the canopy. The dip in PAR interception near noon for the youngest leaves is caused by their proximity to the vertical, and the greater interception of PAR by the oldest leaves at that time is caused by their proximity to the horizontal (Fig. 5.8A).

The predicted PAR distribution (Fig. 5.8A) over the rosette of *A. deserti* is similar to that measured (Fig. 5.8B,C) for agaves of quite different morphology: *A. fourcroydes* (Fig. 1.7) and *A. lechuguilla* (Fig. 1.8). Specifically, the PAR intercepted by the youngest leaves tends to be greater than for the oldest leaves during the early and the late parts of the day and less near the middle of the day, with leaves of intermediate age exhibiting intermediate PAR interception patterns (Fig. 5.8). The average value of the instantaneous PAR is also of interest. For the three species considered, the average instantaneous PAR at the three canopy positions considered never exceeds 600 μmol m^{-2} s^{-1}, a level at which CO_2 fixation by the C$_3$ pathway approaches PAR saturation. Thus, the rosette geometry not only tends to distribute the PAR uniformly, but also the average levels tend to be below saturation, and hence the PAR can be used efficiently.

Before assuming that the distribution of PAR over the rosettes of agaves is uniform, we should indicate certain asymmetries in PAR interception. For instance, the lower (abaxial) side of an east-pointing leaf tends to intercept more PAR at the beginning of the day, and the upper (adaxial) side of the same leaf tends to intercept more PAR near midday; yet the total daily PAR, which can be related to nocturnal CO_2 uptake in the CAM mode (Fig. 5.2), can be similar for the two surfaces. Actually, the lower leaf surface receives 38% less total daily PAR than the upper surface for the rather open rosettes of *A. deserti*, 40% less for *A. lechuguilla*, and 68% less for the rather dense rosettes of *A. fourcroydes*. At an equinox for *A. fourcroydes* at 21°N and *A. lechuguilla* at 25°N, the average total daily PAR intercepted by leaves in the southern quadrant is 27% less than for leaves in the other cardinal directions, which is a consequence of leaf orientation relative to the sun's trajectory (Nobel, 1985a; Nobel and Quero, 1986). Also, at a point on a leaf one-sixth of the distance from the base to the tip, PAR interception averages 36% less than at five-sixths of the distance. Nevertheless, average PAR, both instantaneous and total daily, is remarkably uniform over the leaf surfaces of the agave rosettes (Fig. 5.8).

As we have just indicated, the rosette geometry of agaves leads to a rather uniform distribution of PAR over their leaves. This tends to minimize the fraction of leaves below PAR compensation, which would have a net CO_2 loss, as well as the fraction of leaves near PAR saturation, for which an increase in the photon flux density would lead to no increase in net CO_2 uptake, so any additional photons would not be effective for photosynthesis. Productivity for many agricultural plants is maximized when PAR is distributed fairly uniformly over the leaves (Loomis and Williams, 1969). We note that the common and highly productive *Beta vulgaris* (sugar beet) has vertical leaves toward the top of its canopy and progressively more horizontal leaves toward the bottom of its canopy, thereby distributing PAR more uniformly and hence increasing productivity (Hodáňová, 1979). The leaf orientation and overall canopy architecture achieved by the rosettes of agaves are actually one of the most efficient PAR interception strategies with respect to maximizing CO_2 fixation.

PRUNING

Older plants of *Agave tequilana* (Fig. 1.5A) are commonly "pruned," because farmers believe pruning increases the size of the piña (Fig. 1.5B) and hastens its maturity, although data on such effects have apparently not been published. Pruning also prevents the moth *Acenthroneme hesperiarsis* from laying its eggs near the base of leaf terminal spines. In practice, pruning involves using a machete to cut off the distal, terminal-spine-bearing quarter of leaf length for leaves projecting between the rows of plants, which facilitates passage along the rows, and also cutting the upper part of the central cone of folded leaves. Such removal of leaf material will affect the PAR distribution for the remaining canopy and is fundamentally different from severing the inflorescence of *A. salmiana* (Fig. 1.4A), a practice that causes its stem to swell as the carbohydrates and other material destined for the inflorescence are diverted to the base of the stem.

To examine the consequences of pruning for nocturnal acid accumulation by *A. tequilana*, three treatments were considered for seven-year-old plants: (1) no pruning (the control); (2) pruning 20–30 cm from the distal portions of leaves projecting between the rows plus the distal 50 cm from the central cone, done eighteen months before acidity measurement; and (3) same as (2) except that the leaves projecting between rows plus an additional 30 cm from the central cone are repruned six months before measurement. At the time of measuring nocturnal acid accumulation during a wet period in late summer, the single pruning had removed 8% of the leaf area and the double pruning had removed 20% (A. G. Valenzuela and P. S. Nobel, unpublished observations). For four plants under each treatment, the nocturnal acid accumulation was measured at the middle of the leaf surface area on the upper side of leaves pointing in the four cardinal directions from the upper, middle, and lower parts of the canopy (forty-eight measurements under each treatment). Compared with the control, nocturnal acid accumulation averaged 9% higher for one pruning and 21% higher for two prunings. Pruning apparently raised the PAR available for the remaining leaves, with a corresponding increase in nocturnal acid accumulation (incident PAR was below 15 mol m^{-2} day^{-1}, so nocturnal acid accumulation was approximately proportional to PAR; Fig. 5.3A). However, because of the decrease in leaf area by pruning, nocturnal acid accumulation per plant was similar for the three treatments. Clarification of other possible effects of pruning on stem growth, such as changes of hormonal balances or the types of carbohydrates and their allocation patterns, must await future research.

The influences of pruning were also examined for *Agave salmiana* at various sites in the Mexican states of Hidalgo, Mexico, and Tlaxcala. Such pruning is done to prevent theft of the cuticle, which is stripped off the upper surface of leaves still folded about the central cone (Fig. 5.9A) and can damage the whole plant if crudely done. The large pieces of cuticle so removed (Fig. 5.9B) are used to wrap rabbit, chicken, and other meats, forming a delicacy termed "mixiote" that is especially popular during the Christmas season. Removal of the distal 40 cm by pruning (Fig. 5.9C), which is approximately 30% of the leaf length and 17% of the leaf surface area, did not in-

Figure 5.9. *Agave salmiana* growing in central Mexico near Zinguilucan, Hidalgo. **(A)** Inner leaves have been pulled back (note crease marks on lower surface) to allow removal of cuticle from upper surface of leaves still folded about the central spike for seven-year-old plants. **(B)** Cuticle removed from a leaf in **(A)**. **(C)** Pruning of distal portion of leaves [on farm 12 km southwest from that in **(A)**]. **(D)** Edmundo García Moya clipping tips of leaves as part of a collaborative project on productivity.

fluence nocturnal acidity increases at various locations in the canopy that received the same total daily PAR. For instance, for leaf surfaces facing east or west at midcanopy, the nocturnal acid accumulation averaged 930 mmol m^{-2} for four pruned plants and 920 mmol m^{-2} for four control plants during a period of clear days in late summer when the soil was wet and the minimum nighttime temperature was 12°C (total daily PAR was 25 mol m^{-2} day^{-1}; P. S. Nobel and E. García-Moya, unpublished observations). The pruning decreased the number of newly unfolding leaves on sixty plants during the wet period from May through September by 12%, compared with sixty control plants (Fig. 5.9D) for which a total of 319 leaves unfolded. The average number of newly unfolding leaves per plant over the same period was reduced by 30% for twenty plants of *A. salmiana*, whose leaf surface area was reduced 40% by pruning. Thus, pruning clearly decreases the photosynthetic surface area that can intercept PAR and hence decreases the productivity of these plants.

Platyopuntias

Of all the groups of cacti, perhaps none has been studied as intensively as the platyopuntias (Table 1.2). The orientation of their cladodes has important consequences for the interception of PAR and hence for productivity. We shall see that the direction of the ambient PAR also has important influences on cladode orientation. The morphology of certain platyopuntias has already been presented, such as that of *Opuntia basilaris* (Fig. 1.2D), *O. ficus-indica* (Fig. 1.13), *O. rastrera* (Fig. 1.21), and *O. stricta* (Fig. 2.15). Here we will quantify the orientation of cladodes, develop an analytic framework for interpreting the consequences of orientation, and then indicate a mechanism for the observed orientation tendencies. The influences of topographical features and seasonal changes in the sun's trajectory on cladode orientation will also be evaluated.

ORIENTATION WHEN UNSHADED

We begin by considering the orientations observed for terminal cladodes that are unshaded by other cladodes throughout the day.

Later we will consider situations where mountains and other features of the landscape change the PAR from that expected for an exposed, level site.

Because of the highly variable orientation of individual cladodes, many cladodes must be examined before an unambiguous pattern can be determined for a particular species. In addition, the measured orientations should be organized into categories to help recognize patterns. To satisfy these criteria, a large number (660) of cladodes were routinely measured at each site. The observed orientations are then sorted into eighteen angle classes of 10° width each (for a planar object whose two surfaces are equivalent, a clockwise rotation of 10° from true north is the same as a rotation of 190°, so only eighteen classes are needed). Only essentially vertical cladodes, which is typical for platyopuntias, are considered.

Using sufficiently large numbers of cladodes and taking necessary precautions to avoid effects of shading, three orientation patterns have been observed for platyopuntias on level, unobstructed sites: (1) no orientation tendency, (2) a tendency to face east-west, and (3) a tendency to face north-south. For certain platyopuntias in the central Sonoran Desert in Arizona, the tendency is to face east-west (Fig. 5.10); for instance, almost four times as many cladodes face within 5° of east-west as within 5° of north-south for *Opuntia chlorotica* and *O. phaeacantha*. Considering all the terminal cladodes, the ratio of those facing within 45° of east-west to those facing within 45° of north-south is 2.25 for *O. chlorotica* and 1.93 for *O. phaeacantha* at sites in Arizona (Fig. 5.10). On the other hand, cladodes of these two species can tend to face north-south at other sites. Specifically, the ratio of cladodes facing within 45° of east-west to those facing within 45° of north-south is 0.92 in the southern Mojave Desert in California for *O. chlorotica* (Fig. 5.10A) and 0.53 for *O. phaeacantha* in Israel (Fig. 5.10B). These latter two sites have predominantly winter rainfall, an important consideration that we will return to shortly.

We next summarize the orientation data for the more than twenty species of platyopuntias that have been investigated (Table 5.1). In nearly all cases the cladodes tend to face east-west. This tendency is not significant

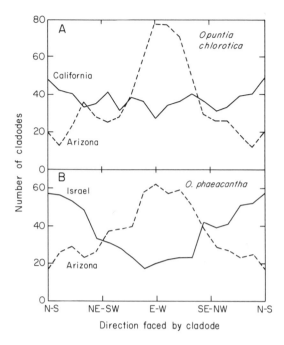

Figure 5.10. Orientation of terminal, unshaded cladodes of *Opuntia chlorotica* (**A**) and *O. phaeacantha* (**B**). Data for *O. chlorotica* were obtained at a southern Mojave Desert site (35°N) in California with mainly winter rainfall (Nobel, 1981b) and a central Sonoran Desert site (31°N) in Arizona with both winter and summer rainfall (Nobel, 1982b). Data for *O. phaeacantha* were obtained at a central Sonoran Desert site (32°N) in Arizona and at a nonirrigated experimental garden (31°N) in Israel, where rainfall occurs predominantly in the winter (Nobel, 1982b). Angle classes for the 660 cladodes examined at each site are 10° wide centered on the indicated direction faced (same as the direction in which the normal to the surface points).

for *O. inamoena* from northeastern Brazil, a species of small stature with only two or three cladodes joined end to end (Fig. 5.11A), but it is for taller species with more cladodes along a branch, such as *O. palmadora* (Fig. 5.11B) and *O. quimilo* (Fig. 5.11C). The greatest orientation tendency is found for *O. stricta* in Florida (Fig. 2.15A), where fourteen to eighteen cladodes often join end to end along the branches. However, at certain sites, the terminal, unshaded cladodes tend to face north-south for several species. Besides the examples already discussed (Fig. 5.10), cladodes of *O. hyptiacantha, O. stricta*, and *O. ficus-indica* tend to face north-south in Israel (Table 5.1), where about 80% of the annual precipi-

tation occurs in the winter (Nobel, 1982c); however, terminal cladodes of *O. ficus-indica* tend to face east-west at irrigated sites in Israel (Table 5.1). Certain studies have claimed that cladodes of platyopuntias such as *O. ficus-indica* (Konis, 1950), *O. humifusa* (Abrahamson and Rubinstein, 1976), and *O. phaeacantha* (Gibbs and Patten, 1970) are randomly oriented, but no actual data were presented. Other studies have stated that the terminal cladodes of *O. humifusa* typically face east-west (Baskin and Baskin, 1973), in agreement with detailed studies (Table 5.1). To help interpret the observed orientations, we next quantitatively consider the interception of PAR.

PAR INTERCEPTION FOR VARIOUS SEASONS AND LATITUDES

To understand PAR interception properly, we need to consider the relation between the direction of sunlight and the orientation of a particular surface. To project the solar beam onto a surface, as is necessary for calculating incident PAR, we need to determine the cosine of the angle between the direct solar beam and a line perpendicular to the surface (its *normal*). We can then predict how total daily PAR changes for various situations.

Radiation relations

In Chapter 4, we considered the direct shortwave irradiation on a particular surface, S^{direct} (Eq. 4.1). Because of absorption and scattering by the earth's atmosphere as well as the position of the sun in the sky, S^{direct} is less than the shortwave radiation incident on the atmosphere for a surface perpendicular to the solar beam, S^{atm}. We note that S^{atm} is referred to as the *solar constant*, which, because of the eccentricity of the earth's orbit and variations in solar activity, such as for sunspots, actually varies by about ±4% from its mean value of 1,360 W m^{-2} (Gates, 1980; Nobel, 1983a).

The fraction of S^{atm} transmitted through the earth's atmosphere depends on intrinsic properties of the atmosphere embodied in its *transmittance* τ (the fraction transmitted when the sun is directly overhead) and on the path length through the atmosphere. The path length depends on the sun's *altitude* or angle

Table 5.1. *Summary of the orientation of terminal, unshaded, vertical cladodes of platyopuntias*

Species	Observation latitude	Site location	East-west facing / North-south facing	Reference
Opuntia basilaris	34°N	California	1.20	Nobel (1980a; 636 cladodes measured)
	34°N	California	1.21	Nobel (1982b) (A)
O. caracasana	10°N	Venezuela	1.90	A
O. chlorotica	35°N	California	0.92	Nobel (1981b) (B)
	34°N	California	0.71	A
	33°N	Arizona	1.78	B
	31°N	Arizona	2.25	A
O. cordobensis	31°S	Argentina	1.72	A
O. echios	0°	Ecuador	1.60	B
	1°S	Ecuador	1.22	B
O. ficus-indica	33°N	Israel	0.49	Nobel (1982c) (C)
	33°N	Israel (irrigated)	2.28	C
	32°N	Israel (irrigated)	2.00	C
	31°N	Israel	0.40	C
	31°S	Argentina	1.77	A
	34°S	Chile (irrigated)	2.19	C
O. galapageia	0°	Ecuador	1.50	A
O. humifusa	42°N	Ontario (Canada)	1.30	A
	26°N	Florida	1.76	A
O. hyptiacantha	31°N	Israel	0.63	A
O. inamoena	8°S	Brazil	1.08	A
O. littoralis	34°N	California	1.21	A
	34°N	California	2.29	Yeaton et al. (1983; 626 cladodes measured)
O. megasperma	1°S	Ecuador	1.58	A
O. melanosperma	2°S	Ecuador	1.18	A
O. palmadora	8°S	Brazil	1.42	A
O. phaeacantha	32°N	Arizona	1.93	A
	31°N	Israel	0.53	A
O. quimilo	30°S	Argentina	1.28	A
O. salagria	30°S	Argentina	1.32	A
O. stricta	31°N	Israel	0.47	A
	26°N	Florida	3.34	A
	25°N	Florida	4.10	B
	13°N	Thailand	1.98	(P. S. Nobel, unpublished observations) (D)
	8°S	Indonesia	2.24	D
	34°S	New South Wales (Australia)	1.60	A
O. sulfurea	31°S	Argentina	1.40	A
	33°S	Argentina	1.22	A
O. utkilio	31°N	Argentina	1.19	A
O. violacea	32°N	Arizona	2.44	A
O. wentiana	11°N	Venezuela	1.38	A

Note: The angular orientation was determined within 1° for 660 cladodes at each site, except where indicated. Data are presented as the number of cladodes facing within 45° of east-west divided by the number facing within 45° of north-south (the remaining cladodes). Studies cited more than once are assigned a boldface letter the first time they occur, this letter being used for subsequent citations.

Figure 5.11. Platyopuntias of various forms. (A) Small-statured *Opuntia inamoena* in the caatinga of northeastern Brazil. (B) *O. palmadora* with many cladodes facing in the same direction, also in northeastern Brazil. (C) The tall *O. quimilo* at the southern part of the chaco in northern Argentina.

above the horizon, γ (Fig. 5.12), which leads to the following relation:

$$\begin{array}{l}\text{Shortwave irradiation on} \\ \text{a surface perpendicular to} \\ \text{solar beam at the earth's} \\ \text{surface on a clear day}\end{array} = S^{\text{atm}} \tau^{1/\sin \gamma}$$

(5.1)

When the sun is directly overhead (the case of maximum transmission), γ is 90°, $\sin \gamma$ is 1, and so the fraction transmitted is $\tau^{1/1}$ or τ, as just indicated. Depending on the presence of water vapor, dust, and other particles, τ can vary from about 0.5 to 0.8 on a clear day (Gates, 1980), with values of 0.74 to 0.76 being

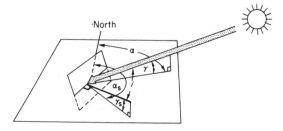

Figure 5.12. Angles involved in calculating the radiation incident on a surface. The cosine of angle *i* is used to project the direct solar beam (stippled arrow) onto the surface (normal indicated by heavy line). The direction of the solar beam is indicated by its altitude γ above the horizontal and azimuth α clockwise from north. The normal to the surface occurs at altitude γ_s above the horizontal and azimuth α_s clockwise from north.

used for the calculations in this book (Geller and Nobel, 1984; Garcia de Cortázar, Acevedo, and Nobel, 1985; Nobel, 1986b).

Besides altitude, we need another angle to define the position of the sun in the sky, namely its *azimuth*, α. Azimuth is the compass angle from north to the projection of the direct solar beam onto the horizontal (Fig. 5.12). The two angles, α and γ, vary with time of day, time of year, and latitude (ϕ). We are assuming that clock time has been converted to true solar time, in which case the sun reaches its highest point in the sky at noon, so that our equations are independent of longitude (see List, 1963). Using solar time and converting a 24-h day to an angle basis (360°/24 h or 15°/h), we obtain an hour angle *h* to express the time of day:

$$h = 15(t - 12) \qquad (5.2)$$

where *t* is true solar time in hours. At solar noon, *t* equals 12 h, so *h* is 0° by Equation 5.2.

The effect of the time of year on solar altitude and azimuth is handled by the solar declination δ:

$$\delta = -23.5° \cos\left[(D + 10)\frac{360°}{365.25}\right] \qquad (5.3)$$

where *D* is the day of the year (January 1 = 1), $-23.5°$ represents the tilt of the earth's axis relative to the plane of the earth's orbit, and 365.25 recognizes that there are an average of $365\frac{1}{4}$ days per year (Eq. 5.3 is only approxi-

mate, mainly because the earth's orbit is non-circular). On the spring equinox, which is about day 81 or March 22, the argument of the cosine in Equation 4.3 is (81 + 10)(360°)/ (365.25) or 90°. Hence, δ is then $-23.5°$ cos 90°, or 0°, and the sun rises due east and sets due west.

Using Equations 5.2 and 5.3 and relations from spherical trigonometry, we can express the sun's altitude γ and azimuth α in terms of the hour angle, the solar declination, and the latitude ϕ (Gates, 1980):

$$\sin \gamma = \sin \phi \sin \delta + \cos \phi \cos \delta \cos h \qquad (5.4a)$$

$$\sin \alpha = -\frac{\cos \delta \sin h}{\cos \gamma} \qquad (5.4b)$$

Although these expressions appear rather formidable, they are important for predicting the PAR incident on a surface. We next need to consider angles that indicate the orientation of the absorbing surface.

Just as for the sun's position, we use altitude (γ_s) and azimuth (α_s) to define the orientation of a surface, where γ_s is the angle that the normal to the surface makes with the horizontal (Fig. 5.12) and α_s is the angle clockwise from north into which the surface faces. To project the shortwave radiation onto the surface, we multiply the expression in Equation 5.1 by the cosine of the angle *i* between the direct solar beam and the normal to the surface (Fig. 5.12). The relation we need to calculate cos *i* is (Gates, 1980)

$$\cos i = \cos \gamma_s \cos \gamma \cos(\alpha - \alpha_s)$$
$$+ \sin \gamma_s \sin \gamma \qquad (5.5a)$$

which for vertical surfaces ($\gamma_s = 0°$) becomes

Vertical: $\cos i = \cos \gamma \cos(\alpha - \alpha_s) \qquad (5.5b)$

For applications to platyopuntias, we will consider vertical surfaces facing in the four cardinal directions, which leads to the following further simplifications:

North: $\quad \cos i = \cos \gamma \cos \alpha \qquad (5.6a)$

East: $\quad \cos i = \cos \gamma \sin \alpha \qquad (5.6b)$

South: $\quad \cos i = -\cos \gamma \cos \alpha \qquad (5.6c)$

West: $\quad \cos i = -\cos \gamma \sin \alpha \qquad (5.6d)$

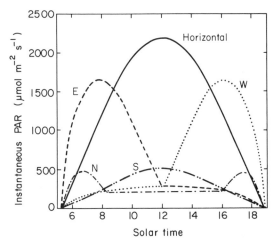

Figure 5.13. Predicted instantaneous PAR for clear days on the summer solstice at 34°N for vertical surfaces facing in the indicated cardinal direction or for a horizontal surface. Both direct and diffuse components are included. Used by permission from Nobel (1986b). © 1986 by Cambridge University Press.

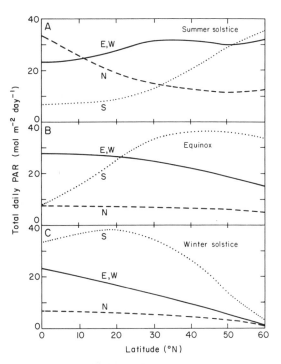

Figure 5.14. Predicted total daily PAR for clear days and unshaded vertical surfaces facing in the indicated cardinal direction at various latitudes. (A) Summer solstice. (B) Equinox. (C) Winter solstice. For southern latitudes, N and S must be interchanged throughout the figure. Used by permission from Nobel (1986c). © 1986 by Cambridge University Press.

Predictions of PAR

Using Equations 5.1 through 5.6, we can predict instantaneous values of shortwave irradiation on clear days for surfaces of any orientation. We can convert shortwave irradiation to a photon basis, using a conversion factor (for instance, 1.00×10^6 J/2.02 mol photons; Varlet-Grancher et al., 1981) or empirical results based on field measurements that take into consideration the sun's altitude (Geller and Nobel, 1984). The instantaneous PAR values during the day can be integrated to obtain the total daily PAR, as we indicated previously.

The predicted instantaneous PAR for vertical surfaces facing in the four cardinal directions is given in Figure 5.13 for the Agave Hill site (Figs. 1.1, 1.2, and 3.20) at 34°N for the summer solstice. The sun rises at a solar time of about 5 h, and initially the greatest PAR is on an east-facing surface. However, by a solar time of 8 h the direct PAR on the east-facing surface begins to decline, and after solar noon only diffuse PAR is incident on that surface (Fig. 5.13). The total daily PAR for the east-facing surface is 32 mol m^{-2} day^{-1}, which is just less than half of the 66 mol m^{-2} day^{-1} incident on a horizontal surface. For the clear day conditions employed, the west-facing surface intercepts the same total daily PAR as the east-facing surface, the north-facing surface then receives 14 mol m^{-2} day^{-1}, and the south-facing surface receives 15 mol m^{-2} day^{-1} (Nobel, 1986b). In this case, vertical surfaces facing in the four cardinal directions receive an average of only 35% as much PAR as a horizontal surface, which has important implications for PAR interception by the sides of vertical stems of cacti.

We will next use the model to predict total daily PAR on vertical surfaces for clear days at various latitudes at the two solstices and an equinox (Fig. 5.14). For low latitudes of 0–20°N, east- and west-facing surfaces generally have a total daily PAR of at least 20 mol m^{-2} day^{-1}, as does a north-facing surface at the summer solstice and a south-facing surface at the winter solstice. Thus, CO_2 uptake by the parts of cactus stems facing in such directions would be essentially PAR saturated (Fig. 5.2). On the other hand, south-facing surfaces at the summer solstice and north-facing surfaces at the winter solstice and at an equinox receive less than 10 mol m^{-2} day^{-1} (Fig. 5.14), and CO_2 uptake by surfaces with such orientations is markedly PAR limited. For midlatitudes of 20–40°N, the north-facing surfaces receive even less total daily PAR than at low latitudes, and south-facing surfaces receive more, surpassing the total daily PAR of east- or west-facing surfaces at an equinox and the winter solstice; these trends continue at high latitudes of 40–60°N (Fig. 5.14). Averaging over a year by weighting an equinox twice as heavily as each solstice and averaging over surfaces in the four cardinal directions, we find that the total daily PAR on clear days is 21 mol m^{-2} day^{-1} for low latitudes (0–20°), 21 mol m^{-2} day^{-1} for midlatitudes (20–40°), and 17 mol m^{-2} day^{-1} for high latitudes (40–60°), or about 20 mol m^{-2} day^{-1} overall.

Considering the dependency of CO_2 uptake on total daily PAR (Fig. 5.2) and the total daily PAR at various times and latitudes (Fig. 5.14), we can now see the consequences of the observed cladode orientations for platyopuntias (Fig. 5.10, Table 5.1) on potential CO_2 uptake. Indeed, cladodes offer an opportunity for observing orientation effects that is not available for cylindrical stems of cacti. As before, we will consider two orientations for vertical cladodes – north-south and east-west. Because both surfaces of an unshaded cladode facing east-west will generally intercept a total daily PAR of 20 mol m^{-2} day^{-1} or more (Fig. 5.14), such a cladode will have nearly maximal net CO_2 uptake at all latitudes on the summer solstice (Fig. 5.15A) and at all except high latitudes for an equinox (Fig. 5.15B). Net CO_2 uptake by a cladode facing north-south is much less for these conditions, because one of the surfaces generally has a total daily PAR below 10 mol m^{-2} day^{-1} (Fig. 5.14A,B). Although predicted net CO_2 uptake tends to be lower near the winter solstice than at other times of the year, it is still higher for an east-west facing cladode than a north-south facing one within 27° of the equator (Fig. 5.15C). At latitudes further from the equator than 27°, a north-south facing cladode has a higher net CO_2 uptake than an east-west facing one at the

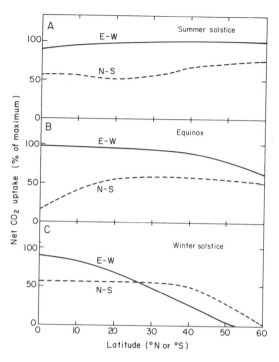

Figure 5.15. Net CO_2 uptake over 24-h periods by vertical cladodes facing east-west or north-south for the summer solstice (**A**), an equinox (**B**), and the winter solstice (**C**). Data were calculated for wet conditions and approximately optimal temperatures from the PAR levels in Figure 5.14 and the response of CO_2 uptake to total daily PAR in Figure 5.2. Used by permission from Nobel (1986c). © 1986 by Cambridge University Press.

winter solstice because of the large amount of total daily PAR intercepted by the equatorially facing side (Fig. 5.14C).

The seasonal and latitudinal variation in PAR (Fig. 5.14), and hence variation in net CO_2 uptake (Fig. 5.15), provides a straightforward explanation for the observed orientation patterns of platyopuntias. In nearly all cases, the tendency for the cladodes is to face east-west (Fig. 5.10, Table 5.1), which under most conditions maximizes net CO_2 uptake (Fig. 5.15). Indeed, east-west facing cladodes of *Opuntia phaeacantha* intercept more radiation than north-south facing ones near the spring equinox and summer solstice at 32°N (Gibbs and Patten, 1970). Also, east-west-facing cladodes of *O. amyclaea* and other platyopuntias in central Mexico intercept more PAR and have up to twice the productivity as north-south facing ones, including the production of

Figure 5.16. Platyopuntias occurring where topographical features affect PAR availability. **(A)** *Opuntia erinacea* in the Providence Mountains near Mitchell Caverns Nature Preserve in southeastern California. **(B)** *O. littoralis* growing on coastal bluffs near Malibu, California.

more fruit (Becerra Rodríguez, Barrientos Pérez, and Diaz Montenegro, 1976; Barrientos Pérez, 1983). Where north-south orientation occurs, such as in the Mojave Desert and unirrigated sites in Israel (Fig. 5.10, Table 5.1), the latitudes are more than 27° from the equator, and most rainfall, together with temperatures favorable for growth, tends to occur near the winter solstice. Even so, most of the natural populations of platyopuntias observed at latitudes greater than 27° from the equator have cladodes that tend to face east-west (Table 5.1), suggesting that winter is not the predominant season for initiating new cladodes. We will discuss the mechanism for orientation shortly; here we again note that its consequence is to lead to more PAR interception at times of the year favorable for growth and that PAR is on the verge of being limiting for net CO_2 uptake at all latitudes and seasons (Figs. 5.14 and 5.15).

TOPOGRAPHICAL AND SEASONAL EFFECTS

Cladode orientation can be influenced by topographical features, and the orientation

patterns of newly initiated cladodes can vary seasonally. We next consider how such effects are related to PAR interception.

Topographical features

In the previous discussion, we have avoided situations in which the terminal cladodes are shaded by vegetation or in which the direct solar beam is intercepted by mountains or other topographical features. We now relax these constraints and see what happens to the orientation of terminal cladodes of platyopuntias.

Many species of platyopuntias occur on hilly or mountainous terrain. For instance, *Opuntia erinacea* occurs in the mountains of inland California (Fig. 5.16A), and *O. littoralis* can occur on coastal bluffs facing the Pacific Ocean (Fig. 5.16B). The presence of such topographical features can block the direct PAR for part of the day and thus affect the orientation of the terminal, unshaded cladodes (Fig. 5.17). When mountains block PAR from the west and northwest, the cladodes of *O. erinacea* tend to face southeast-northwest (Fig. 5.17), an orientation that maximizes PAR in-

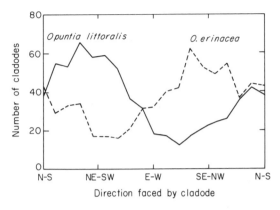

Figure 5.17. Influence of topographical features on orientation of terminal, unshaded, vertical cladodes of two platyopuntias. Mountains subtended angles of 24° above the horizontal to the west and 36° to the northwest for *O. erinacea*, and *O. littoralis* occurred on a west-facing hill with a mean slope of 32°. Sites are illustrated in Figure 5.16. Data are presented as for Figure 5.10 and are adapted from Nobel (1982b).

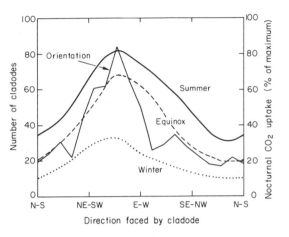

Figure 5.18. Orientation of terminal cladodes for 100-year-old plants of *Opuntia ficus-indica* in Til Til, Chile, and simulated CO_2 uptake at various times of the year. Trees and mountains affected the availability of PAR, especially to the west and northwest. After allowing for such effects, PAR was simulated for vertical surfaces on clear days, and the total daily PAR was used to determine the net CO_2 uptake, expressed here as a percentage of maximum (Fig. 5.2). Modified from Nobel (1982c).

terception. Similarly, when mountains block PAR from the east and northeast, the cladodes of *O. polyacantha* tend to face southwest-northeast (Nobel, 1982b). For *O. littoralis* occurring on a west-facing hillside (Fig. 5.16B), the tendency is to face approximately west-southwest-eastnortheast, which again maximizes PAR interception over the course of a year. Yet *O. littoralis* on a nearby horizontal site with no blockage by topographical features has cladodes that tend to face east-west (Table 5.1), indicating that orientation can change within a population as the availabiltiy of PAR changes. This is also observed for *Opuntia chlorotica* in the Granite Mountains of California, where the ratio of cladodes facing within 45° of east-west to those facing within 45° of north-south is 0.92 at an unobstructed site (Table 5.1) but becomes 0.50 when in a north-south canyon (Nobel, 1980a, 1982d). The canyon walls block much of the direct PAR coming from the east and west, which greatly enhances the tendency for the cladodes to face north-south at this site in the southern Mojave Desert.

An interesting case occurs for 100-year-old individuals of *Opuntia ficus-indica* in central Chile. The site was chosen because of the age of the plants, which are the oldest cultivated locally (Nobel, 1982c). The orientation pattern is at first puzzling, because the peak clearly occurs at about 70° instead of at 90° (Fig. 5.18). Inspection of the site shows that mountains and surrounding trees have a considerable influence on the available PAR, blocking the sun up to an altitude angle of 38° above the horizontal to the northwest, which has a major influence on direct PAR in the southern hemisphere. Angular changes from the horizontal for the effective horizon were noted for all directions and used to calculate the available PAR at three times of the year. In turn, PAR interception by vertical surfaces can be converted to CO_2 uptake, because the CO_2 dependency on total daily PAR is known for *O. ficus-indica* (Fig. 5.2). Calculations show that when the topographical features and surrounding vegetation are taken into account, net CO_2 uptake is also maximal for surfaces facing near an angle of 70° for the summer solstice, an equinox, and the winter solstice (Fig. 5.18). Thus, the cladodes of *O. ficus-indica* tend to face in the direction that maximizes PAR interception and hence net CO_2 uptake.

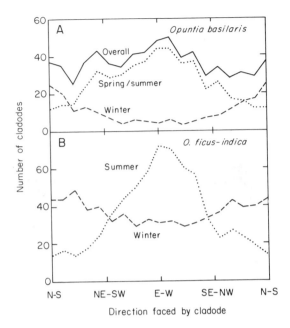

Figure 5.19. Seasonal influences on the orientation of newly developed, terminal, unshaded, vertical cladodes. (A) Orientation of cladodes initiated in the winter (December–January) or in the spring/summer (March–September) was determined for *Opuntia basilaris* near Palm Desert, California, where 660 cladodes were measured overall (no new cladodes were initiated in October–November; Nobel, 1982b). (B) Orientation of two- to four-week-old cladodes initiated near the summer solstice or the winter solstice was measured on irrigated *O. ficus-indica* near Fillmore, California (Nobel, 1982c). Both sites were at 34°N.

Seasonal variations

Cladodes can be initiated at different times of the year, so an overall orientation pattern may actually represent the summation of patterns that vary seasonally. For instance, the overall orientation pattern for *Opuntia basilaris* at 34°N in California shows a slight tendency for the cladodes to face east-west (Fig. 5.19A, Table 5.1). In this case, about 27% of the cladodes are initiated in the winter, when the tendency is to face north-south (Fig. 5.19A), an orientation that leads to maximal interception of PAR and hence to maximal net CO_2 uptake at that time of the year (Figs. 5.14 and 5.15). The rest of the cladodes are initiated in the spring through the summer, when the tendency for the cladodes is to face east-west (Fig. 5.19A). Because most (73%) of the clad-

odes are initiated at times when facing east-west is favored, the overall pattern shows a tendency for the cladodes to face east-west.

Similar seasonal effects are observed for newly initiated cladodes on cultivated *O. ficus-indica* (Fig. 5.19B). In this case, 660 new cladodes were examined in the winter, when the tendency is to face north-south, as might be expected at the latitude of 34°N. On the other hand, cladodes initiated on these plants near the summer solstice have a strong tendency to face east-west (Fig. 5.19B), which again is the direction that maximizes PAR interception and hence net CO_2 uptake at that time of the year (Figs. 5.14 and 5.15). Therefore, the orientation of cladodes on the same plants depends on the season of initiation.

ORIENTATION MECHANISM

The mechanism underlying the orientation patterns involves an interaction between development, morphology, and physiology. In this regard, phototropism can influence the orientation of cladodes, as we discussed in Chapter 1, at least until their orientation becomes fixed early in development.

With respect to morphology, there is a tendency for newly initiated or "daughter" cladodes of many platyopuntias to face in the same general direction as the underlying or "mother" cladodes. Thus, a tendency to face in a favorable direction will tend to be perpetuated. This effect is seen for cultivated *Opuntia ficus-indica*, which are planted by placing detached mature cladodes vertically in the ground with about one-third to one-half of their area below the soil surface. Because planted cladodes are closely aligned along the field axis at Til Til, Chile (Fig. 1.13A), most of the cladodes that develop during the first year also tend to be aligned along the field axis. Even after five years when each plant has sixty to seventy cladodes whose preferred orientation tends to be east-west (Table 5.1), traces of the original planting direction can still be seen. In particular, about twice as many cladodes face within 5° of north-south on these five-year-old plants when the initially planted cladode is aligned on a north-south axis compared with along an east-west axis (Nobel, 1982c).

Figure 5.20. *Stenocereus gummosus* at the two extremes of its range in Baja California. (**A**) In the north near Ensenada, Baja California Norte, where the surrounding sparse vegetation is short, and the tallest stems of *S. gummosus* average 0.7 m in height. (**B**) In the south near San Antonio, Baja California Sur, in a thorn-scrub forest, where stems of *S. gummosus* are often over 4 m tall (note 3-m pole in foreground).

Goebel (1895) long ago noted that newly initiated cladodes of *Opuntia leucotricha* tend to become flattened in a direction perpendicular to the light beam, thereby intercepting more PAR. When vertical cladodes of *O. ficus-indica* are exposed to a horizontal PAR beam making an angle of 45° with the cladode surface, the newly developing cladodes are rotated toward the light beam by an average of 16° from the plane of the mother cladodes (Nobel, 1982c). This phototropic response of the newly developing cladodes leads to more PAR interception than if the new cladodes were in the same plane as the mother cladodes.

Cladodes oriented so as to intercept more PAR tend to have more net CO_2 uptake (Fig. 5.15) and a higher overall productivity (Becerra Rodríguez et al., 1976). The increased productivity of cladodes intercepting more PAR causes them to produce more new cladodes, which also tend to face in the preferred direction. Thus, plants tend to develop cladodes facing in the direction receiving the most PAR, which under most circumstances is facing east-west (Table 5.1). At latitudes further from the equator than 27° and where growth is favored in the winter, north-south orientations can dominate. In any case, clad-odes initiated in the winter can tend to face north-south and those initiated at other times of the year on the same plant can tend to face east-west (Fig. 5.19).

To summarize, we can recognize three aspects that account for the observed orientation tendencies of platyopuntias: (1) daughter cladodes tend to occur in the same plane as mother cladodes; (2) phototropism allows for some rotation toward the predominating PAR direction; and (3) favorably oriented cladodes produce more cladodes, which also tend to be favorably oriented. If we assume that the first (basal) cladode is randomly oriented, then orientation tendencies would not be expected to be very pronounced if only a few cladodes develop along a branch, which is the case for *O. inamoena*. On the other hand, orientation tendencies might be more pronounced when many cladodes occur along a branch, as for *O. stricta*. One can observe such reinforcement of preferred tendencies (Table 5.1), paying appropriate regard to seasonal and topographical effects on PAR interception. Because PAR generally limits net CO_2 uptake, even on clear days for unshaded terminal cladodes, the preferred orientations are advantageous for maximizing CO_2 uptake by platyopuntias.

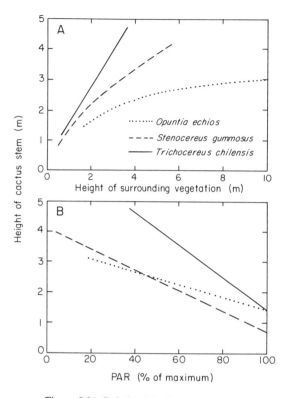

Figure 5.21. Relationships between the height of surrounding vegetation (**A**) or the available PAR (**B**) and the height of the tallest stems of three species of cacti. Data are for thirty adjacent cacti per site at six sites for *Opuntia echios* over a 16-km transect in the Galápagos Islands, Ecuador (Nobel, 1981b); fifteen adjacent cacti per site at twenty-nine sites for *Stenocereus gummosus* over a 1,300-km transect (actually, more of a series of sites than a true transect) in Baja California, Mexico (Nobel, 1980a); and thirty-two adjacent cacti per site at eight sites for *Trichocereus chilensis* over a 12-km transect in central Chile (Nobel, 1981b). PAR data represent the average PAR received from the four cardinal directions (at 1 m above the ground for *S. gummosus* and at 2 m for the other two species) divided by the analogous PAR in an exposed region where the sun's rays were not blocked by any vegetation.

Other cacti

Now that we have presented the dependency of net CO_2 by cacti on the total daily PAR and given a detailed consideration of the interaction between PAR and cladode orientation for platyopuntias, we are ready to consider interactions between morphology and PAR interception for other cacti. As we have already indicated, optimal utilization of PAR occurs when as little surface area as possible

is either below PAR compensation, where a net CO_2 loss occurs, or close to PAR saturation, where the photons are not efficiently used for CO_2 fixation (see Fig. 5.2). In this section we shall begin by considering the importance of stem height for PAR interception, where the key is actually the height of the surrounding vegetation. For instance, *Opuntia fragilis* is the first vascular plant to colonize certain rock outcrops in Manitoba, Canada, but it disappears at later stages of plant succession, presumably because it is outcompeted for PAR by taller grasses and shrubs (Frego and Staniforth, 1986). We will examine how spines and ribs influence the PAR reaching a unit area of the stem surface. PAR interception is also affected by the branching and tilting of the stems as well as by the overall orientation of the outermost branches of various cacti.

HEIGHT

The height of a particular species of cactus at maturity can vary considerably across its geographical distribution. This is dramatically apparent for *Stenocereus gummosus*, which varies from a short, bramblelike shrub in the northern part of its distribution in Baja California Norte, Mexico (Fig. 5.20A), to a tall, treelike plant in the southern part of its distribution in Baja California Sur (Fig. 5.20B). Indeed, the maximal height and physical appearance of this "species" changes continuously over its 1,300-km range in Baja California (Nobel, 1980a). The change in height of *S. gummosus* in Baja California is associated with changes in the height of the surrounding vegetation (Fig. 5.21A). At the northern part of its distribution, *S. gummosus* occurs in open habitats with no overtopping vegetation, and toward the southern part of its distribution it competes for light with many trees in a thorn-scrub forest (Figs. 5.20 and 5.21A). Thus, the height of *S. gummosus* changes with habitat and associated vegetation.

Analogous changes in stem height with changes in height of the surrounding vegetation occur for *Opuntia echios* in the Galápagos Islands of Ecuador (Fig. 5.22A) and *Trichocereus chilensis* in central Chile (Fig. 5.22B). The height of *O. echios* doubles as the mean height of the surrounding vegetation increases

Figure 5.22. Two species of cacti whose mean stem heights can vary over rather short distances (Fig. 5.21): **(A)** *Opuntia echios* on Santa Cruz Island in the Galápagos Archipelago, Ecuador; and **(B)** *Trichocereus chilensis* near Olmne in central Chile.

from 2 m to 10 m along a 16-km transect, and the height of *T. chilensis* quadruples as the height of the surrounding vegetation increases from 0.6 m to 3.6 m along a 12-km transect (Fig. 5.21A). Changes in height of other cacti also occur in concert with changes in the height of surrounding vegetation in Sonora and Sinaloa, Mexico. For instance, the height of the tallest stems of *Stenocereus alamosensis* (Fig. 1.14C) triples from 28°N to 25°N, *Stenocereus thurberi* (Fig. 1.14D) doubles in height from 31°N to 26°N, and *Pachycereus pecten-aboriginum* more than doubles in height from 28°N to 25°N (Nobel, 1980a). As the surrounding vegetation increases in height, the cactus stems become more shaded, meaning that PAR is reduced, an aspect that we consider next.

The increases in stem height can be related to decreases in available PAR caused by the surrounding vegetation (Fig. 5.21B; the PAR measurements along a transect for a particular species were all made at the same height above the ground). Compared with the relatively short stems that occur in open habitats, where PAR at 1 or 2 m above the ground is maximal because of lack of shading by surrounding vegetation, *O. echios* increases twofold in height, *S. gummosus* sixfold, and *T. chilensis* fourfold as the PAR becomes less where the surrounding vegetation is taller (Fig. 5.21A). These three species have typical responses of nocturnal acid accumulation to total daily PAR, 90% saturation occurring at about 20 mol m^{-2} day^{-1} (Fig. 5.3B). The sides of their unshaded vertical stems are generally exposed to a total daily PAR averaging about 20 mol m^{-2} day^{-1} on clear days (Fig. 5.14), so they are on the verge of PAR limitation for net CO_2 uptake. Potential productivity is therefore sensitive to shading. The response of these cacti to increased shading is to become taller, thereby enabling the stems to intercept more PAR.

Of the three species considered (Fig. 5.21), the greatest overtopping by surrounding vegetation occurs for *O. echios*, which has a mean stem height of 3 m where the height of the surrounding vegetation averages 10 m (Fig. 5.21A). *Opuntia echios* is absent where the PAR on vertical surfaces 2 m above the ground is reduced 92% by the surrounding vegetation, although it is present where PAR is reduced by 82% (Nobel, 1981b). We have already indicated that daily net CO_2 uptake for cacti is negative below a total daily PAR of about 3 mol m^{-2} day^{-1} (Fig. 5.2). Such a compensation level corresponds to an approximately 85% reduction in available PAR. Thus, where shading reduces PAR by more than 85%, cacti are not expected to survive, in agreement with the observations on *O. echios* in the Galápagos Islands.

What mechanism is responsible for the observed increase in height of cactus stems as the total daily PAR is reduced? Two factors may be involved: (1) genetic variation within a species and (2) *etiolation* responses, as occur for other plant taxa, in which low light levels lead to the development of longer stems. Genetic variation is plausible for *S. gummosus* over its 1,300-km range; indeed, plants from its two distributional extremes are morphologically distinct (Fig. 5.20). Because of the differences in size, *O. echios* is given different varietal names (subdivisions of a species) at the two ends of the transect considered (Wiggins and Porter, 1971), although the height variation is continuous with distance (Nobel, 1981b) and different varietal names may not be warranted (Dawson, 1965). When the variations in height occur over a geographically short distance for which the plants are genetically essentially identical, possible etiolation responses must be considered. In this regard, cladodes of *Opuntia humifusa* and *O. leuchotricha* are longer when development occurs in the dark (MacDougal, 1903). Also, *O. humifusa* is shorter for a given number of cladodes in an open field than under a dense canopy (Abrahamson and Rubinstein, 1976). Indeed, etiolation responses may be common for cacti.

Factors other than light have also been proposed to influence the height of cacti. For instance, changes in nutrient levels in the soil may cause cacti and the surrounding vegetation to change in concert. Also, grazing on the fruits and cladodes of *O. echios* by land iguanas and giant tortoises has been proposed to have increased its height in the Galápagos Islands on an evolutionary time scale (Stewart, 1911; Dawson, 1965, 1966). However, these animals can forage over the entire transect considered for *O. echios* (Nobel, 1981b), and competition for PAR with the surrounding vegetation seems a much more plausible explanation than grazing for the changes in height of this species (Racine and Downhower, 1974). In any case, research is needed on the etiolation responses of cacti as well as on the genetic variability that can occur over the distribution of a particular species.

SPINES

Properties of spines vary considerably among different species of cacti; moreover, spines can have different biological effects. We have already discussed the raising of min-

Figure 5.23. *Opuntia bigelovii* growing in Joshua Tree National Monument in southeastern California. The spiny stem segments are readily detached and are notorious for becoming attached to the clothing or skin of passersby, leading to its common name of "jumping cholla," although it is also more charitably referred to as "teddy-bear cholla."

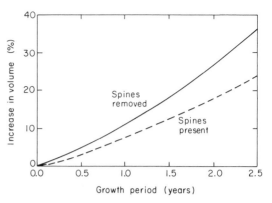

Figure 5.24. Growth of *O. bigelovii* with and without spines. Two control plants initially with six stem segments each were maintained in the laboratory with a mean ambient total daily PAR of 9 mol m^{-2} day^{-1}; two similar plants whose spines were removed every two months were also maintained at the same total daily PAR (Nobel, 1983b).

imal stem temperatures by spines, especially in the apical region, thereby allowing certain cacti to extend their distributional boundaries to colder regions (Chapter 4). Spines can also benefit a species by discouraging herbivory by most vertebrates (Gibson and Nobel, 1986; Janzen, 1986). However, two "costs" of spines must be considered: (1) they represent an investment of carbohydrates and other materials, ranging up to 38% of the total stem dry weight for seedlings of *Ferocactus acanthodes* (Nobel, 1983b); and (2) they reduce the PAR incident on the stem, an effect that we will consider here.

The percentage shading of the stem surface by spines depends on the species. Many species have essentially no spines, including a spineless variety of *Opuntia ficus-indica* that is cultivated worldwide. At the other extreme, spines can prevent over 90% of the incoming PAR from reaching the stem surface of other species, such as near the apex of ecotypes of *F. acanthodes* occurring in regions with prolonged subzero temperatures (Nobel, 1980e). An intermediate situation occurs for the cylindropuntia *O. bigelovii* (Fig. 5.23), whose spines reduce the PAR incident on the stem surface by 30% to 40%; actually because of spine geometry and reflectance properties, the fractional PAR attenuation by the spines depends on the angle between the direct solar

beam and the normal to the stem surface (Nobel, 1983b).

The presence of spines reduces the nocturnal acid accumulation at a particular ambient PAR level compared with the nocturnal acid accumulation when the spines are removed (Table 5.2). The effect is greater for *F. acanthodes* in a region of the stem where PAR is reduced 74% by spines than for an individual of *O. bigelovii* whose stem is shaded 30% by spines. Because nocturnal acid accumulation is approximately proportional to total daily PAR up to 5 mol m^{-2} day^{-1} for these two species (Fig. 5.3B), the percentage decrease in nocturnal acid accumulation at 5 mol m^{-2} day^{-1} is similar to the percentage shading of the stem by the spines (Table 5.2). At higher ambient PAR levels, the percentage reduction in nocturnal acid accumulation by a certain percentage of spines is less, falling to a 7% reduction in nocturnal acid accumulation for *O. bigelovii* at a total daily PAR of 20 mol m^{-2} day^{-1} (Table 5.2); this lowering in the fractional reduction by spines of the nocturnal acid accumulation at higher total daily PAR levels is a consequence of the shape of the PAR response curve (Fig. 5.3B). Indeed, if the ambient PAR could be raised high enough, shading by spines would have little effect on nocturnal acid accumulation, although such high PAR levels generally do not

Table 5.2. *Influence of spine removal on nocturnal acid accumulation by various species of cacti*

Species	Shading of stem surface by spines (%)	Ambient total daily PAR (mol m^{-2} day^{-1})	Increase in nocturnal acid accumulation upon spine removal (%)	Reference
Coryphantha vivipara	—	5–20	24	Norman and Martin (1986)
Ferocactus acanthodes	74	5	74	Nobel (1983b)
		10	71	
		15	66	
		20	57	
Opuntia bigelovii	30	5	32	Nobel (1983b)
		10	27	
		15	18	
		20	7	

occur in nature for these two species, especially *F. acanthodes*.

The reduction of nocturnal acid accumulation (Table 5.2) and the associated nocturnal net CO_2 uptake by spines should reduce the growth of cacti. This phenomenon has been investigated for *O. bigelovii* (Fig. 5.23) over a two-and-one-half-year period (Fig. 5.24). When spines are removed, the plants grow faster, as measured by increases in stem volume. After two and a half years, plants with the spines removed, resulting in an approximately 40% higher PAR level on the stem surface, increase in stem volume 50% more than do plants with the spines present (Fig. 5.24). Spines clearly reduce the already limiting PAR available to the stem surface and therefore reduce net CO_2 uptake and hence cactus growth.

RIBS

Increases in stem surface area, such as are caused by the projecting tubercles of *Mammillaria dioica* (Fig. 1.2C) and *Coryphantha vivipara* (Fig. 1.19) or the tubercles fused into ribs for many species of cacti, can decrease the PAR per unit surface area and hence affect net CO_2 uptake. Here we focus on the influence of ribs, which have a simpler geometry and are therefore more amenable to quantitative evaluation than individual projecting tubercles.

To characterize ribs, we define the *fractional rib depth*, which equals the rib depth (radial distance from the rib trough or base to the midpoint of a line between two adjacent rib crests) divided by the radial distance from the stem axis to a rib crest (Fig. 5.25). Fractional rib depths for mature plants in the field are 0.14 for *Carnegiea gigantea* (Fig. 1.12A), 0.15 for *Ferocactus acanthodes* (Fig. 1.2A), and 0.35 for *Lophocereus schottii* (Fig. 1.14A), which has five to eight deep ribs (Geller and Nobel, 1984). Ribbing was initially considered to be rather neutral with respect to CO_2 uptake by cacti (Nobel, 1980a). In particular, ribbing increases the surface area but decreases the PAR per unit surface, so little effect was expected to occur per unit stem height. Actually, for deep ribs with a fractional rib depth of 0.44, many of the rib surfaces on a sixteen-ribbed plant are close to or below PAR compensation, so net CO_2 uptake per unit plant height is predicted to be slightly lower than for shallow ribs in this case (Nobel, 1980a).

More detailed simulations have been performed that include the effects of fractional rib depth, rib number, and reflection from one rib surface to another for both the sunlit and the shaded portions (Geller and Nobel, 1984). In addition, a new term was defined that includes effects of both rib number and fractional rib depth – the *perimeter ratio*; it equals the total

A

B

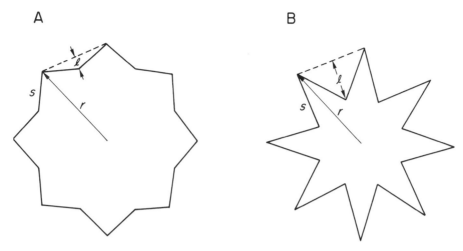

Figure 5.25. Variations in fractional rib depth (l/r) and perimeter ratio (here 16 $s/2r$). As the ribs become deeper, fractional rib depth increases from 0.15 (**A**) to 0.45 (**B**), and perimeter ratio increases from 3.3 (**A**) to 4.7 (**B**).

distance or perimeter from rib crest to trough to the next crest along all ribs around a stem divided by the stem diameter (Fig. 5.25). For a cylinder, the perimeter ratio is $\pi d/d$ or π (3.14); the presence of ribs tends to increase the perimeter ratio, depending mainly on the fractional rib depth. Using the perimeter ratio, we can examine the influence of a wide range in values of rib number and fractional rib depth together (Fig. 5.26). In particular, nocturnal CO_2 uptake for unshaded stems is maximal at a perimeter ratio near 6.5 on the summer solstice, 6.0 on an equinox, and 5.5 at the winter solstice (Fig. 5.26A). The perimeter ratio for maximal nocturnal CO_2 uptake also decreases with shading. For instance, at an equinox the optimal perimeter ratio decreases from 6.0 to 5.5 to 4.0 to 3.1 as the shading increases from 0% (no shade, Fig. 5.26A) to 10% to 30% to 50% (Fig. 5.26B).

Perimeter ratios determined in the field are often slightly above 4. In particular, the perimeter ratio is 4.1 for *C. gigantea* and *L. schottii* and 4.4 for *F. acanthodes* (Geller and Nobel, 1984). Such perimeter ratios favor nocturnal net CO_2 uptake near the equinox if spines, surrounding vegetation, topographical features, or cloudiness reduce PAR 20–30% from the clear-day, unshaded values (Fig. 5.26B). Compared with a cylindrical stem, the presence of ribs thus usually tends to increase net CO_2 uptake; for a perimeter ratio of 4.2, the increase averages about 14% on clear days

throughout the year for unshaded stems and 3% on an equinox with 30% stem shading (Fig. 5.26).

For certain cacti, rib spacing and rib depth vary around the stem. For instance, ribs have been reported to be closest together on the south side of *C. gigantea* (Spalding, 1905) and *Ferocactus wislizenii* (MacDougal and Spalding, 1910), although other studies indicate they are closest on the southwest side for these two species (Walter and Stadelmann, 1974). Ribs of *C. gigantea* and *F. acanthodes* are about 20% closer on the south than the north side, and the fractional rib depth is 24% greater for *C. gigantea* and 5% greater for *F. acanthodes* where the ribs are closest, between south and southwest (Geller and Nobel, 1984). The greater rib surface area per unit circumference occurs for regions of the stem where the total daily PAR tends to be higher. Partially shading the south side or causing the PAR distribution to be uniform in other ways results in the ribs of *C. gigantea* becoming equidistant (Walter and Stadelmann, 1974). In any case, net CO_2 uptake by a stem with asymmetric rib spacing and depth is only a few percent greater than when rib depth and spacing are uniform (Geller and Nobel, 1984).

BRANCHING

Considerable variation in branching with respect to number, position, and size occurs among cacti. In this section we consider the

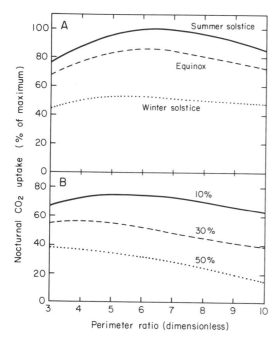

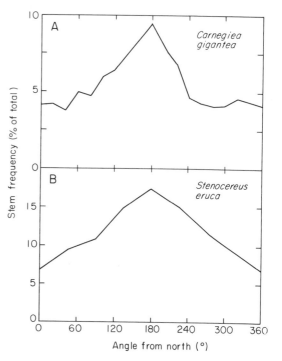

Figure 5.26. Simulated influence of perimeter ratio on net nocturnal CO_2 uptake per unit height. (A) Simulations in which rib number was varied from 3 to 34 and fractional rib depth from 0.0 to 0.6 for unshaded stems of constant diameter at the indicated times of the year. (B) Simulations for a 16-ribbed plant on an equinox with the indicated percent shading. Modified from Geller and Nobel (1984).

Figure 5.27. Frequency of angular stem position measured clockwise from north (azimuth). (A) Branches on *Carnegiea gigantea* in the Tucson Mountain Unit of Saguaro National Monument, Arizona, where about 700 branches on 200 plants were measured. (B) Direction pointed by apex of essentially horizontal stems of *Stenocereus eruca* near San Carlos, Baja California Sur, where 467 stems were measured. Angle classes were 20° wide for *C. gigantea* and 45° wide for *S. eruca*. Modified from Geller and Nobel (1986 and unpublished observations).

consequences of such morphological variation for PAR interception and CO_2 uptake, recognizing that branches, or multiple stems, can also have other effects. For instance, branching can increase the numbers of flowers, fruits, and seeds produced by *Carnegiea gigantea* (Fig. 1.12A,B; Steenbergh and Lowe, 1977, 1983; Yeaton, Karban, and Wagner, 1980; Cody, 1984).

Let us first consider branch position around a main trunk. For *C. gigantea*, the branch frequency is about twice as great on the south side compared with the north side (Fig. 5.27A), in agreement with other studies (Yeaton et al., 1980). Also, *Pachycereus pringlei* (Fig. 1.14B) in Baja California has about twice as many branches on the south side as on the north side of the main trunk (Geller and Nobel, 1986). The angular location of branches can be affected by topographical features that block direct PAR. For instance, when mountains to the east cause the angle to the horizon

to be 28° above the horizontal in that direction, the position for maximal branch frequency for *C. gigantea* shifts toward the southwest (Geller and Nobel, 1986). Although not branches, the horizontal stems of *Stenocereus eruca*, commonly known as the "creeping devil," are also affected by light direction; in particular, nearly three times as many of its stems tend to point south compared with north (Fig. 5.27B).

A model has been developed to predict PAR on the main trunk and a single branch of *C. gigantea* (Geller and Nobel, 1986). On a clear day at an equinox, PAR and net CO_2 uptake per unit stem area are lowest when the branch occurs on the east or the west side of the trunk, which maximizes interstem shading. Compared with having the single branch on the north side, net CO_2 uptake per unit height on

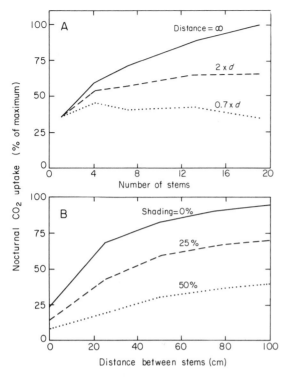

Figure 5.28. Predicted influence of number of stems (**A**) and distance between stems (**B**) on nocturnal net CO_2 uptake for clear days at an equinox. For panel **A**, stem length and stem diameter (d) were varied by the same factor such that total plant volume was kept constant as stem number changed; interstem distance is indicated in multiples of d. For panel **B**, simulations were performed at the indicated percentage shadings for nineteen stems that were 15 cm in diameter. After calculating the total daily PAR on the various stem surfaces, Figure 5.2 was used to estimate nocturnal net CO_2 uptake (maximum CO_2 uptake occurs for the nineteen stems when they are infinitely spaced). Modified from Geller and Nobel (1986).

the trunk where the branch occurs is about 10% less if the branch is on the east or the west side and 2% greater if it occurs on the south side for the summer solstice; these differences are about half as great on an equinox and are very small for the winter solstice. Thus, only a small advantage accrues in terms of PAR distribution and net CO_2 uptake when a single branch of *C. gigantea* is on the south side.

If stems were infinitely spaced, then no shadows would be cast from one stem to another, and for plants of constant volume net

CO_2 uptake would increase with the number of stems (Fig. 5.28A). In this case, the stem surface area increases with the cube root of stem number, so net CO_2 uptake would similarly increase with stem number. When the separation between stems is 0.7 times the stem diameter, the increase in stem area as the number of stems increases is accompanied by increased self-shading, and overall net CO_2 uptake is little affected (Fig. 5.28A). Because infinite stem spacing is obviously impossible for a particular plant and stems separated by less than their diameter can lead to no increase in CO_2 uptake, an intermediate spacing of twice the diameter is considered (Fig. 5.28A). In this case, nocturnal CO_2 uptake on clear days at the equinox nearly doubles (at constant plant volume) as the stem number per plant increases from one to nineteen. In this regard, the separation between stems divided by the stem diameter is about 1.1 for *P. pringlei*, 2.4 for *C. gigantea*, 2.6 for *Stenocereus thurberi*, and 3.2 for *Lophocereus schottii*, the latter two averaging fifteen to eighteen stems per plant (Geller and Nobel, 1986).

For a plant with nineteen stems that are 15 cm in diameter, nocturnal CO_2 uptake nearly doubles as initially appressed stems are separated by about 15 cm and doubles again at a separation of 100 cm (Fig. 5.28B). PAR reduction due to shading from spines, other vegetation, or topographical features as well as clouds leads to an even greater fractional reduction in nocturnal net CO_2 uptake, reflecting the nonlinear relation between total daily PAR and net CO_2 uptake (Fig. 5.2). For instance, at a 50-cm separation between 15-cm-diameter stems, shading that reduces PAR by 25% reduces net nocturnal CO_2 uptake by 30%, and 50% shading reduces CO_2 uptake by 65% (Fig. 5.28B). In summary, considered in terms of PAR interception, optimal separation between stems on multiple-stemmed columnar cacti should be at least twice the diameter of the stems, and such plants would be expected to occur where the available PAR is not reduced more than about 25% by external factors.

Columnar cacti tend to have a much higher area/volume (A/V) ratio than do spheres, for which A/V is minimized. Thus, increasing the photosynthetic area of columnar cacti is apparently favored over maximizing the volume for water storage (see Chapter 3). For *C. gigantea*, branching with its accompanying increase in A/V is more common in wetter areas (Yeaton et al., 1980; Steenbergh and Lowe, 1983). Branches tend to occur higher off the ground when the surrounding vegetation is taller for *Pachycereus pectenaboriginum* and *Stenocereus montanus*, especially the first branch (Cody, 1986b). Therefore, changes in net CO_2 uptake as the number of stems increases (Fig. 5.28A) can interact with the availability of water and of PAR.

TILTING

Many different aspects of tilting affect the PAR relations of cacti, including the slope of the ground. For instance, PAR may be limiting for *Carnegiea gigantea* on north slopes (Niering et al., 1963), just as it is for *Agave deserti* (Fig. 5.7). For nearly all cases in which ten species of cacti were considered on slopes (greater than 8°) in the Mojave and the Sonoran deserts, the number of plants per unit ground area is lower on slopes facing north than on slopes facing east, west, or south, although PAR levels were not determined (Yeaton and Cody, 1979). Here, we will emphasize the effects of tilting by stems of barrel and columnar cacti and by the whole canopy of certain cylindropuntias.

The stems of columnar cacti such as *Carnegiea gigantea* (Fig. 1.12A), *Lophocereus schottii* (Fig. 1.14A), *Stenocereus thurberi* (Fig. 1.14D), and *Trichocereus chilensis* (Fig. 5.22B) show no tendency to tilt, even when presented with essentially horizontal PAR during development (Table 5.3). The lack of tilting for these columnar cacti is not surprising, because gravity would exert a considerable bending moment (tendency to cause rotation) on their massive stems if they were tilted. On the other hand, the solitary stems of barrel cacti tend to tilt toward the equator (Fig. 1.10). For three species of *Copiapoa*, the tilt angle in the field averages 42° (Table 5.3). The equatorial tilt averages 13° for *Eriosyce ceratistes* in Chile and the morphologically similar species of *Ferocactus* in the southwestern United States. When grown in the laboratory with unidirectional radiation 10° above the horizontal, the average tilt from vertical for the barrel cacti is 33°, which is 5° greater than for the

same species in the field (Table 5.3). Actually, equatorial tilts of *F. wislizenii* toward the directions of higher PAR have long been recognized (Humphrey, 1936), and equatorial tilts of various species of barrel cacti have been used as "compasses" (Chapter 1).

Equatorial tilting does not affect the PAR incident on the east or west sides of unshaded barrel cacti, but has seasonally dependent effects on the north and south sides. In particular, equatorial tilting reduces PAR interception on the north plus south sides for all latitudes at the winter solstice and an equinox (Nobel, 1981b). At the summer solstice at 34°N or S, an equatorial tilt of 8° decreases PAR on the north plus south surfaces by 21%, but a tilt of 40° increases it by 78% (Nobel, 1981b). Averaged over the whole stem, the observed equatorial tilt of *Copiapoa cinerea* at 25°S (Table 5.3) leads to 22% less PAR interception than does random orientation (Ehleringer et al., 1980) and 4% less than vertical orientation (Geller and Nobel, 1987) over the course of a year. In summary, tilting of the longitudinal stem axis toward the sun's trajectories generally leads to less PAR annually incident on the stems and hence does not appear to be advantageous for enhancing PAR interception and thus CO_2 uptake by the whole plant (Nobel, 1981b). However, such phototropic responses could be of benefit in terms of PAR interception for plants in shaded microhabitats, because tilting would then result in growth toward regions of higher PAR (certain thermal aspects of tilting are discussed at the end of the previous chapter).

Stem orientation toward the light for cylindropuntias is very pronounced (Fig. 5.29); the average angle from the vertical for four species is 52° in response to a unidirectional laboratory PAR source 10° above the horizontal (Table 5.3). In the field, the orientation of terminal cladodes of *Opuntia echinocarpa* (Fig. 5.29B) is radially outward in each part of the canopy, meaning toward the east on the east side, toward the south on the south side, and so forth (Geller and Nobel, 1987). Such orientation decreases the average PAR intercepted by the stem sides, but can result in a more even distribution of PAR over the entire plant canopy. Additionally, the tops of the dense canopies of *Opuntia acanthocarpa* (Fig.

Table 5.3. *Orientation of cylindrical stems of various cacti*

Species	Field orientation			Laboratory orientation	Reference
		Longitudinal axis			
	Site latitude	Angle clockwise from north (°)	Declination from vertical (°)	Declination from vertical (°)	
Columnar cacti					
Carnegiea gigantea	31°N	—	0	3	Nobel (1981b) (**A**), Geller and Nobel (1987) (**B**)
Lophocereus schottii	31°N	—	0	8	**A,B**
Stenocereus thurberi	31°N	—	1	3	**A,B**
Trichocereus chilensis	34°S	—	1		**A**
Barrel cacti					
Copiapoa alticostata				47	**B**
C. cinerea	25°S	11	38		Ehleringer et al. (1980) (**C**)
	26°S	12	37	48	**A,B**
C. gigantea				48	**B**
C. haseltoniana	25°S	11	48	50	**C**
C. lembckei	27°S	10	43		**B**
Eriosyce ceratistes	33°S	9	9		**A**
Ferocactus acanthodes	34°N	171	8	17	**A**
	37°N	185	34[a]		**A,B**
					Ehleringer and House (1984)
F. covillei	32°N	208	12	27	**A,B**
F. viridescens	33°N	204	8	13	**A,B**
F. wislizenii	33°N	197	17	14	**A,B**
Cylindropuntias					
Opuntia acanthocarpa				55	**B**
O. bigelovii				38	**B**
O. echinocarpa				58	**B**
O. ramosissima				56	**B**

Note: Angles were measured for an average of about fifty plants in the field. In the laboratory, fourteen plants were studied for which unidirectional radiation 10° above the horizontal provided 600 $\mu mol\ m^{-2}\ s^{-1}$ normal to the beam for 14 h day^{-1} (a total daily PAR of 30 mol m^{-2} day^{-1}). Mean angles were generally obtained using circular statistics (Batschelet, 1981). Studies cited more than once are assigned a boldface letter the first time they occur, this letter being used for subsequent citations.

[a] Plants on south-facing slopes.

Figure 5.29. Cylindropuntias with dense, southward-tilting canopies. (**A**) *Opuntia acanthocarpa* at 36°N in the Granite Mountains near Kelso, California. (**B**) *O. echinocarpa* at 34°N in the University of California Philip L. Boyd Deep Canyon Research Center near Palm Desert, California. Rods along which PAR sensors were mounted are shown in panel **B** (photographed by Gary N. Geller).

5.29A) and *O. echinocarpa* (Fig. 5.29B) tend to tilt southward by 30° (Geller and Nobel, 1987). When sunlight from the west is blocked by hills, the canopy tilt for *O. acanthocarpa* is to the southeast, similar to the cladode orientation effects induced by topography for platyopuntias.

A PAR model indicates that changing the canopy of a highly branched cactus from horizontal to 30° southward tilting at 34°N increases PAR on the stem surfaces averaged over the whole plant by 26% at the winter solstice, 12% at an equinox, and 0% at the summer solstice (Geller and Nobel, 1987). The increased total daily PAR is a result of less interstem shading and leads to increases in net CO_2 uptake at night. The mean angles at the plant base from the soil surface to the lower canopy edge are over 20° greater on the north side than the south side and are intermediate on the east or west sides (Fig. 5.29). Thus, individuals of both *O. acanthocarpa* and *O. echinocarpa* appear more open when viewed from the north side because of the relative lack of branches there compared with the other three sides. Just below the canopy edge, the average total daily PAR is only about 3 mol m^{-2} day^{-1}, which is the PAR compensation value at which no net CO_2 uptake takes place (Fig. 5.2). In addition, 77% of newly initiated stems on unilaterally illuminated stems of *O. echinocarpa* occur on the illuminated side (Geller and Nobel, 1987). The overall can-

opy shape of cylindropuntias with dense canopies (Fig. 5.29) apparently is in response to the combined effects of phototropism, lack of growth below PAR compensation, and maximum growth in regions of high ambient PAR.

Other aspects of PAR

We begin this section with a consideration of PAR attenuation down through cactus canopies (PAR incident on agave leaves in various parts of the canopy has already been discussed). We next consider certain details of photon absorption. Although CO_2 uptake by agaves and cacti is generally PAR limited, detrimental effects can occur at very high PAR levels, especially if a plant suddenly goes from days with low total daily PAR to days with high levels. Besides the direct effects of PAR on CO_2 uptake and nocturnal acid accumulation, daylength or photoperiod can influence various plant responses. Nurse plants and other neighboring plants also affect PAR availability.

CANOPY PAR ATTENUATION

The canopies of cacti vary considerably, from dense ones for species like *Echinocereus engelmannii* (Fig. 5.30A), where very little PAR penetrates to the base, to the moderately open canopy of *Opuntia acanthocarpa* (Fig. 5.29A) to the very open canopies of species like *O. ramosissima* and *O. versicolor* (Fig. 5.30B). For *E. engelmannii* and *O. echino-*

Figure 5.30. Variations in canopy architecture. (A) Closely appressed stems of *Echinocereus engelmannii* at Agave Hill (Figs. 1.1 and 1.2), showing sensor wires for PAR determination. (B) *Opuntia versicolor*, resembling a dancer, near Guaymas, Sonora, Mexico.

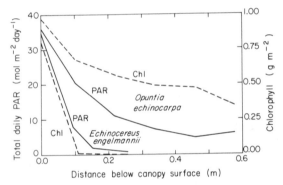

Figure 5.31. Average total daily PAR near an equinox and stem chlorophyll level (Chl) measured at different depths throughout the canopies of *Echinocereus engelmannii* (Fig. 5.30A) and *Opuntia echinocarpa* (Fig. 5.29B). Modified from Geller and Nobel (1987).

carpa (Fig. 5.29B), spines and shadows from the stems halve the total daily PAR at about one-quarter of the way from the top of the canopy to the base (Fig. 5.31). Near the equinox on clear days, the PAR compensation level occurs just over halfway down through the canopy for *E. engelmannii*, whereas the more open canopies of *O. echinocarpa* with considerable diffuse PAR near the plant base generally do not exhibit a reduction of PAR to the compensation level except near the lower canopy edge.

The amount of chlorophyll per unit stem area can also decrease down through the canopy. The fractional decrease of chlorophyll with distance is somewhat greater than the fractional PAR decrease for *E. engelmannii* and somewhat less for *O. echinocarpa* (Fig. 5.31). For *O. echinocarpa*, as chlorphyll per unit stem area decreases from 0.95 g m^{-2} to 0.35 g m^{-2} down through the canopy, the nocturnal acid accumulation at a low total daily PAR of 7 mol m^{-2} day^{-1} is little affected. However, the maximum nocturnal acid accumulation at a high total daily PAR of 30 mol m^{-2} day^{-1} is only 45% as great at the lower chlorophyll level (P. S. Nobel and T. L. Hartsock, unpublished observations). Thus, the chlorophyll level is matched to the ambient PAR for these two species (Fig. 5.31). In any case, the decrease in the availability of PAR down through the canopy of various cacti decreases the potential for CO_2 uptake at the lower parts of the canopy.

ABSORPTANCE – PHOTON REQUIREMENT

When discussing the absorption of shortwave radiation in Chapter 4, we introduced the shortwave absorptance *a*, which indicates the fraction of impinging shortwave energy that is absorbed (Eq. 4.1); for plants, *a* generally varies from 0.4 to 0.7. The PAR absorptance, which equals the fraction of incident photons constituting the photosynthetically active radiation (400–700 nm) that is absorbed, tends to be higher than the shortwave absorptance, primarily because of the absorption by photosynthetic pigments. Chlorophyll absorbs strongly in the blue and the red regions, and carotenoids absorb strongly in the blue and the green regions (Nobel, 1983a).

The PAR absorptance is 0.82 for the upper side of the leaves of *Agave americana*, 0.86 for the lower side, 0.88 for the stem of *Carnegiea gigantea*, and 0.73 for *Opuntia phaeacantha* (Gates, 1980). In another study, the PAR absorptance was 0.86 for *C. gigantea*, 0.77 for *Echinocereus engelmannii*, 0.84 for *Ferocactus viridescens* and *Mammillaria dioica*, and averaged 0.71 for nine species of *Opuntia* (range of 0.61 to 0.82; Ehleringer, 1981). Thus, the PAR absorptance averages 0.76 for the agave and the cacti that have been studied, which is a fairly typical value for plants in general. However, in addition to reflection of an average of about 24% of the incident PAR by the leaf or stem, not all the photons absorbed are actually absorbed by photosynthetic pigments in the chlorenchyma. For instance, for *Opuntia ficus-indica* only about 57% of the incident PAR is transmitted through the cuticle, epidermis, and hypodermis (Fig. 1.34A) and thus is available for absorption by chlorophyll and other photosynthetic pigments (Nobel and Hartsock, 1978). Because the chlorenchyma of agaves and cacti is thick and has a high amount of chlorophyll per unit area, only about 1% of the incident PAR passes through it (Nobel, 1977b, and unpublished observations; Nobel and Hartsock, 1983).

Most of the absorbed photons are used for photosynthesis, which in the absence of oxygen has a minimum theoretical requirement of 8 photons per CO_2 fixed into a carbohydrate (Eq. 2.1). In the presence of normal atmospheric levels of oxygen (21% O_2), some of the carbohydrate is oxidized, leading to CO_2 production by respiration and by a light-dependent process called *photorespiration*. Consequently, the minimum theoretical requirement becomes about 11 photons absorbed per net fixation of one CO_2 (Nobel, 1983a). However, plants rarely reach this theoretical minimum in the field, where C_3 plants at 21% O_2 generally require at least 15 to 20 absorbed photons per CO_2 fixed (Rabinowitch, 1951). In

the laboratory in the near absence of O_2, the minimum measured requirement is about 11 photons per CO_2 fixed for C_3 plants, and in the presence of 21% O_2 the requirement is 15 to 19 absorbed photons per CO_2 fixed (Ehleringer and Björkman, 1977; Ehleringer and Pearcy, 1983). Using a somewhat indirect method, about 11 photons are found to be required per CO_2 fixed during the daytime for a species (or intergeneric hybrid) of *Heliocereus* in the absence of photorespiration, similar to values for other CAM plants and also for C_3 plants in the absence of photorespiration (Adams, Nishida, and Osmond, 1986).

For agaves and cacti operating in the CAM mode, photons are absorbed during the day, but most of the net CO_2 uptake occurs at night (Chapter 2). Because some of the energy stored in carbohydrates is consumed in other processes with the accompanying release of CO_2, such as the active transport of malate into the vacuoles at night and respiration required for general plant growth and maintenance, the minimal photon requirement per CO_2 fixed tends to be high. Specifically, the number of photons absorbed per CO_2 fixed at night equals 68 for *Ferocactus acanthodes* under mild water stress (Nobel, 1977b), 46 for *Agave deserti* (Nobel and Hartsock, 1978), 30 for *Cereus validus* (Nobel et al., 1984), 25 for *Opuntia ficus-indica* (Nobel and Hartsock, 1983), and 22 for *O. chlorotica* (Nobel, 1980a) and *Trichocereus chilensis* (Nobel, 1981b). The lowest number of absorbed photons during the day per CO_2 fixed at night for cacti is thus 22, which is at the high end of the range of absorbed photons per CO_2 fixed during the daytime for C_3 agricultural crops under optimal conditions (Rabinowitch, 1951).

PHOTOINHIBITION

Although PAR is generally limiting for net CO_2 uptake by agaves and cacti, situations can occur in which the PAR level is too high. Excessive PAR levels can lead to a bleaching of the leaves or stems as the green color is lost because of partial destruction of chlorophyll. Sometimes more subtle damage is done to the component reactions of photosynthesis, again resulting in a decrease in net CO_2 uptake. Such light-dependent damage to the photochemical processes in chloroplasts is termed *photoinhi-*

bition; the primary site of damage is often in the reaction center of Photosystem II (Chapter 4), thereby adversely affecting the electron transport reactions of photosynthesis (Osmond, 1981). Darbishire (1904) long ago suggested that spines might benefit *Mammillaria elongata* and other cacti by protecting them from high light levels, especially in the region of developing tubercles, and thereby preventing the destruction of chlorophyll in high light.

Photoinhibition tends to occur when all the light cannot be processed by normal photosynthetic reactions, such as might occur during drought and its accompanying low rates of photosynthesis. CAM plants such as agaves and cacti actually have an advantage over C_3 plants in such situations, because CAM plants tend to maintain considerable photosynthetic activity using internally recycled CO_2, even when net CO_2 exchange with the environment is minimal (Ting, 1985; Gil, 1986). Specifically, CO_2 can be released internally by respiration and incorporated into organic acids at night; during the day, such organic acids are decarboxylated and the internally released CO_2 is fixed behind closed stomata by C_3 photosynthesis (Chapters 2 and 3). The electron transport reactions are kept active even during drought, and photoinhibition is thus minimal.

Photoinhibition can occur when plants are suddenly exposed to much higher PAR levels or are exposed to very high PAR levels. For instance, photosynthetic electron transport is initially somewhat disrupted when *Opuntia stricta* is transferred from 50% to 100% of full sunlight, but recovery occurs the next day (Osmond, 1982). Inhibition of electron transport is noted for the south-facing sides of cladodes of *O. basilaris* at 37°N compared with the north-facing sides, which receive less total daily PAR. Specifically, one CO_2 can be fixed for 18 photons absorbed on the north side, but 32 photons are required on the south side during a period of extended drought, which can exacerbate photoinhibition (Adams, Smith, and Osmond, 1987). For *Cereus validus* growing in the laboratory at a total daily PAR of 10 mol m^{-2} day^{-1}, both tissue chlorophyll levels and nocturnal acid accumulation are decreased about 50% when the plants are placed under a total daily PAR of 25 mol m^{-2} day^{-1}, and the decreases are even

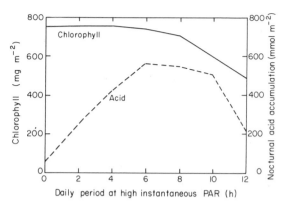

Figure 5.32. Effect of daily period at a relatively high instantaneous PAR of 1,200 μmol m^{-2} s^{-1} on chlorophyll content and nocturnal acid accumulation for *Opuntia ficus-indica*. Plants previously maintained for three months at 560 μmol m^{-2} s^{-1} in the planes of the cladodes for 12 h each day were shifted to 1,200 μmol m^{-2} s^{-1} for the indicated number of hours each day for one week (a background PAR of 40 μmol m^{-2} s^{-1} was provided for 12 h each day to keep the photoperiod constant). Modified from Nobel and Hartsock (1983).

greater if the plants are additionally subjected to a salt stress (Nobel et al., 1984). Photoinhibition in cacti may be more severe at higher tissue temperatures (Adams et al., 1987). The inverse may also be true; for instance, the greater sensitivity to high temperatures for the aerial compared with the subterranean parts of a large cladode of *Opuntia ficus-indica* inserted into the ground (Nobel et al., 1986) may result from damage to the aerial part caused by photoinhibition.

When *O. ficus-indica* maintained in the laboratory at a total daily PAR of 24 mol m^{-2} day^{-1} (an instantaneous PAR of 560 μmol m^{-2} s^{-1} for the 12-h photoperiod) is shifted to 36 mol m^{-2} day^{-1} or higher, nocturnal CO_2 uptake and nocturnal acid accumulation decline. For instance, after one week at a total daily PAR of 49 mol m^{-2} day^{-1}, both processes are reduced over 75% compared with the values at 24 mol m^{-2} day^{-1} (Nobel and Hartsock, 1983). Also, transferring the plants from 560 to 1,200 μmol m^{-2} s^{-1} causes a progressive loss of chlorophyll as the amount of time at the higher instantaneous PAR is increased (Fig. 5.32). Seven or more hours at 1,200 μmol m^{-2} s^{-1} results in the destruction of chlorophyll accompanied by a decrease in nocturnal acid accumulation, even though the total daily PAR is increasing (Fig. 5.32). However, under field conditions the instantaneous PAR level never exceeds 1,200 μmol m^{-2} s^{-1} for 7 h on the vertical sides of stems, and the total daily PAR only rarely exceeds 36 mol m^{-2} day^{-1} on vertical surfaces (Fig. 5.14). Therefore, photoinhibition may not be a severe limitation for cacti in the field, except when other stresses, such as drought or temperature extremes, make them more sensitive, or when the total daily PAR changes very rapidly from low to high levels.

PHOTOPERIOD

Photoperiod affects the daily pattern of gas exchange, longer photoperiods tending to favor daytime CO_2 uptake at the expense of nighttime CO_2 uptake for CAM plants (Kluge and Ting, 1978). The length of the light period can also affect other plant responses, such as the levels and activities of various enzymes involved in carbon metabolism in CAM plants (Kluge and Ting, 1978; Queiroz and Brulfert, 1982). Because use of the CAM pathway can result in a higher water-use efficiency (Chapters 2 and 3), and for some habitats droughts occur at a predictable time of the year, changes in photoperiod could allow certain CAM species to anticipate a seasonal drought and thus shift from daytime to nighttime CO_2 uptake patterns. We have already indicated that shortening the photoperiod, which happens naturally as winter approaches, has very little influence on the low-temperature tolerance of an agave and two species of cacti (Chapter 4). Also, the high-temperature tolerance is not affected by photoperiod.

As would be expected, changes in the photoperiod affect flowering, but the effects vary considerably with species. Increasing the photoperiod from 8 h to 14 h leads to the formation of more flower buds for an unidentified species of *Echinopsis* only if receiving nutrients (Pushkaren et al., 1980). Also, longer photoperiods after a chilling period lead to the induction of flower buds for *Mammillaria longicoma* (Rünger, 1968). Induction of flower buds on *Schlumbergera truncata* occurs for shorter photoperiods at higher temperatures; buds form on most terminal stem segments at

daylengths up to 8 h at 30°C, 15 h at 20°C, and continuous light at 10°C (Rünger and Führer, 1981; but see Ho, Sanderson, and Williams, 1985). Low temperatures and short days are optimal for flower buds and flowering of *Rebutia marsoneri*, but long days at 20°C are better after three months at low temperatures (Rünger, 1969). Longer photoperiods increase the number of new stem segments and total plant dry weight for *Opuntia microdasys*, increase the number of new stem segments for *Cereus silvestrii*, and have little effect on *Mammillaria elongata* (Sanderson et al., 1986).

Growing *Opuntia ficus-indica* under short photoperiods with wet conditions and then changing the plants to long photoperiods with dry conditions leads to typical dark CO_2 fixation and plant survival. However, initially growing the plants under long photoperiods with wet conditions and then changing them to long photoperiods with drought can lead to plant death (Queiroz and Brulfert, 1982; day/night temperatures were 27°C/17°C throughout). When the initial short photoperiod (wet condition) is 8 h, enzyme levels during the subsequent long photoperiod (dry condition) are about the same as when the initial long photoperiod is 16 h, suggesting that readjustment of the levels of enzymes in the CAM pathway is not the key to survival for *O. ficus-indica* (Brulfert, Guerrier, and Queiroz, 1984). When the initial day/night temperatures under wet conditions are 22°C/12°C and the subsequent temperatures under dry conditions are 40°C/30°C, no CO_2 uptake is observed during the drought for plants initially subjected to the long photoperiod, but those initially subjected to the short photoperiod have some net CO_2 uptake. However, the activity of PEP carboxylase in the plants initially under long photoperiods is maintained (Brulfert et al., 1987). Further experimentation is obviously needed to understand fully the influence of photoperiod during wet periods on the survival of cacti during subsequent drought.

NURSE PLANTS

Nurse plants can have many influences on the environmental biology of agaves and cacti. In Chapter 4 we considered the moderation of both low- and high-temperature extremes by nurse plants. The interception of radiation involved in temperature moderation also reduces the PAR available to the associated seedling. We will consider such influences of nurse plants on PAR availability, as well as certain other effects, including some mediated by animals (Yeaton and Cody, 1976; Serey and Simonetti, 1981; McAuliffe and Janzen, 1986).

The interception of shortwave radiation by the bunchgrass *Hilaria rigida* (Fig. 1.1) can be crucial for protecting seedlings of *Agave deserti* from high-temperature damage (Chapter 4), but it also leads to slower growth for the seedling. For instance, after thirteen years of being protected by *H. rigida*, an individual of *A. deserti* was only 16 cm tall (Fig. 3.20). For a small seedling 4 cm tall (Fig. 1.1), the seasonally averaged PAR attenuation in the planes of the leaf surfaces caused by the nurse plant can be up to about 70%, decreasing to about 40% attenuation for the 16-cm-tall plant (Nobel, 1984c, and unpublished observations; Nobel and Geller, 1987; Franco and Nobel, 1988). Under unshaded conditions on clear days, the total daily PAR annually averages about 15 mol m^{-2} day^{-1} in the planes of the leaves for mature plants of *A. deserti* (Woodhouse et al., 1980) but about 25 mol m^{-2} day^{-1} for the more open canopies of small seedlings (Franco and Nobel, 1988). If we take cloudiness into consideration, the total daily PAR for the seedlings considered here would annually average about 20 mol m^{-2} day^{-1}, and so 70% and 40% reductions in PAR would lead to total daily PAR values of about 6 and 12 mol m^{-2} day^{-1}, respectively. Based on the nonlinear CO_2 response to total daily PAR for *A. deserti* (Fig. 5.2), its net CO_2 uptake is one-seventh of maximal when the total daily PAR is 6 mol m^{-2} day^{-1} and one-half of maximal at 12 mol m^{-2} day^{-1}. Therefore, PAR attenuation by the nurse plant *H. rigida* can cause the expected growth rate of seedlings of *A. deserti*

to be two to seven times lower than for an exposed location where no shading occurs.

Common nurse plants for *Carnegiea gigantea* (Fig. 1.12A) are *Cercidium floridum* (blue paloverde) and *Cercidium microphyllum* (foothill paloverde). Even when nearly leafless on the winter solstice, these trees can reduce the shortwave irradiation and PAR at ground level by 31% (Lowe and Hinds, 1971). The cactus ultimately overtops the nurse plant, which eventually dies; the cycle can begin again when *C. gigantea* dies (Vandermeer, 1980; McAuliffe, 1984a). Cyclical replacement of one species by another can also occur for the widely distributed desert shrub, *Larrea tridentata* (creosote bush), and *Opuntia leptocaulis*. Seeds of the cactus are dispersed by birds that use *L. tridentata* as a perch and by rodents that use the shrub as a refuge (Yeaton, 1978). Eventually *O. leptocaulis* overtops *L. tridentata* as well as successfully competes for the limited rainfall by having shallow roots, and the nurse plant dies. Upon death of *O. leptocaulis, L. tridentata* recolonizes the area and the cycle can begin again. Seed densities of *C. gigantea* are greater under the trees destined to become its nurse plants than in the open, reflecting dispersal aspects or perhaps differential seed predation (Hutto, McAuliffe, and Logan, 1986). Animals are involved in another way when the treelike *Opuntia fulgida* serves as an apparent nurse plant for *Echinocereus engelmannii* (Fig. 5.28A) and *Mammillaria microcarpa*. The easily detachable stem segments of *O. fulgida* fall to the ground, where their spines deter mammalian herbivores and thus favor the establishment of the other two cactus species (McAuliffe, 1984b). In summary, in addition to the widely assumed competition for water by roots of desert plants, such as between a nurse plant and its associated seedling, competition for PAR may occur between a seedling and its nurse plant, which can also have other effects on desert succulents that are mediated by animals.

6 Nutrient relations

A full understanding of the influences of water (Chapter 3), temperature (Chapter 4), and PAR (Chapter 5) on the biology of agaves and cacti is difficult. Yet a comparable understanding of the nutrient relations is even more elusive, partly because of the many mineral elements that are involved. *Macronutrients* (elements required in large amounts from the soil or the "watering" solution, leading to high levels in the tissues) include nitrogen (N), phosphorus (P), potassium (K), calcium (Ca), magnesium (Mg), and sulfur (S), each of which generally comprises at least 0.1% (1 g kg^{-1}) of the dry weight of plants. *Micronutrients* (elements that are also necessary, but at low levels), such as manganese (Mn), copper (Cu), zinc (Zn), iron (Fe), boron (B), cobalt (Co), molybdenum (Mo), and chlorine (Cl), generally range up to 200 parts per million by plant dry weight (abbreviated ppm, which numerically equals mg kg^{-1}). We will also consider sodium (Na), an element not required by most plants, and hydrogen (H), whose concentration in aqueous solutions we specify by the pH.

Besides the sheer number of mineral elements required by plants, other factors also contribute to the complexity of nutrient relations. Photosynthesis, growth, and productivity respond nonlinearly to nutrient level. For instance, no response may occur below a low or "threshold" nutrient level in the soil or plant tissue; responses may be linear with somewhat higher levels and then level off, followed by inhibition at even higher nutrient lev-

els. Such nonlinearity can be exacerbated by interactions between nutrients, such as between those of similar charge. In addition, availability of nutrients to plants is affected by the soil pH. We also note that the roots of agaves and cacti tend to be shallow, so the mineral nutrients are obtained from the upper part of the soil, which is different than for certain other plants.

Let us begin by briefly reviewing some element responses for CAM plants other than agaves and cacti. Adding NaCl to the watering solution can substantially increase the osmotic pressure and can induce a shift from C$_3$ to CAM for *Mesembryanthemum crystallinum* (Winter and von Willert, 1972), *Portulacaria afra* (Ting and Hanscom, 1977), and other succulents (Kluge and Ting, 1978). Such NaCl-induced shifts do not seem to be very pertinent to mature agaves and nonleafy cacti, most of which already predominantly exhibit CAM instead of C$_3$. Growth of most CAM plants is increased by nitrogen fertilization (Nose et al., 1985; Winter, 1985); N is required for proteins, nucleic acids, and other categories of molecules vital to plant function. Potassium is critical for cell osmotic regulation, such as that involved with stomatal opening, and indeed K is required for stomatal opening by CAM plants (Willmer and Pallas, 1973; Loucks and Ownby, 1978; Jewer, Incoll, and Howarth, 1981). The ratio of K to Ca in tissues of *Kalanchoe tubiflora* and *Sedum telephium* exceeds unity when CAM predominates but is just less than unity when C$_3$ metabolism predominates (Mathur, Natarella, and Vines, 1978; Mathur and Vines, 1982); however, K levels can be manyfold lower than Ca levels in other CAM plants (Phillips and Jennings, 1976; Rössner and Popp, 1986).

Mineral elements are taken up from the soil and occur in the plant in forms carrying a positive or a negative charge. For instance, potassium occurs as a positively charged ion, or *cation*, K^+; phosphate can occur as $H_2PO_4^-$, an *anion* carrying a single negative charge. Certain elements occur in multiple forms, such as Mn^{2+} and Mn^{4+} for manganese or ammonium (NH_4^+) and nitrate (NO_3^-) for nitrogen (in keeping with convention, we will not always indicate the charge when presenting a specific atomic name or symbol). To maintain charge balance or *electroneutrality* for plant cells (Nobel, 1983a), the total charge of the cations must closely match the total charge of the anions. Thus, when a cation is taken up from the soil by a root, an anion is generally also taken up or a cation is released. For instance, K^+ uptake can be accompanied by Cl^- uptake or H^+ release; such H^+ release affects soil pH in the immediate vicinity of the root surface, which can influence the uptake of other ions. Electroneutrality also affects the ionic relations of individual chlorenchyma cells of agaves and cacti. For instance, the accumulation of calcium (Ca^{2+}), which can greatly exceed the potassium level in the chlorenchyma, presumably occurs in concert with the accumulation of the anion oxalate ($^-OOCCOO^-$). Because nutrients move toward a root or its associated mycorrhizae in the soil water, nutrient relations and water relations are intertwined. Moreover, nutrients move throughout a plant primarily in the xylem, so xylary flow generated by transpiration also plays a role in the nutrient relations of agaves and cacti.

We will first indicate the levels of various nutrients and Na in agaves and cacti. Element levels can affect metabolism; for instance, we shall see that chlorenchyma N levels are correlated with nocturnal acid accumulation. Likewise, element levels in the soil or watering solution can affect the growth of seedlings and adult plants. Out of all the macronutrients and micronutrients tested, we will identify nitrogen as the soil nutrient most limiting for the growth of agaves and cacti. We will also examine the effects of soil salinity, pH, and particle size in the nutrient relations of agaves and cacti.

Tissue element levels

Here we will survey the levels of eleven elements in the chlorenchyma of agaves and cacti and compare such levels with element levels in the leaves (mainly chlorenchyma) of agronomic plants. Element levels in other tissues of agaves and cacti will also be summarized. We will relate tissue element levels to the nutritive value of agaves and cacti. To help understand the metabolic implications of different element levels, the levels of each element will be correlated with nocturnal acid accumulation in the field for various species of agaves and again for various species of cacti. Although this technique cannot be considered an "experiment," nocturnal acid accumulation is an easily measured metabolic process that can be related to CO_2 uptake. The approach can suggest whether certain mineral elements are limiting for productivity of agaves and cacti and thus help provide a perspective on the growth responses to various elements considered in subsequent sections.

CHLORENCHYMA AND PARENCHYMA LEVELS

Because of the large amount of water-storage parenchyma in the photosynthetic organs of agaves and cacti and the 2- to 3-mm thickness of the chlorenchyma, the two tissues can be readily separated. Thus, element levels in the chlorenchyma and the parenchyma can be separately determined for most agaves and cacti. Also, element levels in the shoots can be compared with those in the roots.

Element levels in the chlorenchyma have been determined in the field for eight species of agaves and eleven species of cacti (Table 6.1). The ranges in level of all elements examined overlap for the two taxa, and the mean element levels are similar. The major differences are at least twofold higher average levels of Na and Mn in cacti. In addition to the similarity between agaves and cacti, another point of interest is the comparison with representative values for agronomic plants. The only major differences are levels of Na, which av-

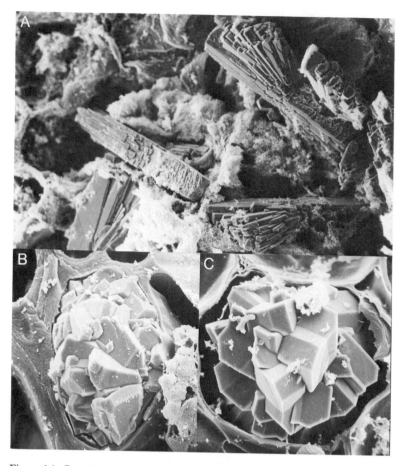

Figure 6.1. Crystal aggregates of calcium oxalate known as druses. **(A)** Flat crystals in *Cephalocereus polygonus*. **(B)** Crystals filling an outermost hypodermal cell of *Opuntia schottii*. **(C)** Crystals filling a chlorenchyma cell of *Pereskia sacharosa*. The druses are about 100 μm across. Used by permission from Gibson and Nobel (1986). © 1986 by Harvard University Press.

erage sevenfold lower for agaves and cacti, and Ca, which average twofold higher (Table 6.1). Other seemingly aberrant element levels compared with agronomic plants are the low B levels for *Opuntia phaeacantha* from Coahuila, Mexico, and the extremely low chlorenchyma P levels (below 500 ppm) for three species of cacti.

Let us next consider the high levels of Ca. Calcium can be above 6% of the chlorenchyma dry weight for both agaves and cacti (Table 6.1). The Ca level can increase from 2% for a two-week-old cladode of *Opuntia ficus-indica* to 6% for a one-year-old cladode (Nobel, 1983d). Such high levels of Ca are mostly associated with the accumulation of calcium oxalate in agaves and cacti (Darling,

1976; Gibson and Nobel, 1986), which can form crystals, referred to as druses (Fig. 6.1). The crystal aggregates can nearly fill certain specialized cells in the chlorenchyma, although druses tend to occur primarily in the hypodermis for the cactus subfamily Opuntioideae. Druses may act as a deterrent to herbivory, a storage for calcium or organic matter, and a reservoir for otherwise toxic material, hypotheses that require further testing.

For *O. ficus-indica*, as well as other cacti, high Ca levels can also be associated with cells that store and secrete mucilage (Trachtenberg and Mayer, 1981, 1982a). Mucilage is a sticky, fibrous polysaccharide, each molecule generally composed of over 30,000 sugar subunits (mainly, arabinose, galactose,

Table 6.1. *Element levels in the chlorenchyma of agaves and cacti*

Species	Site	N (%)	P (ppm)	K (%)	Na (ppm)	Ca (%)	Mg (%)	Mn (ppm)	Cu (ppm)	Zn (ppm)	Fe (ppm)	B (ppm)	Maximal nocturnal acid accumulation (mol m^{-2})
Agave americana	Mexico (details unknown)	1.53	1,280	1.78	46	3.94	0.59	50	4	49	53	34	0.80
A. deserti	Palm Desert, California	1.08	2,760	1.65	10	3.57	0.62	8	2	34	43	20	0.52
A. fourcroydes	Mérida, Yucatán, Mexico	—	2,800	—	—	3.6	0.6	21	6	26	43	—	—
		1.69	2,800	1.46	29	4.64	0.61	10	2	33	38	39	0.78
A. lechuguilla	Saltillo, Coahuila, Mexico	1.14	1,220	1.27	45	6.11	0.40	14	7	36	77	18	0.66
A. salmiana	Salinas de Hidalgo, San Luis Potosí, Mexico	1.10	1,790	2.26	46	4.37	0.59	8	4	7	118	26	0.66
	Tanzania	0.8	1,300	2.1	—	1.8	1.0	35	6	23	50	—	—
A. sisalana	Nairobi, Kenya	0.6	2,600	1.5	—	1.2	0.32	41	7	15	148	13	—
A. tequilana	Tequila, Jalisco, Mexico	1.47	3,300	2.97	62	5.33	1.32	53	3	30	155	22	0.70
A. utahensis	Clark Mountain, California	0.89	1,450	1.31	66	2.30	0.51	18	1	14	34	19	0.24
Agave mean		1.15	2,100	1.81	43	3.70	0.66	26	4	26	77	24	
Carnegiea gigantea	Buckeye, Arizona	2.48	1,180	1.18	332	1.69	0.60	26	4	21	117	23	0.46
Coryphantha vivipara	Mercury, Nevada	1.52	3,820	1.02	92	1.92	0.58	134	3	33	178	54	0.24
Ferocactus acanthodes	Palm Desert, California	1.62	1,700	1.95	315	4.62	0.62	122	9	22	161	62	0.38

Species	Location												
F. wislizenii	Tucson, Arizona	1.92	1,830	1.68	248	2.58	0.60	242	9	29	155	32	0.56
Opuntia basilaris	Palm Desert, California	1.19	444	1.70	121	3.10	1.25	31	2	7	133	13	0.58
O. bigelovii	Palm Desert, California	1.00	1,220	1.52	282	4.98	1.34	46	6	14	219	35	0.19
O. chlorotica	Granite Mountains, California	2.06	292	0.63	18	6.10	1.75	498	5	28	204	52	0.76
O. echios	Santa Cruz Island, Galápagos Islands, Ecuador	1.58	1,720	1.58	484	3.14	1.68	209	3	25	102	18	0.32
O. ficus-indica	Fillmore, California	2.61	3,320	1.18	31	6.33	1.43	54	15	52	88	109	0.81
	Til Til, Chile	1.53	1,160	1.97	30	3.90	1.31	371	13	52	79	35	0.65
O. phaeacantha	Kingsville, Texas	2.11	1,970	3.69	179	3.81	1.84	92	4	31	73	23	—
	Coahuila, Mexico	0.90	370	1.48	78	9.66	1.05	28	4	11	61	4	—
Trichocereus chilensis	Cuesta la Dormida, Chile	1.23	2,620	1.67	215	5.92	0.78	478	14	44	203	44	0.26
Cactus mean		1.67	1,670	1.63	187	4.44	1.14	179	7	28	136	39	
Agronomic plants (mean)		2	3,000	2	1,000	2	0.7	70	8	40	150	30	—

Note: Means are presented in % or ppm on a dry-weight basis for six to nine samples from mature field plants. Data are from Nobel and Berry (1985) for agaves and Nobel (1983d) for cacti, except Sprague et al. (1978) for the first entry for *A. fourcroydes*, Lock (1962) for the first entry for *A. sisalana*, and Pinkerton (1971) for the second, P. S. Nobel (unpublished observations) for *A. tequilana*, and Nobel et al. (1987) for *O. phaeacantha*. Data for agronomic plants are calculated from Epstein (1972) and Larcher (1980).

rhamnose, and xylose together with galacturonic acid; Gibson and Nobel, 1986). Because H^+ can dissociate from the galacturonic subunits of the mucilage, leading to a negative charge, Ca^{2+} can be electrostatically attracted to mucilage. This binding can account for about 10% of the Ca in many cacti.

Averaged for eleven species of cacti, Na is threefold lower, Ca just over 50% higher, and Mn threefold higher in the chlorenchyma than in the parenchyma (Nobel, 1983d; Nobel et al., 1987). The higher Mn level in the chlorenchyma may be related to the presence of certain enzymes and photochemical reactions that require Mn. The higher level of Ca in the chlorenchyma is due to the occurrence of druses (Fig. 6.1) in this region for most cacti. The low level of Na in the chlorenchyma is the result of the general exclusion of this element from the photosynthetic tissue of cacti.

Four of the species of cacti considered have 2% or more N in the chlorenchyma, similar to values for the chlorenchyma of agronomic plants (Table 6.1). On a dry weight basis, the N level in the water-storage parenchyma of cacti averages only about 5% less than the chlorenchyma N (Nobel, 1983d; Nobel et al., 1987). The water-storage parenchyma contains large cells with thin walls and relatively few organelles, so its dry weight per unit fresh weight is lower than that of the chlorenchyma. For instance, for fully hydrated cladodes of *Opuntia phaeacantha*, dry weight corresponds to 11% of the fresh weight of the chlorenchyma but only 6% of that of the water-storage parenchyma (Nobel et al., 1987). On a fresh weight or a volume basis, the N level can thus be about twice as high in the chlorenchyma as in the water-storage parenchyma.

Nitrogen levels for *Opuntia amyclaea* and *O. ficus-indica* are 0.8% to 1.5% on a whole cladode basis (Monjauze and Le Houérou, 1965). The stems of two platyopuntias (*O. ficus-indica* and *O. phaeacantha*) and stems as well as fruits of two cylindropuntias (*O. fulgida* and *O. spinosior*) average 1.1% N by dry weight (Vinson, 1911). Nitrogen levels tend to be higher for young compared with mature cladodes: 95% higher for *O. cantabrigiensis*, 25% higher for *O. phaeacantha* (Martinez Davila, 1980), and 45% higher for *O. ficus-indica* (Nobel, 1983d). The N level can

also vary with leaf age for agaves; for *Agave sisalana*, the N level averaged over the whole leaf is 1.0% of the dry weight for recently unfolded leaves, decreasing to 0.7% for the oldest leaves (Lock, 1962).

Besides N, the levels of other elements can also change with age of the tissue. For leaves of *A. sisalana*, P decreases from 1,300 ppm for recently unfolded leaves to 700 ppm for the oldest leaves, and K decreases from 2.2% to 0.7% (Lock, 1962). On the other hand, Ca increases from 0.7% for leaves still folded about the central spike to 1.3% for recently unfolded ones to 2.0% for the oldest leaves, indicating the relative immobility of this element in the plant (element levels are again averaged over the whole leaf). The accumulation of Ca and the exclusion of Na is also seen for data averaged over the whole stem for five species of cacti from the southwestern United States; the dry weight of Ca in the mature stems is double that of K and fifty times higher than that of Na (Griffiths and Hare, 1906). When the Na level is 608 ppm in the roots of *O. ficus-indica*, it is 71 ppm in the lowermost (oldest) two cladodes and 21 ppm in the fifth cladode along a branch. Indeed, little Na enters most healthy cacti, and that which enters does not accumulate in the distal portions, where there is a further exclusion of Na from the chlorenchyma.

NUTRITIVE VALUE

Agaves and cacti are consumed by both wild and domesticated animals in arid and semiarid regions. Various species of *Opuntia* can be the principal food of birds and various mammals, including wild pigs and deer (Arnold and Drawe, 1979; Everitt et al., 1981). *Ferocactus diguetii* and *Pachycereus pringlei* have served as stock feed during drought in Baja California (Lindsay, 1952). In Monterrey, Mexico, over 70% of the cattlemen feed opuntia cladodes to dairy cattle, and in nearby Saltillo, about 50 kg of cladodes is fed per cow during the winter drought season (de la Cruz et al., 1983). The pulp remaining after the fiber has been extracted from leaves of *Agave fourcroydes* makes up 80% of the feed given to penned dairy cattle in Yucatán. After natural or human-caused prairie fires singe the spines off of *Opuntia polyacantha* in southern Alberta, Canada, pronghorn antelopes feed on the cladodes, especially during periods of drought (Stelfox and Vriend, 1977).

The levels of various elements in agaves and cacti have important implications with respect to such usage by animals. For instance, the P levels near 1,400 ppm in the fruits of *Opuntia leptocaulis* and *O. lindheimeri* can be below the nutritional requirements of birds and mammals, as are the levels of Na, whereas the levels of K, Ca, and Mg are fully adequate (Everitt and Alaniz, 1981). Compared with nine other range species including five grasses, *O. lindheimeri* tends to be low in N and P, although these elements are higher in the cactus than in three of the grass species examined (Meyer and Brown, 1985). For sheep, the major nutrient limitation of cactus fruits appears to be N; a diet of about 85% cactus fruit by dry weight and 15% hay leads to weight gain (Vinson, 1911), but a diet consisting solely of platyopuntias leads to weight loss (Terblanche, Mulder, and Rossouw, 1971). When using cacti as a forage or a fodder, best results are achieved when N exceeds about 1.5% by dry weight. Indeed, the levels of N for some cacti can exceed 2% (Table 6.1), even when averaged over the chlorenchyma and the water-storage parenchyma, in which case no N deficiency would be expected for animals fed these species.

When the daily ration of hay pellets given to cattle is supplemented by an additional 40% dry weight of opuntia cladodes with the spines singed, the daily weight gain is 72% higher than that obtained with hay pellets alone (Shoop, Alford, and Mayland, 1977). Although the cactus has nearly the same protein content as the hay pellets, it has 1,000 ppm P, which is only about one-third of the level required in cattle diets. Also, the cactus contains 3.6% Ca, leading to a much higher Ca:P ratio than considered optimal for cattle. The low Na content of the cactus can be compensated for by supplying salt, and no other elements appear to lead to mineral deficiency or toxicity for the cattle (Shoop et al., 1977). Thus, although the cactus would not be sufficient by itself, it considerably increases the daily weight gain when given in addition to the basal ration of hay pellets.

Because of CAM, cactus tissue can have large daily oscillations in malate concentration and in pH. Indeed, the low pH in the early morning for *Opuntia ficus-indica* can apparently cause diarrhea in cattle (Samish and Ellern, 1975). Timing of feeding during the day may alleviate some of the pH problems. The N level in many species is below that required by cattle, so the cladodes often primarily supply carbohydrates and water (Griffiths, 1905; Gonzalez and Everitt, 1982). The problems of low N and low P may be overcome by research with different species of cacti, some of which may be better adapted to the local combination of soils and animal nutrient needs. Also, fertilizer can be applied to change the element levels in the cladodes to more acceptable levels, such as raising N or P levels. Actually, cladodes of various platyopuntias are already successfully used as cattle fodder in many areas of the world (Monjauze and Le Houérou, 1965; also see Fig. 1.22).

CORRELATION WITH NOCTURNAL ACID ACCUMULATION

To help determine whether certain elements are limiting metabolic processes or perhaps acting in a toxic manner, we can correlate element levels in the chlorenchyma of agaves and cacti with the maximal nocturnal acid accumulation observed in the field under essentially optimal conditions of water, temperature, and PAR. Maximal nocturnal acid accumulation varies randomly with levels of most elements for both agaves and cacti (Table 6.1). Thus, only elements that show significant correlations will be considered here.

For agaves growing under similar environmental conditions, most of the interspecific variation in nocturnal acid accumulation is directly correlated with variations in chlorenchyma nitrogen level ($r^2 = 0.70$, meaning that 70% of the overall variation in nocturnal acid accumulation is accounted for by variations in N; Nobel and Berry, 1985). In particular, nocturnal acid accumulation is low for *Agave utahensis* with only about 0.9% N by dry weight in the chlorenchyma, attaining maximum values for species in which N exceeds 1.5% (Table 6.1). The second highest positive correlation is with B ($r^2 = 0.43$). The only significant negative correlation between element level in the chlorenchyma and nocturnal acid accumulation by agaves is with Na ($r^2 = 0.13$; Nobel and Berry, 1985). Similar patterns occur for cacti (Table 6.1), in that the highest positive correlation is again with tissue N ($r^2 = 0.39$; Nobel, 1983d). The second highest correlation is negative and with Na ($r^2 = 0.32$). Thus, for both agaves and cacti, chlorenchyma N tends to be positively correlated and chlorenchyma Na negatively correlated with an important metabolic process, nocturnal acid accumulation (the negative correlation may simply mean that more Na gets into those cells that are not as metabolically competent). These correlations have served as a useful guide for subsequent experimental studies on the element responses of agaves and cacti.

Macronutrients

Much of the ecological and agronomic interest in the nutrient relations of plants centers on macronutrients. Indeed, nitrogen, phosphorus, and potassium form the basis of the fertilizer industry; for instance, a 6-12-6 fertilizer means one containing 6% N by weight, 12% P_2O_5, and 6% K_2O. Based on practical experience gained from growing succulent plants, including results such as those discussed in the previous section, most of our emphasis will be on N because this element nearly always has a major influence on the growth of agaves and cacti. For example, field trials with *Agave deserti* show that application of nitrogen leads to a doubling of the number of leaves unfolding in a particular year, phosphorus application causes a slight increase, and potassium and calcium have very little effect (Table 6.2). The relative influence of various macronutrients is also apparent for *Rhipsalis gaertneri*, for which element omission from the watering solution for plants growing in peat leads to the following decreases in dry-weight accumulation: N, 83%; P, 34%; K, 32%; Ca, 26%; and Mg, 6% (Penningsfeld, 1972). It is instructive to consider the amount of various nutrients removed from a site when leaves of *Agave sisalana* containing 1 metric ton (1,000 kg) of fiber by dry weight are harvested. In particular, for the acidic red soils of Tanzania, such a harvest leads to the following removal of macronutrients: about 30 kg N, 5 kg P, 80 kg K, 70 kg Ca, and 40 kg Mg (Lock, 1962).

Table 6.2. *Influence of various fertilizer levels on the number of leaves unfolding over a one-year period for* Agave deserti *at Agave Hill (Fig. 1.1)*

Macronutrient	Level applied (g m^{-2})				
	0	3	10	30	100
N	58	87	113	103	—
P	59	66	62	60	—
K	60	54	58	59	58
Ca	56	60	60	55	58

Note: Nutrients were applied to groups of twenty plants on 22 April 1986, and the number of leaves newly unfolding per group during the ensuing year was determined on 21 April 1987. Nitrogen was added as $Ca(NO_3)_2$, phosphorus as $CaHPO_4$, potassium as K_2SO_4, and calcium as $CaSO_4$. The application region extended about 1 m beyond the roots of plants in a particular group, and the levels indicated are based on the element mass (1 g m^{-2} = 10 kg ha^{-1}). Data represent unpublished observations of P. S. Nobel and T. L. Hartsock.

NITROGEN

Nitrogen is required for proteins (16% N by dry weight), chlorophyll (6% N), nucleic acids (15% N), and many other crucial cellular components. Thus, nutrient studies for many plants often begin with a consideration of N, and agaves and cacti are no exception. Quantitative research is difficult to do using natural fertilizers such as manure or *bagasse* (the organic residue from an agroindustrial process, such as the extraction of fibers from agave leaves or the fermentation of agave stems), because these fertilizers contain many elements in addition to N. On the other hand, we now have fairly detailed information on the specific growth responses to N. Fortunately, both agaves and cacti can be grown "hydroponically" in aqueous solutions. The use of hydroponics facilitates nutrient studies, because element levels can easily be varied in such solutions. Moreover, balanced solutions containing proper ratios of all the elements for optimal growth have been devised, so deviations from normal growth attributable to nutrient deficiency or toxicity are readily detected (Hoagland and Arnon, 1950).

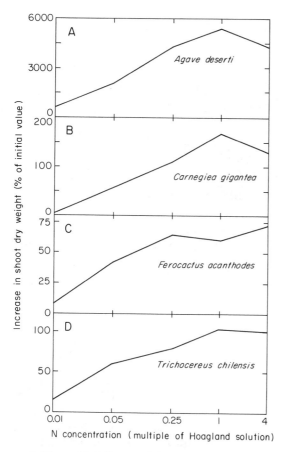

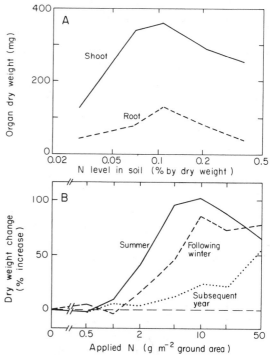

Figure 6.2. Influence of the nitrogen concentration in a hydroponic solution on the growth of seedlings of four species: **(A)** *Agave deserti*, **(B)** *Carnegiea gigantea*, **(C)** *Ferocactus acanthodes*, and **(D)** *Trichocereus chilensis*. Under each condition, six to eight seedlings were grown for five to six months at the indicated nitrogen concentration with other nutrients at the level in 0.25 × Hoagland solution (Number 1, supplemented with micronutrients), which at full strength contains 15 mM NO_3^- (Hoagland and Arnon, 1950). The seedlings of *A. deserti* were initially ten days old, so their percentage increases are greater than for the cacti, which were initially one year old. Data are expressed on a logarithmic scale for the abscissa, which is the case for many of the nutrient responses presented. Modified from Nobel (1983d) and Nobel and Hartsock (1986b).

Figure 6.3. Influence of nitrogen on the growth of *Agave deserti* in soil. **(A)** Root and shoot dry weight of seedlings maintained for six months in soil from Agave Hill (Fig. 1.1) whose N level was adjusted by preconditioning with KNO$_3$. **(B)** Dry-weight enhancement during the summer, the following winter, and the entire next year for mature plants at Agave Hill that received the indicated application of Ca(NO$_3$)$_2$ to the soil surface just prior to the summer rainfall of the first year (dry-weight increase is relative to the natural condition). Note that 0.1% by dry weight equals 1 g kg^{-1}. Modified from Nobel and Hartsock (1986b and unpublished observations).

As the nitrate (NO_3^-) level is increased in hydroponic solutions, growth of seedlings of *Agave deserti*, *Carnegiea gigantea*, *Ferocactus acanthodes*, and *Trichocereus chilensis* is enhanced (Fig. 6.2). In all cases, maximum growth occurs at or near 15 mM, the nitrate level in full-strength Hoagland solution (Hoagland and Arnon, 1950). From 0.01 to 1 times the N level of Hoagland solution, tissue N approximately doubles for all four species (Nobel, 1983d; Nobel and Hartsock, 1986b). For *Harrisia tortuosa* and *Mammillaria elegans* growing in peat-perlite (1:1 by volume), increases in plant dry weight over a three-month period are maximal when the watering solution contains 7 mM NO_3^- (Stefanis and Langhans, 1980).

The importance of nitrogen for the growth of seedlings of *A. deserti* can also be observed for plants growing in soil (Fig. 6.3A). Both root growth and shoot growth increase up to a soil N level of about 0.1% (1 g N kg^{-1} soil by dry weight) and then decrease at higher N levels. For mature plants of *A. deserti* in the field, application of 10 g N per square meter of ground area leads to the maximum increases in dry weight during the summer after application and the following winter (Fig. 6.3B; 10 g m^{-2} = 100 kg N per hectare, the conventional unit of land area for agricultural studies; 1 hectare = 1 ha = 10,000 m^2 = 2.47 acres; we also note that 100 kg ha^{-1} $\cong$ 100 lb acre^{-1}). The application of 10 g N m^{-2} increases the N level in the leaves by nearly 40% (from 1.1% to 1.5% by dry weight; P. S. Nobel and T. L. Hartsock, unpublished observations). When more than about 10 g N m^{-2} is applied, growth of *A. deserti* decreases slightly during the first year (Fig. 6.3B). During the ensuing winter, however, higher initially applied levels are required to give a particular enhancement for *A. deserti* and the maximum enhancement is about 20% less (Fig. 6.3B). Some enhancement of dry weight is also in evidence during the subsequent year, especially for the higher initial application levels of N (Fig. 6.3B).

Similar results occur for *Agave lechuguilla*, for which applications of 2 and 11 g N m^{-2} enhance growth by 16% and 39%, respectively (Quero and Nobel, 1987). For *Agave fourcroydes* in Cuba, an application of 10 g N m^{-2} increases the yield of fiber by 60–70% for eight- or nine-year-old plants, but little further enhancement of growth occurs for applications of 20 or 30 g N m^{-2} (Carrion, 1981). For *Agave sisalana* on plantations in Tanzania, growth in soils with less than 0.1% N is reduced, but 0.15% N in the soil is fully adequate for growth; application of 10 g N m^{-2} annually is generally recommended for the first two or three years (Lock, 1985). Of course, the optimal dose of fertilizer depends on the soil and harvest conditions, 100 g N m^{-2} applied in increments of 20 g m^{-2} year^{-1} for five years proving best for *A. sisalana* in soils depleted of nutrients (Lock, 1962). Also, planting *Pueraria phaseoloides* and other legumes, which sequester atmospheric nitrogen by means of bacteria in special root nodules, as intercrops between rows of *A. sisalana* can increase the yield of the agave, especially on soils deficient in N (Hopkinson, 1969).

Figure 6.4. Commercial planting of *Opuntia ficus-indica* near Fillmore, California, which is periodically harvested for young cladodes of sizes similar to the smallest ones illustrated. About 20 g N m^{-2} (200 kg N ha^{-1}) is applied in the form of ammonium sulfate as fertilizer each year, such as from the bag held by Arthur C. Gibson.

Nitrogen application (Fig. 6.4) also stimulates the growth of cacti in the field. As the soil N level at three nearby sites increases from 0.07% to 0.09% to 0.14%, the chlorenchyma N level of *Opuntia chlorotica* increases from 1.3% to 1.7% to 2.0%, which should lead to greater nocturnal acid accumulation and greater net CO_2 uptake (Nobel, 1983d). For *Opuntia phaeacantha* and *O. rastrera* (Fig. 1.21) in Coahuila, Mexico, increases in plant dry weight are positively correlated with soil N levels (Nobel et al., 1987). In particular, as soil N increases from 0.05% to 0.30%, the annual dry-weight productivity of these platyopuntias increases about twentyfold. Applying 5 and 10 g N m^{-2} to *O. cochenillifera* in northeastern Brazil results in a productivity increase of 43% and 80%, respectively (Metral, 1965). Applying 20 g N m^{-2} increases the productivity of *O. ficus-indica* by about 30% in Mexico, Mexico (Flores Valdez and Aguirre Rivera, 1979). In one year the number and the wet weight of cladodes of *O. polyacantha* in Colorado both increase about 20% for the control and 50% when 15 g N m^{-2} is applied (Dodd and Laurenroth, 1975). As the supplemental level of N applied to *O. phaeacantha* in the field in south Texas is increased up to 16 g m^{-2}, new cladodes increase by 55% in number, 70% in area, and 73% in dry weight (Table 6.3).

Table 6.3. *Influence of applied nitrogen and phosphorus on growth of* Opuntia phaeacantha *in Kingsville, Texas*

Quantity	Control	Applied N (g m^{-2})		Applied P (g m^{-2})	
		4	16	2	8
Number of new cladodes per plant	27.3	31.4	42.3	30.1	35.6
New cladode area (both sides) per plant (m^2)	1.12	1.31	1.90	1.49	1.60
Increase in dry weight per plant (kg)	0.71	0.84	1.23	0.96	1.05

Note: Morphology and dry weight were determined at the time of nutrient application and then one and a half years later. Nitrogen was added as NH_4NO_3 and phosphorus as $Ca(H_2PO_4)_2$. Before fertilizer application, the soil in the root zone contained 330 ppm N by dry weight and 23 ppm P. Modified from Nobel et al. (1987).

Although further research on the influence of N on the growth of agaves and cacti is clearly needed, certain generalizations can be made concerning this important nutrient. Metabolic activity is promoted when N is about 2% of chlorenchyma dry weight, just as for representative agronomic plants (Table 6.1). Of all the elements tested for about twenty species, nocturnal CO_2 uptake and nocturnal acid accumulation are positively and most highly correlated with tissue N levels. Plant growth increases with increasing N up to about 0.1% N by dry weight in the soil. The ecological implications of a growth optimum near 0.1% N in the soil for agaves and cacti warrant further field studies of soil nutrient levels compared with the success of these succulents in competition with other plants. Related to this, growth of agaves and cacti in the field can be substantially enhanced as N applied to the soil surface is raised up to 10 g m^{-2} (100 kg ha^{-1}), with little further enhancement at higher doses. Of course, the benefit of fertilizer applications of 10 g N m^{-2} to agaves or cacti for agronomic purposes must be judged in the appropriate economic context, a topic that needs much more field research.

PHOSPHORUS

Phosphorus is taken up from the soil in the form of phosphate (generally as $H_2PO_4^-$ or HPO_4^{2-}) and becomes an important ionic constituent of plant cells, where it is incorporated into the energy currency ATP, other molecules directly involved in metabolism, nucleic acids, and membranes. Indeed, the second most important element with respect to limitations of soil fertility on plant growth is usually P, although much lower P levels occur in the soil and in plant tissue (Table 6.1) than N levels. For instance, soil from Agave Hill contains 600 ppm N by dry weight but only 45 ppm P (for comparison, K occurs at 140 ppm, Na at 40 ppm, Ca at 1,800 ppm, and Mg at 150 ppm in soil from Agave Hill; P. S. Nobel, unpublished observations).

Compared with representative levels in agronomic plants, the levels of P in the chlorenchyma of agaves and cacti are on average relatively low and tend to be quite variable (Table 6.1). The responses of these two taxa to added P also tend to be varied. For *Opuntia phaeacantha* in the field in Texas, both the number of new cladodes and total cladode area are enhanced by an application of 2 or 8 g P m^{-2} to the soil surface (Table 6.3). At the higher application level, plant dry weight increases 48% over a one-and-a-half-year period compared with no P addition (Nobel et al., 1987). For *Opuntia cochenillifera* in northeastern Brazil, applying 5 and 10 g P m^{-2} increases productivity by 62% and 101%, respectively (Metral, 1965). On the other hand, applying 20 g P m^{-2} does not enhance productivity of *O. ficus-indica* in Mexico, Mexico (Flores Valdez and Aguirre Rivera, 1979). Also, *Agave sisalana* generally does not respond to P fertilization in Tanzania, although

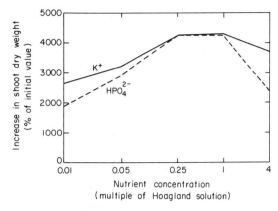

Figure 6.5. Influence of potassium and phosphorus concentrations in a hydroponic solution on the growth of seedlings of *Agave deserti*. Conditions are as for Figure 6.2, except that K and P (as HPO₄²⁻) were varied from their concentration in full strength Hoagland solution (10 mM and 1 mM, respectively). Modified from Nobel and Hartsock (1986b).

an application of 2 to 4 g P m^{-2} at the time of transplanting can help root growth (Lock, 1985).

Relatively little influence of P level on growth is noted when one-year-old seedlings of certain cacti are grown hydroponically for six months. In particular, raising the P level from 0.01 to 1 times the level in Hoagland solution (1 mM HPO₄²⁻) increases the stem dry weight an average of only 20% for *Carnegiea gigantea*, *Ferocactus acanthodes*, and *Trichocereus chilensis* (Nobel, 1983d); increasing N over the same range causes an average increase of 100% for the stem dry weight of these species (Fig. 6.2B–D). When P in a hydroponic solution is similarly varied from 10 μM to 1 mM for seedlings of *Agave deserti* that are initially much smaller and hence have less stored P than do the cactus seedlings, shoot dry weight is enhanced just over twofold (Fig. 6.5); in comparison, increasing N from 0.01 to 1 times the level in full strength Hoagland solution for *A. deserti* increases the gain in shoot dry weight about ninefold (Fig. 6.2A). Thus, agaves and cacti tend to respond much less to P fertilization than to N fertilization.

POTASSIUM

Potassium is the main cation in most plant cells. Although it does not take part in

biochemical reactions as a reactant or a product, K has a major influence on the ionic milieu in which such reactions take place as well as on specific enzymes. Also, an increase in the concentration of K$^+$ in guard cells caused by metabolically driven active transport increases the osmotic pressure in the guard cells, thereby lowering their water potential ($\Psi = P - \pi + \rho_w gh$, Eq. 3.1). Water then diffuses in from the surrounding epidermal cells toward the lower Ψ, causing the turgor pressure in the guard cells to increase and hence the stomatal pores to open (Jarvis and Mansfield, 1981; Nobel, 1983a).

Growth of seedlings of *Agave deserti* in hydroponic solution can be enhanced by increasing the level of K, but the effects are even smaller than for P and much smaller than for N (Fig. 6.5). Specifically, as K is increased from 0.01 to 1 times its level in full strength Hoagland solution (10 mM), the gain in shoot dry weight over about a six-month period is 61% greater. Smaller enhancement occurs for one-year-old seedlings of *Carnegiea gigantea*, *Ferocactus acanthodes*, and *Trichocereus chilensis* grown for six months in hydroponic solution (see Fig. 6.2B–D for N effects on these seedlings). Specifically, their shoot dry weight is enhanced an average of only 10% as K is increased from 0.1 mM to 10 mM (P. S. Nobel, unpublished observations).

Applying 5 or 10 g K m^{-2} to the soil surface leads to no significant change in productivity for *Opuntia cochenillifera* in northeastern Brazil (Metral, 1965). Also, applying K to *Agave sisalana* in Tanzania generally does not increase productivity for most plantations. However, the most common nutrient deficiency disease observed there for *A. sisalana*, "banding disease," is attributable to a lack of K (Lock, 1962, 1985). In the incipient stage of banding disease, small pale green spots about 2 mm in diameter develop on the underside of the leaves, eventually leading to a band of necrotic tissue about 10 cm wide across the base of the leaves. Leaves containing 0.6% or more K by dry weight do not exhibit banding disease, but it always occurs when K is less than 0.4%, a very low leaf K level (Table 6.1). Banding disease can be avoided by applying about 4 g K m^{-2} to the soil surface each year, which can increase productivity of *A. sisalana*

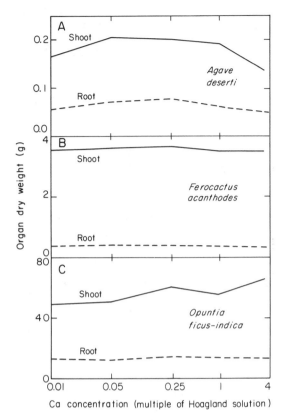

Figure 6.6. Influence of calcium concentration on plant growth in washed sand for three species: **(A)** *Agave deserti*, **(B)** *Ferocactus acanthodes*, and **(C)** *Opuntia ficus-indica*. Under each condition, three plants were grown for six months and watered weekly at the indicated Ca concentration with other nutrients at the level in 0.25 × Hoagland solution, which at full strength contains 4 mM Ca^{2+}. *Agave deserti* was grown from seed (6 mg each), *F. acanthodes* was grown using seedlings that were initially one year old and 3 cm tall (total plant dry weight of 0.75 g), and *O. ficus-indica* was grown using cladodes that were initially 10 cm in length (dry weight of 6.4 g). Modified from Berry and Nobel (1985) and Nobel and Berry (1985 and unpublished observations).

by about 60% for soils initially deficient in K (Lock, 1962, 1985).

CALCIUM

Not only can calcium be the second most concentrated cation in plant cells, where it is required for the functioning of certain enzymes, but also it can occur in an insoluble form. In addition to being incorporated into druses (Fig. 6.1), Ca is an important compo-

nent of cell walls, where it facilitates the binding of one cell to another. Calcium can be applied to plants in the form of lime (CaO), which is usually used to raise the pH of acid soils. On the other hand, many soils are already very high in Ca, such as those derived from limestone (CaCO$_3$, generally coming from the fossil remains of sea animals). *Pediocactus sileri* appears to be restricted to soils in northern Arizona and possibly southern Utah that are high in Ca, presumably in the form of gypsum (CaSO$_4$·2H$_2$O; Benson, 1982). *Echinocactus platyacanthus* also tends to occur on soils high in Ca (Trujillo Arguenta, 1982), but *Ferocactus histrix* occurs in soils low in Ca (del Castillo Sanchez, 1982), although again the physiological reasons have not been identified.

Over the 200-fold range in calcium concentration employed, from 60 μM to essentially saturation (with respect to precipitation of calcium sulfate), Ca^{2+} exerts relatively little influence on the growth of three desert succulents in sand culture. Specifically, for *Agave deserti* grown from seeds (Fig. 6.6A), one-year-old seedlings of *Ferocactus acanthodes* (Fig. 6.6B), and rooted cladodes of *Opuntia ficus-indica* (Fig. 6.6C), neither shoot dry weight nor root dry weight deviate by more than about 10% from the mean value for Ca concentrations from 0.1 to 5 mM. At the lowest and the highest Ca concentrations, growth of *A. deserti* is slightly lower than the mean (Fig. 6.6A), and at the highest Ca concentration, growth of *O. ficus-indica* (Fig. 6.6C) is slightly greater than the mean. Such minor changes can hardly be considered as responses for major nutrient deficiency or mineral toxicity for these plants. In agreement with the lack of correlation between nocturnal acid accumulation and tissue Ca levels for agaves and cacti (Table 6.1) and the lack of correlation between productivity and soil Ca levels for *Opuntia phaeacantha* and *O. rastrera* in Coahuila, Mexico (Nobel et al., 1987), we conclude that Ca is generally not present in the soil in limiting or excessive levels for these taxa. Indeed, as a group, agaves and cacti occur natively in soils with widely differing Ca levels, although some species have more specific Ca requirements.

Field trials with *Agave sisalana* in Tanzania indicate that Ca application can increase fiber yield. In particular, adding 230 g Ca m^{-2} increases fiber yield by 27%, and adding 460 g Ca m^{-2} increases yield by 41% (Lock, 1962). These very high dosages of Ca are in the form of limestone (CaCO$_3$), and the responses observed may not be solely attributable to Ca. In particular, the soils tend to be acidic with a *p*H near 5 that is raised by adding the limestone, which can then influence the uptake of other nutrients besides Ca.

MAGNESIUM AND SULFUR

Less attention has been directed toward the influence of macronutrients other than N, P, K, and Ca on agaves and cacti. Magnesium occurs in fairly high levels in the chlorenchyma (Table 6.1). It is often the cation second only to K$^+$ in concentration in plant cells and is required by many enzymes for their activity. Moreover, each molecule of chlorophyll contains one atom of Mg. Sulfur also is a crucial constituent of plant cells, occurring in nearly all proteins.

The Mg levels usually tend to be higher for the older leaves and toward the leaf tip (Lock, 1962). Magnesium deficiency for *A. sisalana* can lead to small, brown, deeply sunken necrotic lesions on the upper surfaces of the oldest leaves. Such deficiency symptoms can appear when the Mg that is readily exchangeable from the soil particles drops substantially below 10 ppm by dry weight, which is a very low soil Mg level. Symptoms of sulfur deficiency, such as chlorosis (loss of green color, yellowing), occur for *A. sisalana* when the S content of its leaves is below 400 ppm by dry weight compared with normal S levels of 800 to 1,000 ppm (Lock, 1962). Very little other information is available on the responses of agaves and cacti to soil Mg or S.

Micronutrients

Compared with the number of studies on the responses of agaves and cacti to macronutrients, much less information is available on the responses to micronutrients. The soil often appears to have sufficient levels of these elements for plant nutrition, so no deficiency responses are generally evident. For instance, the more than tenfold variation in dry weight productivity for two species of platyopuntias in Coahuila was not significantly correlated with soil levels of Mn, Cu, Zn, Fe, Co, or Mo (Nobel et al., 1987). Also, dry weight accumulation by *Rhipsalis gaertneri* growing in peat is not significantly decreased by omission of Mn, Cu, Zn, Fe, and Mo from a fertilizing solution, and omission of B leads to a decrease of only 7% (Penningsfeld, 1972).

COPPER AND ZINC

Copper and zinc, each of which is required for the activity of specific enzymes, can both be released upon weathering of soils. Because they are not readily leached under arid conditions, their levels can build up in the soil. Consequently, toxicity because of high levels can become a more important concern for growth than can deficiency. At high soil levels of Cu or Zn, uptake of other micronutrients such as Fe is often depressed, and excess heavy metals can also inhibit the action of certain enzymes.

When young plants of *Agave sisalana* are grown hydroponically for sixteen months in the absence of Cu or Zn, deficiency symptoms appear, including a reduction in the rate of leaf unfolding from the central spike and the production of shorter leaves (Pinkerton, 1971). In particular, the rate of leaf unfolding is reduced 22% in the absence of Cu and 31% in the absence of Zn (Fig. 6.7). For *A. sisalana*, deficiency levels in the leaves are about 2 ppm for Cu and 5 ppm for Zn (Lock, 1962; Pinkerton, 1971); if such levels are pertinent to agaves in general, certain other species may be experiencing Cu deficiency in the field (Table 6.1).

For seedlings of *Agave deserti* grown hydroponically for twelve days, levels of Cu and Zn in the bathing solution can be raised quite high without much inhibition of growth. Shoot growth is relatively uninfluenced from 2 to 20 μM Cu and from 20 to nearly 500 μM Zn (Nobel and Berry, 1985; for comparison, full-strength Hoagland solution contains only 0.3 μM Cu and 0.8 μM Zn). Root growth is affected above 10 μM Cu or 100 μM Zn and is about 90% inhibited at 20 μM Cu or 500 μM Zn.

Copper and Zn can concomitantly reach high levels in the soil in certain arid and semiarid regions, and they appear to have little inhibitory effect on short-term growth of seedlings of *A. deserti* at up to about 100 times their

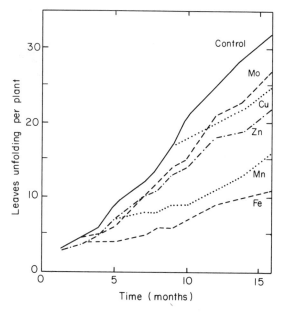

Figure 6.7. Influence of omission of the indicated micronutrients on the unfolding of leaves from the central spike of *Agave sisalana* grown hydroponically. Modified from Pinkerton (1971).

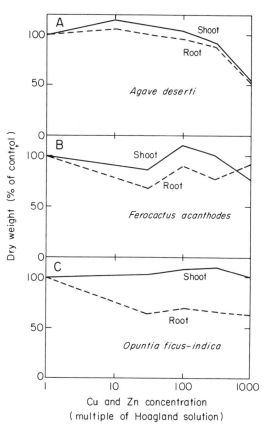

Figure 6.8. Influence of high levels of copper and zinc on shoot and root growth for six months in washed sand for three species: (**A**) *Agave deserti*, (**B**) *Ferocactus acanthodes*, and (**C**) *Opuntia ficus-indica*. Conditions are as for Figure 6.6. Dry weights are expressed relative to the responses to Cu and Zn levels in full-strength Hoagland solution [0.31 μM (0.020 ppm) Cu and 0.76 μM (0.050 ppm) Zn].

concentration in Hoagland solution. For these reasons, Cu and Zn were added together at up to 1,000 times their level in Hoagland solution to determine possible adverse effects on growth over six months in sand culture for *A. deserti* (Fig. 6.8A), *Ferocactus acanthodes* (Fig. 6.8B), and *Opuntia ficus-indica* (Fig. 6.8C). For all three species, the roots tend to be slightly more sensitive than the shoots to increasing concentrations of these heavy metals. However, even when Cu plus Zn concentrations are raised to 1,000 times their level in Hoagland solution, the average inhibition of root and shoot growth is 46% for *A. deserti* (Fig. 6.8A) and only 17% for the two species of cacti (Fig. 6.8B,C). Therefore, these desert succulents, especially the cacti, are presumably quite tolerant of potentially stressful levels of these heavy metals in the soil.

At the higher applied concentrations of Cu and Zn (Fig. 6.8), tissue levels of these heavy metals can become quite high. For 30 μM Cu in the watering solution for plants in sand culture, which is 100 times the concentration in Hoagland solution, the Cu level in the roots exceeds 100 ppm by dry weight for all three species, although the chlorenchyma level averages less than 10 ppm; at 300 μM

Cu, the tissue levels are about fivefold higher (Berry and Nobel, 1985; Nobel and Berry, 1985). For 80 μM Zn in the watering solution, which is again 100 times the level in Hoagland solution, root and shoot Zn levels are similar for *A. deserti*, *F. acanthodes*, and *O. ficus-indica* and average about 100 ppm, increasing about fourfold at 800 μM Zn. Because growth of all three species in sand culture is essentially unaffected when the concentrations of these heavy metals in the watering solution are at 100 times the level in Hoagland solution and is not drastically inhibited at 1,000 times that level (Fig. 6.8), these desert succulents can apparently tolerate high soil and tissue levels of Cu and Zn.

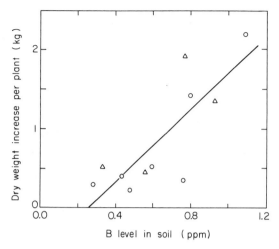

Figure 6.9. Correlation between soil boron level in the root zone (0.1 m below the soil surface) and dry weight increases over three years for *Opuntia phaeacantha* (O) and *O. rastrera* (△) in Coahuila, Mexico. Note that ppm by dry weight equals mg kg^{-1}. Modified from Nobel et al. (1987).

BORON

Boron is required in small amounts for proper plant growth, presumably because of its involvement in meristematic activity and carbohydrate metabolism. Effects of B deficiency have been observed for both agaves and cacti, and nocturnal acid accumulation has a relatively high and positive correlation with chlorenchyma B levels for agaves (Table 6.1). Spraying B on *Echinocactus grusonii* can increase its size and dry weight (El-Meligy, 1980). Boron toxicity for plants in general can occur in many arid and semiarid regions, where soils often have high B levels (Richards, 1954).

Boron deficiency for *Agave sisalana* in the field in Tanzania is often first diagnosed by the presence of a yellow speckling on the upper and the lower surfaces of the distal half of the leaf (Lock, 1962). Next, wrinkling develops on the leaves, and the distal portions become shrunken. Boron deficiency can decrease growth of the terminal spine and can also impair root growth (Pinkerton, 1971). Such symptoms appear when the B level in the leaves falls below 12 ppm. Yet when 0.7 ppm B (as boric acid, H_3BO_3) is added to the nutrient solution, newly unfolded leaves are normal within about two months (Lock, 1962).

An interesting case of apparent B deficiency is suggested when soil levels of B are correlated with growth of *Opuntia phaeacantha* and *O. rastrera* in Coahuila, Mexico (Fig. 6.9). Although many factors differ among the eleven sites sampled, 66% of the more than tenfold variation in productivity observed for these species can be accounted for by variations in soil B level (Nobel et al., 1987). The four highest productivity sites average 0.9 ppm B in the soil and the four lowest productivity sites average just under 0.4 ppm B (Fig. 6.9; for comparison, soil from Agave Hill averages 1.4 ppm B by dry weight). Also, the chlorenchyma B level is only 1 ppm at the two lowest productivity sites, increasing to 8 ppm at two of the four highest productivity sites. The second most important element with respect to explaining the productivity variation between sites is N; regression analysis using both B and N shows that soil levels of these two elements account for 84% of the intersite variation in productivity (Nobel et al., 1987).

What happens when B is raised to very high levels, as can occur for soils in the southwestern United States? Shoot growth of *Agave deserti* is more sensitive to high levels of B (Fig. 6.10A) than is that of *Ferocactus acanthodes* (Fig. 6.10B) and *Opuntia ficus-indica* (Fig. 6.10C), similar to the relative sensitivity of these species to high levels of Cu and Zn. Root growth is more sensitive than shoot growth to increasing B, especially for the two species of cacti. Very little inhibition of shoot growth in sand culture occurs until B exceeds 60 times its level in Hoagland solution for *A. deserti* and over 100 times the Hoagland level for the two cacti (Fig. 6.10). At 150 times the B level in full-strength Hoagland solution (which contains 9 μM B), root and shoot growth of *A. deserti* is completely inhibited but shoot growth of *F. acanthodes* and *O. ficus-indica* is reduced only 30% and root growth just over 50% (Fig. 6.10). Thus, these desert succulents are actually quite tolerant of high levels of soil boron, at least when the B is present in its usual form of borate.

OTHER

Although little experimental work has been done on micronutrients other than Cu, Zn, and B, this does not mean that these other

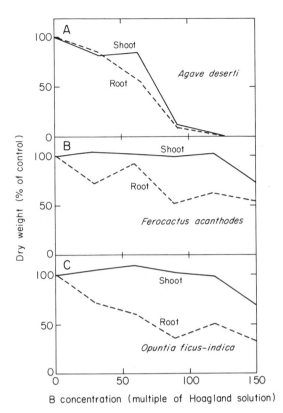

Figure 6.10. Influence of elevated boron (added as borate) concentrations on shoot and root growth for six months in washed sand for three species. (**A**) *Agave deserti*, (**B**) *Ferocactus acanthodes*, and (**C**) *Opuntia ficus-indica*. Conditions are as for Figure 6.6. Dry weights are expressed relative to the responses to B in full-strength Hoagland solution (9.2 μM H_3BO_3), which was the lowest concentration used.

elements are never at deficient or toxic levels in the soil. Again, much of our information comes from *Agave sisalana*, which has been commercially cultivated in more regions than any other agave or cactus except *Opuntia ficus-indica* (Fig. 1.22).

Manganese, which is either a part of or needed for the activation of various enzymes, as well as being part of Photosystem II in chloroplasts, is generally present in adequate quantities in the soil. When Mn is withheld from hydroponically grown *A. sisalana*, root growth is reduced, leaf elongation is retarded, leaves become chlorotic, and the rate of leaf unfolding over a sixteen-month period is approximately halved (Fig. 6.7). Similar symptoms can occur when molybdenum, which is

required for nitrogen metabolism, is withheld; the symptoms develop more gradually than for Mn omission, and the decrease in leaf unfolding over sixteen months is only 15% in absence of Mo compared with normal Mo levels (Fig. 6.7). On the other hand, omission of iron, which is a crucial component for many proteins including the cytochromes involved in electron transport in chloroplasts and mitochondria, has even greater effects with respect to root growth, leaf chlorosis, and leaf unfolding. Specifically, leaf unfolding over a sixteen-month period is reduced 65% by withholding Fe (Fig. 6.7). Even though deficiencies of Mn, Mo, and Fe can greatly reduce growth of *A. sisalana*, growth reductions clearly attributable to these micronutrients have not been identified in the field (Lock, 1962). Growth reductions attributable to low chloride, a micronutrient required for Photosystem II, have also not been identified for *A. sisalana* in the field, or for any other agave or cactus for that matter. In summary, we conclude that micronutrients have so far not been unambiguously found to be limiting for agaves and cacti in the field, with the important exception of B.

Other elements

The distribution and growth of desert succulents can be influenced by elements that are not nutrients. For instance, agaves and cacti tend to be absent from saline soils (Walter and Stadelmann, 1974), suggesting the importance of sodium in the distribution of these desert succulents. The two elements that we will consider here, Na and H, are the two most important nonnutrient elements for most plants; H concentration is measured by pH, which can have many important biological effects.

SODIUM

Sodium, which is often added as sodium chloride and simply referred to as salt – or salinity – can have adverse effects on the growth of desert succulents. For instance, growth of various agaves is markedly stunted at a salinity corresponding to 400 mM NaCl (McDaniel, 1985). Also, salinity induced by 100 mM NaCl plus 100 mM CaCl$_2$ reduces the dry weight of ten-month-old *Agave sisalana* by 40% after five months (El-Gamassy et

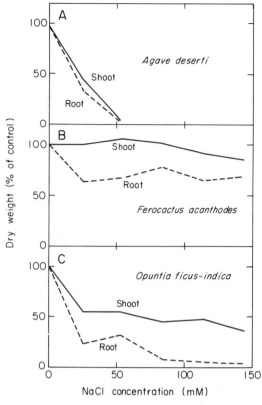

Figure 6.11. Influences of NaCl concentration on shoot and root growth for six months in washed sand for three species: (**A**) *Agave deserti*, (**B**) *Ferocactus acanthodes*, and (**C**) *Opuntia ficus-indica*. Conditions are as for Figure 6.6.

al., 1974). Yet some desert succulents occur in saline areas, or at least areas that become periodically saline in times of drought. A cactus that grows in saline areas of Argentina, *Cereus validus*, has even been proposed as a salt-tolerant crop for the Sonoran Desert (Yensen et al., 1981).

We have already noted that nocturnal acid accumulation is negatively correlated with the sodium level in the chlorenchyma of agaves and cacti (Table 6.1). Thus, we expect that Na will inhibit the growth of these desert succulents. When seedlings of *Agave deserti* are grown hydroponically for twelve days, 50% inhibition of elongation occurs at about 60 mM NaCl for roots and 90 mM for shoots (Nobel and Berry, 1985). This species is even more sensitive to a particular NaCl concentration when grown for six months in sand culture, as both shoot and root growth are in-

hibited over 90% by 50 mM NaCl (Fig. 6.11A). Although *Ferocactus acanthodes* (Fig. 6.11B) and *Opuntia ficus-indica* (Fig. 6.11C) are less sensitive than *A. deserti* to NaCl, root growth is again more inhibited by NaCl than is shoot growth. Of these two cacti, *O. ficus-indica* is more sensitive, shoot growth being about 50% inhibited at 70 mM NaCl. Other studies indicate that shoot growth of seedlings maintained hydroponically for six months is 50% inhibited at 100 mM NaCl for *Trichocereus chilensis* and 120 mM NaCl for *Carnegiea gigantea* (Nobel, 1983d). To summarize these results, we can say that on average the growth of succulents in the presence of NaCl for a relatively long period of six months is noticeably inhibited by 50 mM NaCl (1,100 ppm Na) and severely inhibited by 100 mM NaCl, especially root growth, and that considerable interspecific variation occurs with respect to the sensitivity to NaCl.

CO_2 uptake by *Cereus validus*, a species that is periodically exposed to salt stress in its native habitat, is decreased by adding NaCl to the watering solution (Nobel et al., 1984). As the NaCl concentration in the watering solution increases for nine-month-old plants (Fig. 3.5), the Na concentration in the roots and in the chlorenchyma increases (Fig. 6.12A), a similar increase occurring in the concentration of the anion chloride (Cl^-). Sodium concentrations are higher in the roots than in the chlorenchyma of *C. validus*, similar to the results for *O. ficus-indica* cited previously. The increasing Na concentration in the chlorenchyma is associated with a progressive inhibition of nocturnal accumulation of malate (Fig. 6.12B). At a chlorenchyma Na concentration of 80 mM, as can occur for *C. validus* exposed to 500 mM NaCl (about the salinity of sea water) for two weeks, nocturnal malate accumulation is reduced by 50%. Net CO_2 uptake is even more inhibited than nocturnal acid accumulation, because much of the malate accumulated represents internal recycling of respiratory CO_2. In particular, two weeks at 500 mM NaCl reduces the rate of net CO_2 uptake by the whole plant by 80% (Nobel et al., 1984).

Cereus validus in the laboratory has much higher tissue Na levels than most cacti in the field (Table 6.1). When the concentration data (Fig. 6.12) are reexpressed on a dry-

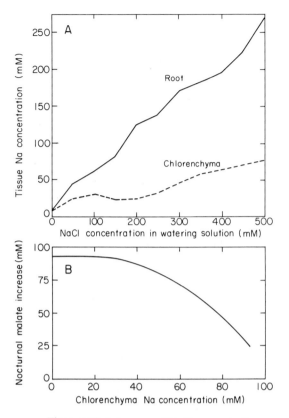

Figure 6.12. Influences of NaCl concentration used in the watering solution for fourteen days on the Na concentration in the roots and in the chlorenchyma of *Cereus validus* (**A**) and the relation between its chlorenchyma Na concentration and nocturnal malate increase (**B**). Tissue Na concentrations are based on the ratio of tissue Na amounts to tissue water volumes; chlorenchyma Na concentrations above 80 mM were obtained by maintaining the plants at 400 mM NaCl for longer than fourteen days. Modified from Nobel et al. (1984).

remain metabolically active during the time of sâlinity stress (Nobel et al., 1984).

Inhibition of growth by NaCl has also been demonstrated for *Opuntia humifusa* over a six-week period (Silverman, Young, and Nobel, 1988). Root growth is halved at about 60 mM NaCl in the watering solution and shoot growth at 80 mM. Compared with no applied NaCl, 150 mM NaCl causes the cladodes to become 30% thinner, with little further decrease in thickness occurring up to 400 mM NaCl. Net CO_2 uptake over a 24-h period after six weeks at 150 mM NaCl is 29% of that in the absence of applied NaCl for plants from the marine strand and only 6% for those from an inland population (Silverman et al., 1988). Thus, this widely occurring platyopuntia is not very tolerant of salinity, especially plants from the inland population, with damage again being especially apparent for the roots.

HYDROGEN – pH

The hydrogen ion concentration is a readily measured property of soils. Moreover, it is biologically important, because the pH of the soil water influences the availability of various soil nutrients. For instance, pH values below 5 increase the solubility of aluminum (Al) and heavy metals such as Fe and Mn, all of which can be toxic to plants, especially Al. Also, at low pH values, Ca, Mg, K, Mo, and especially P can occur in a form difficult for roots to absorb from the soil (Larcher, 1980). At soil pH values above 8.5, many nutrients such as B, Cu, P, and Zn again occur in a form difficult for roots to absorb. Thus, pH studies using hydroponic solutions can show the direct effect of hydrogen ion concentrations; studies varying the pH of soils are ecologically more realistic, because the secondary effects of pH on the availability of various nutrients become manifest; and pH studies using sand culture are intermediate, in that changes in pH affect the binding of ions other than H^+ to the sand particles, but to a lesser extent than the binding of ions to the much smaller particles that occur in soils.

A number of studies have examined the influence of pH on the germination of seeds of agaves and cacti. For *Agave deserti*, germination is insensitive to pH over an extremely wide range of 4.1 to 9.7 (Jordan and

weight basis, the Na levels in the chlorenchyma range from 0.3% in the absence of applied NaCl to 1.6% when 500 mM is applied for two weeks. Moreover, 500 mM NaCl has a water potential of −2.4 MPa at 20°C, indicating that *C. validus* can withstand quite low soil water potentials. Some of its roots are probably lost at continued high salinity, but rapid root growth at the onset of a rainy period should enable the plants to return to normal water uptake then. Meanwhile, during the dry period, nocturnal malate accumulation by CAM would continue recycling the internally produced respiratory CO_2, so the plants would

Table 6.4. *Influence of the* pH *of the watering solution on shoot and root dry weights for plants growing for six months in sand culture*

Species	Organ	pH		
		4.5	6.5	8.5
Agave deserti	Shoot (mg)	204	216	206
	Root (mg)	49	56	51
Ferocactus acanthodes	Shoot (g)	3.45	3.48	3.69
	Root (g)	0.37	0.43	0.40
Opuntia ficus-indica	Shoot (g)	51.6	46.9	47.8
	Root (g)	12.1	13.1	10.9

Note: Conditions are as for Figure 6.6.

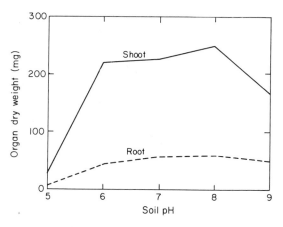

Figure 6.13. Influence of *p*H on root and shoot growth for seedlings of *Agave deserti* in soil. Plants were maintained for six months in soil from Agave Hill (Fig. 1.1); soil *p*H was adjusted by preconditioning with HCl or KOH. Modified from Nobel and Hartsock (1986b).

Nobel, 1979). In agreement with this, when seeds of *A. deserti* are germinated in aerated hydroponic solutions with *p*H values from 4.9 to 8.5, both root and shoot lengths of the seedlings after twelve days vary less than 10% from the mean (Nobel and Berry, 1985). On the other hand, germination of seeds of *Agave lechuguilla* is reduced 50% at *p*H 7.6 compared with *p*H 6.2 (Freeman, 1973). Although seeds of *Agave parryi* germinate more slowly at higher *p*H's, after eight days no significant difference in percentage germination is found from *p*H 7.0 to *p*H 8.5 (Freeman, 1975). Thus, soil *p*H does not appear to be a major factor limiting seed germination for agaves in native habitats, because soil *p*H there is generally in the range suitable for at least 50% germination. However, more research is needed on agaves, as well as on cacti.

For the three species we have been considering in detail throughout this book, *A. deserti* (Fig. 4.20), *Ferocactus acanthodes* (Fig. 1.2A), and *Opuntia ficus-indica* (Fig. 1.12), *p*H has very little effect on growth during a six-month period in sand culture. In particular, neither shoot growth nor root growth varies more than 10% from the mean for *p*H values of 4.5, 6.5, and 8.5 (Table 6.4). Thus, *p*H over a wide range does not appear to have a major direct effect on the dry-weight increases of these desert succulents.

As already indicated, a larger effect of *p*H on growth would be expected for plants

growing in soil compared with hydroponics or sand culture, because of the secondary effects of *p*H on the availability of nutrients. Shoot and root dry weights of *A. deserti* are decreased less than 10% when a hydroponic solution or the watering solution for sand culture (Table 6.4) is slightly below *p*H 5. On the other hand, growth of seedlings of *A. deserti* in soil over a six-month period is greatly reduced when the soil *p*H is 5 (Fig. 6.13); at this *p*H, shoot dry weight is decreased by 88% compared with the average dry weight for *p*H values from 6 to 8 and root dry weight is decreased by 86%. However, this species tolerates relatively high soil *p*H values; shoot growth is inhibited 28% and root growth only 6% at *p*H 9 compared with the average growth for *p*H values of 6 to 8 (Fig. 6.13).

Surprisingly few experiments have been done on the responses of other agaves or cacti to *p*H, although the soil *p*H where these plants occur has often been measured. Satisfactory growth of *Agave sisalana* occurs for soils with *p*H values from 6.0 to 7.6 but not in soils with a *p*H of 5.0 (Lock, 1962, 1985). The *p*H of the latter soils can be increased by adding lime (CaO), which also increases the level of the macronutrient Ca. *Agave fourcroydes* in Yucatán, Mexico, appears to have adequate growth on soils ranging in *p*H from 7.1 to 8.9 (Sprague et al., 1978), and *A. deserti* occurs in the field in soils ranging from pH 6.5 to 7.9 (Nobel and Hartsock, 1986c). The various field observations suggest that agaves tend to occur naturally or to grow well when cultivated for soils whose pH is between about 6 and 8.

Soils

Agaves and cacti tend to thrive in sandy soils and to do less well in clayey soils. We thus surmise that the particle sizes in the soil influence the behavior of these succulents. Indeed, soil particle size plays a crucial role in the nutrient relations of agaves and cacti because of its effect on both the water-holding capacity and the exchangeability of mineral elements. Certain nutrient characteristics of soils can be identified by the soil classification; actually, soils can be classified into a bewildering set of categories. For instance, the influences of N and P fertilization on the growth of *Opuntia phaeacantha* (Table 6.3) were measured for a soil that is a Willacy fine sandy loam in the hyperthermic family of Udic Argiustolls (Nobel et al., 1987). We will briefly introduce some of the textural categories used to define soils, examine the influence of particle size on water-holding capacity, and then consider the soils where agaves and cacti occur.

TEXTURE

Let us begin by considering the size of soil particles. After the larger stones and gravel (soil particles greater than 2 mm in diameter) are removed, we are left with three recognizable categories. Although definitions of the ranges of particle sizes in these categories differ somewhat, a common scheme is as follows: *clay*, particles less than 0.002 mm (2 μm) in diameter; *silt*, particles from 0.002 to 0.05 mm; and *sand*, particles from 0.05 to 2 mm in diameter (Hillel, 1971). Once we have divided the nongravel fraction of soil into these categories, we can then refer to a *textural triangle* to identify the principal soil textural classes (Fig. 6.14). Such textural classes have important influences on the drainage of water and penetration of air into soils. For instance, *Agave deserti* and *Ferocactus acanthodes* tend to occur in well-drained soils such as that at Agave Hill (Figs. 1.1 and 1.2); the nongravel

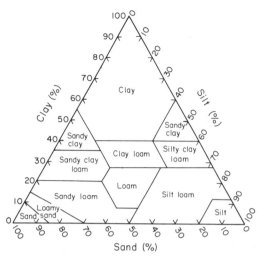

Figure 6.14. Soil textural classes based on particle sizes of the nongravel fraction. Soil particles on a dry-weight basis are divided into three categories based on particle diameter: (1) clay (<0.002 mm), (2) silt (0.002–0.05 mm), and (3) sand (0.05–2 mm). Once the percentages of clay, silt, and sand are known for the soil, a line parallel to the 0% line is drawn for the measured percentage for each category; the intersection of these three lines indicates the textural class. Modified from Hillel (1971).

fraction of this soil is composed of about 77% sand, 17% silt, and 6% clay (Nobel, 1976, 1977b). The soil from Agave Hill is thus close to the border between sandy loam and loamy sand on a textural triangle (Fig. 6.14).

For a given mass of particles of the same density, the surface area from which anions and cations can be exchanged is inversely proportional to the particle radius. Because clay particles are so small in diameter, they have a very large surface area per unit mass. Moreover, clay particles generally have a net charge, and therefore clays tend to dominate the ion exchange properties of soils. The parent material helps determine the amount of surface area and the charge on the soil particles, so the mineralogical identity of the clay becomes important when evaluating various soils from a nutrient point of view. For instance, the clay known as montmorillonite has over ten times more surface area on which cations can bind per unit mass than does the clay kaolinite, primarily because of the way the montmorillonite cyrstals can have exposed internal surfaces (Hillel, 1971). Indeed, montmorillonite has 800 m² of surface area per gram; most of the surfaces are negatively charged, leading to the binding of 1 mmol cations g⁻¹. Besides more cation binding, the anion phosphate can be released more readily from montmorillonite than from kaolinite. Such ion-binding properties affect the availability of nutrients, whose exchangeability can be further influenced by pH.

It has long been recognized that agaves and cacti prefer sandy, well-drained soils. The larger pore sizes in such sandy soils facilitate the diffusion of oxygen to the roots, as was recognized long ago by Cannon (1911, 1925). Yet very few studies have subsequently been done to determine whether roots of succulent plants have especially high rates of respiration and hence would require more O_2. Alternatively, water drainage from the soil may be the key to the preference for porous soils; many species of succulents succumb to fungal root infections if the soil is not periodically dried.

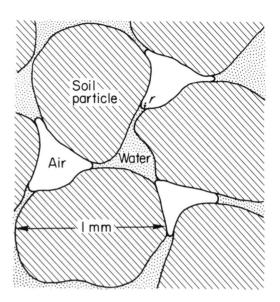

Figure 6.15. Section of a moist, sandy soil showing the three phases: soil particles (cross-hatched), water (stippled), and air (clear). The sizes of the soil particles are typical for those of sand (0.5–2 mm in diameter), and the indicated radius of curvature (r) for the water is about 0.02 mm. The air–water interfaces are concave when viewed from the air side, showing that a negative hydrostatic pressure exists in the water. Modified from Nobel (1983a).

WATER-HOLDING CAPACITY

The sizes of pores in the soil also have important influences on the water-holding capacity and on the water potential of the soil. These properties dictate when water can be taken up by the plants, which in turn influences nutrient relations.

We have already indicated that Ψ^{soil} is negative (Chapter 3), reflecting the negative hydrostatic pressure or tension ($P < 0$) that occurs in the soil ($\Psi = P - \pi + \rho_w gh$, Eq. 3.1). The negative P in the soil water results from effects of surface tension at the many air–water interfaces (Nobel, 1983a). Indeed, we can relate P to the shape of the air–water surfaces, which are generally concave when viewed from the air side (Fig. 6.15). In particular, we obtain

$$P = -\sigma\left(\frac{1}{r_1} + \frac{1}{r_2}\right) \tag{6.1}$$

where σ is the surface tension at an air–water interface, and r_1 and r_2 are the radii of curvature at the interface in planes that are perpendicular to each other (Nobel, 1983; a radius of curvature is the radius of the circle that an arc in the surface corresponds to).

Because of surface irregularities on the soil particles, r_1 and r_2 are considerably less

than the radii of the soil particles (see Fig. 6.15). To be specific, we will let r be 0.02 mm, which is typical for a wet sandy soil (Fig. 6.15). For the value of σ at 20°C (7.28 × 10⁻⁸ MPa m) and letting r_1 equal r_2, using Equation 6.1 we calculate that P is −(7.28 × 10⁻⁸ MPa m)[2/(0.02 × 10⁻³ m)], or −0.007 MPa. If Ψ^{soil} were −0.007 MPa, water uptake by agaves and cacti would be expected, because the water potential of such plants is generally about −0.3 MPa under wet conditions (Table 3.1). On the other hand, the particle size for a clay averages at least 100-fold smaller than for a sand; if the radii of curvature at the air–water interfaces are 100-fold smaller, Ψ^{soil} for the clay would be 100-fold lower than for the sand (see Eq. 6.1), or about −0.7 MPa in the present case. This is a lower water potential than that usually found in agaves and cacti, so such plants would not be able to take up water from a clay under these conditions.

A wide range of pore sizes exists in any soil and the shapes at the air–water interfaces

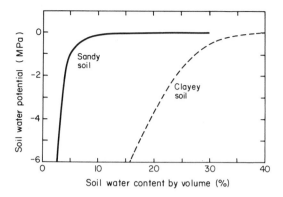

Figure 6.16. Relation between volumetric soil–water content and soil water potential for the sandy soil at Agave Hill and for a clayey soil. Note the flatness and the larger extent of the curve for Ψ^{soil} above -0.3 MPa for the sandy soil compared with the clayey soil, meaning that more water can be held at high water potentials by the sandy soil. Data for the sandy soil are adapted from Young and Nobel (1986, where a printing error exists in the ordinate) and for a clayey soil from Hillel (1971) and Taylor and Ashcroft (1972).

can be quite complicated when considered in three dimensions. Nevertheless, we can still conclude that, although sandy soils tend to drain quite rapidly, they can hold much more water at the high Ψ^{soil} necessary for water uptake by desert succulents than can very clayey soils (Fig. 6.16). Indeed, about 20% of a wet sandy soil by volume can correspond to water that can be released to agaves or cacti as Ψ^{soil} falls from field capacity to -0.3 MPa, which represents two to three times more water available to these desert succulents than would be available from a clay (Hillel, 1971).

In particular, soil from Agave Hill has a volumetric water content of 30% when wet to field capacity ($\Psi^{\text{soil}} \cong -0.01$ MPa) but only 8% at a Ψ^{soil} of -0.3 MPa (Fig. 6.16). Therefore, after a heavy rainfall, 22% of the soil by volume can represent water available to agaves and cacti (whose Ψ^{shoot} averages about -0.3 MPa when wet; Table 3.1), which is three-quarters of the water such soil can hold at field capacity. For comparison, as Ψ^{soil} for a clayey soil decreases from field capacity to -0.3 MPa, water representing only about 8% of the soil volume is released (Fig. 6.16). Thus, even though the clayey soil can hold more water than a sandy soil, the water tends to be

held at a lower water potential because of the smaller radii of curvature at the air–water interfaces (Eq. 6.1) in the clayey soil, so less water is available to succulent plants than for a sandy soil. As the sandy soil dries below about -0.3 MPa, very little water is released for large changes in Ψ^{soil}. For instance, as Ψ^{soil} decreases from -0.3 MPa to -0.6 MPa for soil from Agave Hill, a volumetric water content of only 1% is released (Fig. 6.16). Thus, because of the relatively large particle sizes, sandy soils hold water at the high water potentials (Eq. 6.1) generally required for water uptake by agaves and cacti, a very important aspect of the texture of the soils in which these plants occur.

INFLUENCES ON GROWTH AND DISTRIBUTION

Worldwide studies have shown that *Agave sisalana* grows best on porous, well-drained soils (Smith, 1929). In India, *A. sisalana* does not survive in water-logged clayey soil; in soil with high contents of N and of organic matter, the leaves contain a lower percentage of fiber. The red soils derived from coral limestone in Kenya and Tanzania are very suitable for *A. sisalana*, having appropriately high pH values, much Ca, no micronutrient or P deficiencies, and no heavy-metal toxicities (Lock, 1962). When *A. sisalana* is grown continuously for seventy years at the same location, the major nutrient deficiencies that develop are for K and N (Lock, 1985). The soils used for *A. sisalana* tend to be loams or loamy sands, the sandier soils often being lower in N, other nutrients, and pH. The soil where *Agave fourcroydes* is cultivated in the Yucatán is also derived from limestone (Sprague et al., 1978). Even though the annual rainfall is generally 700 to 1,000 mm, the soil in the Yucatán is very porous, leading to rapid infiltration of water, and the topography is relatively flat; as a consequence rivers are absent. Also, the soils with high frequencies of native cacti tend to be sandy with high porosity. However, exceptions occur, such as the presence of *Opuntia imbricata* on clayey soils (Kinraide, 1978).

Growth and productivity of *A. sisalana* is greatly reduced on saline soils (Lock, 1962), as is generally the case for desert succulents.

Cacti on salt plains in northern Venezuela are exposed to highly saline soil conditions only during the dry season when they have no actively absorbing roots and hence they do not accumulate NaCl; moreover, they tend to occur in mounds from which the salt is readily leached during the rainy periods (Walter and Stadelmann, 1974; Walter, 1985). As we indicated in the previous section, *Cereus validus* can tolerate saline conditions for short periods (Fig. 6.12). On the other hand, irrigation in Arizona with Colorado River water, which can contain 500–900 ppm NaCl (Richards, 1954; United States Department of the Interior, 1985), has caused the salinity to build up in certain cotton fields, preventing the reestablishment of native agaves and cacti until suitable leaching with rainwater occurs.

Soils of arid and semiarid lands in general and of the southwestern United States in particular have certain characteristics that can affect agaves and cacti. Metal-rich deposits occur, leading to potential heavy-metal stresses that can affect plant distribution (Foy, Chaney, and White, 1978). Also, the Cu and Zn released upon weathering of rock are not readily leached under arid conditions. Yet agaves and cacti are quite tolerant of even fairly high levels of heavy metals like Cu and Zn. Arid conditions can lead to high levels of B and Ca in the soil (Richards, 1954), but again agaves and cacti generally appear to be quite tolerant of high levels of these two elements. A possible exception may occur when irrigation waters high in B are used for species with only moderate tolerance of high B levels, such as *Opuntia ficus-indica* (Berry and Nobel, 1985). Low annual rainfall and high rates of evaporation often lead to saline soils, which can have a major impact on the distribution of desert succulents. In particular, growth of agaves and cacti can be substantially reduced by NaCl. Acid soils with pH values near 5 or lower would generally restrict growth, primarily by reducing the availability of certain nutrients, as do very alkaline soils. Although the availability of micronutrients may not be a severe limitation in arid and semiarid regions, soil levels of N in such regions most likely will limit the growth of agaves and cacti (Lock, 1962; Berry and Nobel, 1985; Nobel and Berry, 1985). In conclusion, we have a fairly good overall idea of the responses of desert succulents to soil conditions, but more research is needed on individual elements, selected soils, and particular species.

7 Productivity

Plant productivity is measured in many ways, such as the fresh weight of fruits, the commercial yield of fibers, or the dry weight amassed annually for some region. In all cases, productivity represents the cumulative, integrative, and interactive effects of rainfall, temperature, radiation, and nutrients, in particular as they relate to net CO_2 uptake and carbon storage. It is thus fitting that our final chapter is devoted to productivity. We will begin by indicating the productivities, usually expressed as annual dry-weight gain per unit ground area for aboveground parts of plants, that have been determined for agaves and cacti growing under various conditions in different parts of the world (Nobel, 1988). Yet the belowground parts must also be considered when assessing the fate of carbon from CO_2 taken up by the shoot. We will therefore also consider root respiration and root growth.

For agricultural purposes, generally only part of the plant is harvested, that fraction of total plant biomass often being referred to as the *harvest index*. For *Agave fourcroydes* and *A. sisalana* (Fig. 7.1), the harvest index generally refers to the fibers; commercial fiber productivity is thus the product of aboveground dry-weight productivity, the fraction of the total dry weight represented by the leaves, and the fraction of leaf dry weight represented by fibers. The harvested part of *A. tequilana* is the stems of approximately eight-year-old plants; to estimate commercial productivity in this case, we need to consider the fraction of plant dry weight in the stems times the fraction of stem dry weight represented by fermentable sugars and glucans that can be broken down into such sugars. The harvested portion of platyopuntias can be the fruit or the cladodes, the latter sometimes being multiplied by a "digestibility index" representing the nutritive value for the particular animals involved.

After introducing the characteristics of gas exchange for agaves and cacti (Chapter 2), we considered the influences of water status (Chapter 3), temperature (Chapter 4), and PAR (Chapter 5) on net CO_2 uptake over 24-h periods. Influences of these three environmental factors on productivity were first accurately predicted for *Agave deserti* (Nobel, 1984a). The underlying assumption is that each environmental factor can individually limit CO_2 uptake, so the effect of each factor is multiplicative. We can then simultaneously estimate the effects of water status, temperature, and PAR on productivity in an analytic manner, a substantial improvement over the predilection to focus on a single, limiting factor. Such an approach can be modified when these three environmental factors substantially interact with respect to their influence on CO_2 uptake, or when other factors, such as nutrients, become limiting, as we will also consider in this chapter. Nevertheless, the original approach of simply multiplying limitations of three individual factors on net CO_2 uptake, leading to an environmental productivity index, closely predicts the influences of these environmental factors on the productivity of various agaves and cacti. Using such an ap-

Figure 7.1. (**A**) *Agave fourcroydes* growing near Mérida, Yucatán, Mexico. (**B**) *Agave sisalana* growing near Voi, Kenya.

proach, we will conclude this chapter with a prediction of the potential productivity of these desert succulents under new conditions and in new areas.

A topic related to productivity is growth of individual plants, which is often measured as changes in plant height. Such an index of growth is easier to measure than dry-weight changes per unit ground area because one does not need monospecific stands, knowledge of the ground area explored by the roots, or the sacrifice of the plants for dry-weight deter-

minations. Certain specimens of *Coryphantha* and *Mammillaria* can be less than 10 cm tall and over 100 years old, indicating an average annual growth in height of less than 1 mm. On the other hand, innumerable measurements have indicated that when *Carnegiea gigantea* is over 1 m tall it can increase in height an average of about 10 cm per year (Steenbergh and Lowe, 1977, 1983). For certain barrel cacti, annual growth in height is intermediate in value, about 1.4 cm for *Ferocactus acanthodes* (Nobel, 1977b) and 2.2 cm for *F. wis-*

lizenii (MacDougal and Spalding, 1910). Height increases of platyopuntias are mainly caused by the initiation of new cladodes, which can occur in all directions, so height changes do not accurately measure growth. Plant height annually increases only about 3 mm for *Agave utahensis* but over 10 cm for *A. fourcroydes* and *A. tequilana* (P. S. Nobel, unpublished observations), a very wide range in values. In keeping with agronomic practice and studies of ecosystems, our emphasis here will be on productivity expressed as annual dry-weight gain per unit ground area. We thus shift from changes occurring for individual plants to changes occurring for a particular monospecific stand.

Aboveground

Agave sisalana and *Opuntia ficus-indica* are cultivated worldwide (Fig. 1.22). Many of the productivity studies on agaves and cacti have been performed on such species, namely, agaves commercially exploited for hard fibers or cacti used for animal feed. Before indicating annual productivities of the various agaves and cacti that have been so measured, we will briefly indicate some commercial production values for agave fibers. To place desert succulents into proper perspective, we will also indicate productivities of certain agronomic plants.

AGAVES

In the 1880s, exports of processed fibers of *Agave fourcroydes* from Yucatán to New York averaged about 30,000 metric tons annually; such fibers were used mainly for twine, rope, and sacks. Aboveground productivity for these agaves was estimated at 0.45 kg m^{-2} year^{-1} (4.5 metric tons ha^{-1} year^{-1}; Royal Gardens, Kew, 1898). Because commercial agave species are generally not seriously afflicted by fungi, other pathogens, insects, or vertebrate herbivores, aboveground growth and harvestable material are roughly equivalent. Maximum fiber production for *A. fourcroydes* in Mexico occurred in 1916, when 210,000 metric tons of hard fibers were produced (Camp, 1980). From 1921 to 1940, Mexico annually exported about 100,000 metric tons of fiber; somewhat more fiber was annually exported from East Africa (primarily

from what is now Kenya and Tanzania) and slightly less from the Dutch East Indies (now Indonesia). Even as late as 1963, about 210,000 metric tons of hard fiber were produced annually from agaves in Tanzania, 160,000 in Brazil, 130,000 in Mexico, 90,000 in Mozambique and Angola, 70,000 in Kenya and Uganda, and 65,000 tons in other countries (Gentry, 1982).

The advent of synthetic fibers has severely depressed the natural fiber industry. The nearly 100-year-old plantations of *A. sisalana* north of Nairobi, Kenya (Fig. 1.6), are rapidly being converted to coffee and other crops with higher economic return. Not only are processing factories (Fig. 1.6C) falling into disrepair, but also the agricultural technology for the agaves is being lost, such as the ability to identify visually various nutrient deficiencies (Chapter 6). Even though the relative cost of synthetic fibers will inevitably rise as oil and coal resources become depleted, the economic upsurgence of the agave fiber industry may well rest upon the development of uses for by-products. For instance, the leaf pulp remaining after the fiber is extracted can be fed to cattle and other animals. Of perhaps greater economic potential is the utilization of the steroids present at high concentrations in the leaf sap released when the leaves are crushed to obtain the fiber. Experimentation coupled with industrial ingenuity may take advantage of the powerful effects that various steroids can have on animals, including muscle development as well as control of reproduction and other biological processes.

For obvious economic reasons, commercial productivity of *A. fourcroydes* and *A. sisalana* (Fig. 7.1) has been measured in terms of the dry weight of fiber harvested per year. *Agave sisalana* produces about 220 leaves per plant before "poling" (emergence of the inflorescence) at about seven years of age under usual plantation conditions (Lock, 1985). Leaves can be harvested beginning after two years of age. In what is now Tanzania, a representative productivity for *A. sisalana* over its nearly five years of useful productivity is 0.51 kg fiber (m^2 ground area)$^{-1}$ year^{-1} (Table 7.1). For fiber-producing agaves such as *A. sisalana* and *A. fourcroydes*, the dry weight of the fiber constitutes about one-third

Table 7.1. *Annual aboveground productivities of agaves, cacti, and other plants*

Species	Location	Approximate annual rainfall (mm)	Part	Annual productivity [kg dry weight (m² ground area)⁻¹ year⁻¹]	Reference
Agaves					
Agave deserti	California	430	Leaves + stem	0.71	Nobel (1984a)
A. fourcroydes	Yucatán, Mexico	1,000	Leaf fiber	0.17	Smith (1929) (**A**)
		1,000	Leaves + stem	1.53	Nobel (1985a)
	Tanzania (formerly Tanganyika)	1,300	Leaf fiber	0.33	Lock (1962) (**B**)
A. lechuguilla	Coahuila, Mexico	430	Leaves + stem	0.32	Nobel and Quero (1986)
A. salmiana	San Luis Potosí, Mexico	320	Leaves + stem	1.01	Nobel and Meyer (1985)
A. sisalana	India	900	Leaf fiber	0.17	**A**
	Queensland, Australia	1,200	Leaf fiber	0.25	**A**
	Tanzania	1,300	Leaf fiber	0.51	**B**
	Kenya	1,000	Leaf fiber	0.36	**B**
A. tequilana	Jalisco, Mexico	1,080	Leaves + stem	2.49	Nobel and Valenzuela (1987)
Cacti					
Ferocactus acanthodes	California	150	Stem	0.17	Nobel (1977b)
Opuntia cantabrigiensis and *O. phaeacantha*	Zacatecas, Mexico	300	Stem	0.72	López, Nava, and Gastó (1978)
O. ficus-indica	Argentina	300	Stem	0.25	Braun, Cordero, and Ramacciotti (1979)
		300	Fruit	0.02	

	Location		Part	Value	Reference
	Ceará, Brazil	700	Stem	1.12	Metral (1965)
	Pernambuco, Brazil	1,000	Stem	2[a]	Monjauze and Le Houérou (1965) (C)
	California	450[b]	Fruit	0.12	Curtis (1977)
	Chile	300[b]	Fruit	0.19	Reñasco (1976)
		300[b]	Stem	1.05	Acevedo et al. (1983) (D)
	Sicily, Italy	300[b]	Fruit	0.32	D
	Mexico, Mexico	600	Stem	0.8	C
		700	Stem	0.82	Flores Valdez and Aguirre Rivera (1979) (E)
	Tunisia	700[b]	Stem	1.75[a]	E
		280	Stem	0.60	Le Houerou (1970) (F)
O. phaeacantha	Texas	280	Fruit	0.02	F
O. streptacantha	Argentina	600	Stem	1.5[c]	Nobel et al. (1987)
		300	Stem	0.12	Braun and López (1978) (G)
	Zacatecas, Mexico	300	Fruit	0.02	G
		430	Fruit	0.09	López G. et al. (1977)
Opuntia sp.	Texas	600	Stem	~2	Griffiths (1915a)
	Mexico (central)	700	Fruit	0.11	Hernández Xolocotzi (1970)
Other					
Ananas comosus (pineapple)	Hawaii	1,100	Fruit + leaves	~2	Bartholomew and Kadzimin (1977)
		1,100	Total carbohydrate	1.2[c]	Bartholomew (1982)
Glycine max (soybean)	Illinois	800	Leaves + stem	0.7	Odum (1971) (**H**); Evans (1975) (**I**)
Medicago sativa (alfalfa)	Arizona; California	200–800[b]	Leaves + stem	2.1–3.4	**H**; Loomis, Williams, and Hall (1971)

Table 7.1. *Annual aboveground productivities of agaves, cacti, and other plants (continued)*

Species	Location	Approximate annual rainfall (mm)	Part	Annual productivity [kg dry weight (m² ground area)⁻¹ year⁻¹]	Reference
Oryza sativa (rice)	California; Japan	500[b]	Leaves + stem	1.0–1.6	**I**; Loomis (1983) (**J**)
Saccharum officinarum (sugar cane)	Guayana; Hawaii; Queensland	2,000	Leaves + stem	4–6	**I**
Sorghum vulgare (sorghum)	California	600	Leaves + stem	0.8	**J**
Triticum aestivum (wheat)	Australia; England; Mexico; United States	600–1,000	Leaves + stem	0.4–1.0	Monteith (1977); Fischer and Turner (1978); **J**
Zea mays (maize)	Illinois; Ohio	700–1,000	Leaves + stem	1.1–1.9	**H**; **I**; **J**
Seven broad-leaved trees	—	—	Stem	2.7	Jarvis and Leverenz (1983);**K**
Eleven coniferous trees	—	—	Stem	2.3	**K**

Note: Examples have been selected that indicate the higher reproducible yields. Where necessary, wet weight has been multiplied by 0.12 to obtain an equivalent dry weight. Studies cited more than once are assigned a boldface letter the first time they occur, this letter being used for subsequent citations.

[a] Fertilized.
[b] Also irrigated.
[c] Predicted.

of the dry weight and 4% of the fresh weight of the leaf (Smith, 1929; Lock, 1962). Thus, the dry-weight productivity of leaves of *A. sisalana* in Tanzania is about 1.5 kg m^{-2} year^{-1}, and the corresponding estimate for *A. fourcroydes* there is 1.0 kg m^{-2} year^{-1} (see Table 7.1). The stem of five species of mature agaves has about 16% as much dry weight as the living leaves (Nobel, 1984a, 1985a; Nobel and Meyer, 1985; Nobel and Quero, 1986; Nobel and Valenzuela, 1987). Thus, the aboveground productivity of fiber-bearing agaves in East Africa, where their productivity is reportedly the highest, would be about 1.2 to 1.7 kg m^{-2} year^{-1}. In agreement with this, measured aboveground productivity for six-year-old plants of *A. fourcroydes* cultivated in Yucatán, Mexico, is 1.53 kg m^{-2} year^{-1} (Table 7.1).

How does this productivity of cultivated, fiber-producing agaves compare with the productivity of other agaves? The aboveground dry-weight increase for monospecific stands of *Agave deserti* of various ages under natural conditions at Agave Hill over a wet period in summer and early fall averages 0.11 kg m^{-2} month^{-1} (Nobel, 1984a). Because the mean annual rainfall is only about 240 mm in that region and tends to be seasonal (Fig. 3.3), such a productivity is not sustained throughout the year; nevertheless, during wet periods it is similar to the average monthly productivity of the plantations of *A. fourcroydes* and *A. sisalana*. Under semiarid conditions, the aboveground productivity of *A. lechuguilla* is 0.32 kg m^{-2} year^{-1}, but that of *A. salmiana* is 1.01 kg m^{-2} year^{-1} (Table 7.1). The highest aboveground productivity so far reported for an agave is 2.49 kg m^{-2} year^{-1} measured for *A. tequilana* (Fig. 1.5) growing in Jalisco, Mexico, where the annual rainfall is just over 1,000 mm (Table 7.1). Indeed, this species has the highest productivity so far reported for any CAM plant (Nobel and Valenzuela, 1987).

Besides having a relatively high annual productivity under natural conditions (Table 7.1), *A. salmiana* (Fig. 1.4) can also have a large aboveground biomass (the fresh weight of individual plants can exceed 250 kg; Nobel and Meyer, 1985). In particular, 6% of the monospecific stands of *A. salmiana* in San Luis Potosí exceed 10 kg dry weight of living

material per square meter of ground area (Martinez Morales, 1985; Martinez-Morales and Meyer, 1985). By way of comparison, for monospecific stands under natural conditions, the aboveground biomass on a dry-weight basis can be 4.8 kg m^{-2} for *A. deserti* (Nobel, 1984a), 1.4 kg m^{-2} for *A. lechuguilla* (Nobel and Quero, 1986), 2.2 kg m^{-2} for *A. stricta*, and 1.5 kg m^{-2} for *A. xylonacantha* (González-Medrano et al., 1981). For seven-year-old plants under cultivation, the aboveground biomass is 8 kg m^{-2} for *A. fourcroydes* (Nobel, 1985a), about 7 kg m^{-2} for *A. sisalana* (Lock, 1962, 1985), and 9 kg m^{-2} for *A. tequilana* (Nobel and Valenzuela, 1987). By way of comparison, aboveground biomass for closely spaced large plants is 12 kg m^{-2} for *Opuntia streptacantha* (Lopez G., Gastó C., and Nava C., 1981) and 14 kg m^{-2} for *O. ficus-indica* (P. S. Nobel, unpublished observations). Such biomass figures are much larger than for annual crops but are similar to those for many other perennials.

CACTI

In Chapter 1 we indicated that Burbank in 1911 touted the glories of opuntias. Promises of great productivities were possible at that time because little was actually known, as noted by one of his adversaries, Griffiths: "When information regarding the value of prickly pear began to be demanded some years ago next to nothing was definitely known about the handling of the crop on an economic basis" (Griffiths, 1915a). Griffiths apparently never excavated the root systems and so did not know the ground area explored by the roots of some of the species, which were often planted in two adjacent rows and sometimes even in single rows, which causes further uncertainty when expressing data on a unit ground area basis. There was also confusion with respect to the specific names of the platyopuntias employed as well as dry weight versus fresh weight. Nevertheless, he indicated that the yields of platyopuntias could be high, up to about 2 kg dry weight m^{-2} year^{-1} (Table 7.1).

Despite the widespread interest in cacti and their worldwide cultivation (Fig. 1.22), productivity of relatively few species has been reported. Because of its economic importance, fruit productivity has been quantified in various countries, values ranging from 0.02 to 0.32 kg dry weight m^{-2} year^{-1} for platyopuntias (Table 7.1). For *Opuntia ficus-indica* in Argentina, full fruit production occurs when the plants are seven to eight years old (Braun, Cordero, and Ramacciotti, 1979), but fruit harvests for it and various other platyopuntia species can begin at three years. For fourteen recognizable species and cultivars growing in the Altiplano Potosino-Zacatecano of Mexico, fresh weight of individual fruits ranges from 60 to just over 200 g each (Peralta Mata, 1983), similar to fruit weights of 35 to 210 g for various platyopuntia species in the southwestern United States and northern Africa (Monjauze and Le Houérou, 1965). In Sicily, Italy, two irrigations during the summer increase the quantity and quality of the fruit of *O. ficus-indica* (Barbera, 1984). Individual plants of *O. ficus-indica* (Fig. 1.13), which are started by placing mature cladodes in the ground, can average 170 cladodes after less than eight years of growth in Til Til, Chile (Doussoulin E. and Acevedo H., 1984). Annual productivity of stems can exceed 1 kg m^{-2} year^{-1}, with an aboveground productivity of 1.37 kg m^{-2} year^{-1} reported for *O. ficus-indica* in Chile (Table 7.1). Similarly, dry matter productivity of *O. stricta* in eastern Australia can be 0.003 kg m^{-2} day^{-1}, although such a rate is not sustained for the entire year (Osmond et al., 1979b).

Let us now consider the productivity of *O. ficus-indica* in Chile in more detail to see the pattern of growth throughout the year. Cladode area (totaled for both sides of all cladodes) increases sharply in late spring when the temperatures are favorable and the soil is wet (Fig. 7.2A); the number of cladodes approximately doubles from October through January, when an average of fifteen cladodes are added per plant (an additional thirty-eight cladodes were added during the next year of growth; Acevedo et al., 1983).

A common way to express growth, which is consistent with our productivity values based on dry weight changes, is the *net assimilation rate* (NAR):

$$NAR = \frac{DW_2 - DW_1}{(t_2 - t_1)\overline{A}} \qquad (7.1)$$

where DW_1 and DW_2 are the plant dry weights

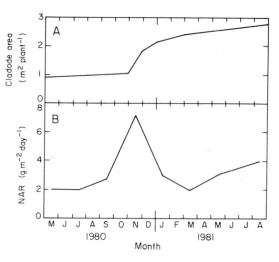

Figure 7.2. Changes in (A) total cladode area and (B) net assimilation rate (NAR) at approximately two-month intervals for *Opuntia ficus-indica* growing in central Chile. The plants, which are started by vertically placing mature cladodes about halfway into the ground, initially were one and a half-years-old and had twelve cladodes each. Plant dry weight is necessary to calculate the net assimilation rate (expressed per unit total surface area of the cladodes) and was obtained as the sum of the dry weights of all the cladodes on the monitored plants; dry weight for a cladode of a particular size was calculated using regression equations developed from twenty cladodes of various sizes harvested from neighboring plants at each sampling date and then oven dried. Modified from Acevedo et al. (1983) and García de Cortázar et al. (1985).

at times t_1 and t_2, respectively, and $\overline{A}$ is the mean photosynthetic surface for the time interval t_1 to t_2 (Larcher, 1980; Nobel, 1983a). NAR is particularly high for *O. ficus-indica* during the late spring because of favorable temperatures and water status (Fig. 7.2B). If the NAR from October through January, which averages about 5 g (m^{-2} stem area) day^{-1}, could be maintained for the entire year, the annual productivity would be 45% higher than measured. Under such optimal conditions, a productivity of about 2.0 kg (m^{-2} ground area) year^{-1} would be expected for *O. ficus-indica* at the fairly wide spacings between plants used in the field in this case.

How does the productivity of agaves and cacti under favorable conditions compare with that of other plants? One of the most produc-

Table 7.2. *Summary of distribution of root and shoot dry weights for mature agaves and cacti*

Species	Total plant dry weight (%)			Reference
	Root	Stem	Leaves	
Agaves				
Agave deserti	10	12	78	Nobel (1984a)
A. fourcroydes	13	9	78	Nobel (1985a)
A. lechuguilla	6	15	79	Nobel and Quero (1986)
A. salmiana	10	7	83	Nobel and Meyer (1985)
A. tequilana	11	17	72	Nobel and Valenzuela (1987)
Cacti				
Ferocactus acanthodes	8	92[a]	0	Nobel (1986a)
Opuntia ficus-indica	12	88[a]	0	P. S. Nobel and T. L. Hartsock (unpublished observations)

Note: Data are expressed as a percentage of the total plant dry weight of living parts.
[a] Including spines.

tive CAM plants is *Ananas comosus* (pineapple), which has an aboveground productivity under wet conditions of about 2 kg m^{-2} year^{-1} (Table 7.1). Both *O. ficus-indica* and *A. tequilana* can have a similar aboveground productivity. In fact, the aboveground productivity of about 2 kg m^{-2} year^{-1} achieved by some CAM plants is higher than that achieved by many agronomic crops (Table 7.1), most of which are grown as annuals. An annual productivity of 2 kg m^{-2} year^{-1} can be exceeded by alfalfa, which generally is multiply harvested, and the acknowledged leader in productivity, sugar cane (Table 7.1). Averaged over periods of a few years, eighteen species of trees can have maximum productivities of 2.5 kg m^{-2} year^{-1} (Table 7.1). Two factors that contribute to the relatively high productivity of certain agaves and cacti are (1) the potential for year-round growth, especially because they generally occur in regions without subzero wintertime temperatures; and (2) CAM, which leads to a relatively high net CO_2 uptake per unit of water transpired (Chapter 2), an important consideration where growth is water limited, as occurs for many crops and other plants except when irrigated.

Belowground

The total productivity of a plant includes both the readily observable aboveground growth as well as that taking place belowground. For agaves and cacti, a relatively small fraction of the total plant dry weight is belowground; in contrast, for many other perennials over half of the plant dry weight can be belowground. Like shoots, roots also exchange gases with the environment. In particular, respiration is required both to maintain the roots in a physiologically active state and for the growth of new roots.

ROOT/SHOOT RATIOS

We will begin by summarizing the percentage of plant dry weight in the root and the shoot for several agaves and cacti (Table 7.2). For five species of agaves and two species of cacti, the roots average only 10% of the plant dry weight. Hence, the root/shoot ratio is only 10%/90% or 0.11 for these desert succulents (Table 7.2).

Even though the root/shoot dry-weight ratio is fairly constant between species and with plant age, at least for mature agaves and cacti, a related matter is how the root system changes as the shoot grows. As the shoot surface area increases for both *Agave deserti* and *Ferocactus acanthodes*, the root length and hence root surface area increase (Fig. 7.3). Shoot area increases slightly faster than does root length, meaning that seedlings of both species have slightly more root length per unit shoot surface area than do the adult plants. For

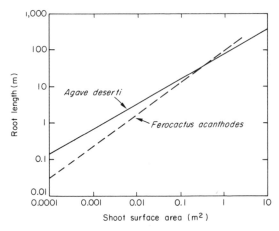

Figure 7.3. Relationship between total root length and total shoot surface area for *Agave deserti* and *Ferocactus acanthodes* of various sizes. Modified from Hunt and Nobel (1987b).

both species, the ground area explored by the roots increases in conjunction with increases in root length; in particular, the root length per ground area explored is approximately constant at 180 m m^{-2} for *A. deserti* and 60 m m^{-2} for *F. acanthodes* (Hunt and Nobel, 1987b). Thus, as the shoot area increases (Fig. 7.3) and hence potential water loss increases, the area from which soil water can be obtained increases approximately as rapidly.

ROOT RESPIRATION

The gas exchange we have discussed so far generally relates to specific parts of the shoot. To understand the physiological basis of productivity, we must consider gas exchange for the whole plant. We must take into account not only the incident PAR and other factors for different locations on the shoot but also respiration by the root system. Actually, there have been very few studies indicating how much of the net CO_2 uptake by the shoot is used for growth respiration and maintenance respiration for the roots of agaves and cacti. Under well-watered laboratory conditions that maximize shoot CO_2 uptake and minimize root turnover, 17% to 19% of total shoot net CO_2 uptake is allocated to the roots of *Agave deserti* and *Ferocactus acanthodes* (Nobel and Hartsock, 1986d). The water status also affects the root respiration rate, which approximately doubles 1 h after droughted plants of *Opuntia decumbens* are watered (Kausch, 1965).

By combining the results of various studies, we can obtain some idea of the respiratory costs associated with root maintenance and root growth in the field. We begin by noting that root surface area is three to four times greater than shoot surface area for medium-sized plants of *A. deserti* and *F. acanthodes* (Table 7.3). Rain roots can be induced three times per year by wet periods, periods which total about five months in duration; rain roots are shed during the ensuing dry periods. Of the total CO_2 annually taken up by the shoot, about 11% of the carbon is incorporated into the established roots or the three pulses of rain roots for *A. deserti*; an additional 21% of the annual shoot CO_2 uptake is expended by root respiration (Table 7.3). The comparable numbers for *F. acanthodes* are 19% for root construction and 27% for root respiration. Thus, 32% to 46% of the annual shoot CO_2 uptake for these two desert succulents is diverted to the roots. Conversely, 54% to 68% of the shoot net CO_2 uptake leads to annual shoot growth under arid conditions in the field. Because the root/shoot ratio is fairly constant at 0.11 for these desert succulents and the turnover of rain roots can be estimated, total productivity for *A. deserti* and *F. acanthodes* can be calculated once aboveground productivity is known.

The foregoing estimates must be viewed with caution, in part because certain rates had to be extrapolated over large portions of the year. Respiration depends markedly on temperature, increasing about 70% for each 10°C rise from 5°C to 45°C for *A. deserti* (Nobel, 1984a). For both *A. deserti* and *F. acanthodes*, the maximal root elongation rate is about 5 mm day^{-1} at 30°C, decreasing an average of 25% at temperatures 10°C higher or lower (Jordan and Nobel, 1984). Also, the root elongation rate is maximal above soil water potentials of -0.6 MPa, decreasing 15% at -1.0 MPa and over 90% at -2.0 MPa. Even with all the approximations, the annual total productivity of *A. deserti* measured as dry weight upon harvesting the plants corresponds to 74% of the total shoot net CO_2 uptake compared with an estimate of 75% when the CO_2 released by root respiration is taken into account (Nobel, 1984a). Root respiration amounts to 22% of annual total shoot CO_2 uptake for *A. salmiana* (Nobel and Meyer, 1985); in more arid habi-

Table 7.3. *Shoot morphology, root morphology, and root respiratory properties*
for two desert succulents

	Agave deserti	*Ferocactus acanthodes*
Shoot		
Dry weight (g)	1320	973
Surface area (m^2)	1.25	0.40
Established roots		
Length (m)	503	182
Major diameter at midlength (mm)	2.9	2.0
Dry weight (g)	163	89
Surface area (m^2)	4.3	1.0
Rain roots		
Length (m)	234	48
Major diameter at midlength (mm)	0.8	0.7
Dry weight (g)	12	3.5
Surface area (m^2)	0.59	0.092
Times induced per year (in 1981 and 1982)	3	3
Annual shoot CO$_2$ uptake (mol m^{-2} year^{-1})	17	12
Annual shoot CO$_2$ uptake incorporated into dry weight of roots (% of total)	11	19
Respiration rate		
Established roots (nmol CO$_2$ g^{-1} s^{-1})	0.51	0.32
Rain roots (nmol CO$_2$ g^{-1} s^{-1})	8.2	9.2
Maintenance and growth respiration		
Established roots (mol CO$_2$ year^{-1})	3.1	0.9
Rain roots (mol CO$_2$ year^{-1})	1.4	0.4

Note: Morphological data refer to a representative plant two weeks after major rainfalls. Respiration rates are averaged over a day for roots of the diameters indicated and at approximately the mean field temperatures. Adapted from Nobel (1976, 1977b, 1984a); Jordan and Nobel (1984); Hunt and Nobel (1987b); Hunt, Zakir, and Nobel (1987); and P. S. Nobel and T. L. Hartsock (unpublished observations).

tats, root plus stem respiration can account for 44% of annual total shoot CO$_2$ uptake for *A. lechuguilla* (Nobel and Quero, 1986). To summarize the results on several desert succulents, at least 20% of the shoot net CO$_2$ uptake is respired by the roots, and ignoring the dry weight of the transient rain roots, belowground productivity is about 11% of aboveground productivity.

Environmental productivity index (EPI)

As we have indicated in Chapters 2 through 5, environmental factors affect net CO$_2$ uptake by agaves and cacti. Influences of water status, temperature, and PAR on CO$_2$ uptake can be most easily measured in the laboratory. On the other hand, in terms of productivity the most pertinent values of environmental factors are those occurring in the field. In this regard, the studies discussed in the rest of this chapter have a common methodology – laboratory measurements of CO$_2$ uptake made when single factors are varied under controlled conditions leading to estimates of net CO$_2$ uptake in the field. When the environmental influences on net CO$_2$ uptake are quantitatively known, productivity can be predicted under a multitude of environmental conditions.

INTRODUCTION

An environmental productivity index (EPI) was introduced in 1984 to provide a simple, quantitative method for evaluating the influence of the primary environmental factors on net CO_2 uptake and hence on productivity (Nobel, 1984a). EPI indicates the ratio of net CO_2 uptake expected over 24-h periods under a particular set of environmental conditions to the net CO_2 uptake under optimal conditions of water status, temperature, and PAR. Thus, EPI ranges from 0.00, when one or more factors abolish net CO_2 uptake, to 1.00, when net CO_2 uptake is maximal (actually, EPI can even be negative). Although secondary effects and interactions between environmental factors on net CO_2 uptake would occur, not even the primary effects of climate on agaves and cacti had previously been predicted accurately, and the primary effects would presumably dominate the net CO_2 uptake over 24-h periods.

As we indicated at the beginning of the chapter, multiplication of the effect of three environmental indices, each based on net CO_2 uptake, was proposed: (1) a *water index*, indicating the effect of soil water potential on net CO_2 uptake by the shoots of agaves and cacti; (2) a *temperature index*, indicating how ambient temperature would influence daytime and nighttime CO_2 uptake; and (3) a *PAR index*, which indicates how the total daily PAR, which varies with position on a particular shoot, affects net CO_2 uptake. The environmental productivity index is defined as follows:

EPI = Water index $\times$

Temperature index $\times$ PAR index   (7.2)

Net CO_2 uptake over 24 h under a particular environmental condition is therefore the maximal net CO_2 uptake observed under optimal conditions times the EPI for that particular water status, temperature regime, and total daily PAR.

In Chapters 3 through 5, we discussed how each of the environmental factors (Eq. 7.2) individually affects net CO_2 uptake over 24-h periods. For instance, we considered each of these factors for three species of agaves: Figure 3.17 shows the effect of drought duration on net CO_2 uptake over 24-h periods, Figure 4.5 shows the effect of particular day/night temperatures, and Figure 5.5 shows the effect of various amounts of total daily PAR. The multiplication of three component indices indicating the limitation by each on net CO_2 uptake is admittedly a simplification. Yet the product, EPI, provides a convenient index for predicting productivity. And such productivity estimates based on CO_2 exchange can be compared with dry-weight estimates of productivity or with measurements of morphological parameters that can be correlated with dry-weight gain.

The shortest time interval over which to consider the influence of environmental parameters like soil water status and total daily PAR on productivity for CAM plants is 24 h. The longest practical time interval would be a season or perhaps a year, but use of average annual temperatures would hide monthly trends of minimum and maximum temperatures. Hence, the time interval chosen for temperature is a month, which also proves convenient when using regional weather records, because daily minimal temperatures and daily maximal temperatures are generally summarized on a monthly basis for a particular weather station.

Total daily PAR on a particular surface varies with location on agaves and cacti, which can have complex, three-dimensional architecture. To calculate the PAR incidence for a whole plant, we could make a series of measurements at various locations on the plant throughout both clear and cloudy days at different times of the year. On the other hand, models have been developed that accurately predict the PAR incident on agaves and cacti (see Chapter 5), and in the last section of this chapter we will introduce an even more powerful modeling technique for calculating total daily PAR. Models are eminently suited for determining PAR at various latitudes, during different seasons, and for different degrees of cloudiness. Moreover, the prediction of net CO_2 uptake by vertically oriented cladodes of platyopuntias at various latitudes and times of the year (Fig. 5.15) is really an application of the EPI approach. In particular, if the data expressing net CO_2 uptake as a percentage of maximum for different latitudes in Figure 5.15 is divided by 100 so that the maximum value

(100%) when PAR is not limiting CO_2 uptake is unity, then the ordinate becomes the PAR index. Just as for the temperature index, expressing the PAR index on a monthly basis has proved to be convenient.

The remaining component index of EPI is the water index, which includes effects on net CO_2 uptake of both the water content of the soil and the water-storage properties of the shoots of these desert succulents (Chapter 3). Using monthly precipitation records to determine the water index can lead to substantial errors, because productivity can be quite different when all the rainfall comes at the end of the month rather than occurring such that the soil remains wet throughout the month. Thus, even though we will average the water index over an entire month, it is calculated on a daily basis taking into consideration the daily values of precipitation. Models have proved useful for relating rainfall amounts to the soil water potential in the root zone, which in turn indicates when water can be taken up by an agave or a cactus. Because of water storage in the succulent tissues, some net CO_2 exchange can occur during drought, defined here as beginning when the soil water potential falls below the root water potential so that no soil water can be taken up by the plant. Indeed, Figure 3.18 is a direct plot of the water index versus drought length for *Ferocactus acanthodes* and *Opuntia ficus-indica*, because it expresses net CO_2 uptake as a fraction of maximal.

Our final topic here is the use of morphological characteristics of growth that can be correlated with EPI. As we have already indicated, an appropriate time interval for expressing the component indices of EPI is a month, so we ideally desire a morphological feature that can be readily, unambiguously, and nondestructively measured on a monthly basis. "Nondestructively" rules out conventional dry-weight measurements, because they require harvesting of plant material, and so the same plants cannot be monitored over consecutive months. "Unambiguously" rules out such straightforward measurements as height and diameter of cactus stems, because both can decrease substantially during periods of drought and so are more an indication of fresh weight. "Readily" is a desirable objective but

obviously subject to interpretation. The method adopted for *O. ficus-indica* involves measuring the area of each cladode for the studied plants and determining dry weight by regression equations obtained for cladodes harvested each month from similar adjacent plants – nondestructive for the plants monitored, fairly unambiguous, but hardly "readily" measured. For *F. acanthodes*, the method adopted involves counting the production of new areoles – again nondestructive, but not really unambiguous or readily measured, as we shall see. But for agaves, the simple expedient of counting the leaves unfolding from the central spike proves to be nondestructive, unambiguous (the leaf tip can be clipped upon unfolding, so that leaf is not counted again), and readily measured.

AGAVES

Beginning with *Agave deserti*, the EPI approach has now been successfully applied to productivity studies on five species of agaves. The component indices for two of these species are based on environmental effects on nocturnal acid accumulation, requiring measurements only at dusk and at dawn compared with the continuous monitoring over 24-h periods necessary to determine net CO_2 exchange. In all five cases, monthly leaf unfolding is highly correlated with monthly EPI.

Agave deserti

We will begin by considering the calculation of EPI in some detail for *A. deserti* over a one-year period. The principles are the same for all the other species considered.

To calculate EPI for a particular site, we need to know the climatic conditions there. First let us consider the soil water status, which means considering all the rainfall events over the period of interest (Fig. 7.4A). The soil water potential in the root zone can be determined by direct measurements, or, more feasibly, Ψ^{soil} can be calculated from models. When the soil locally has a higher water potential than the roots, water can be taken up by the plant and the water index is 1.00. That is, when by this criterion the soil is wet, water is not limiting net CO_2 uptake. As the soil begins to dry, stomatal opening and net CO_2 up-

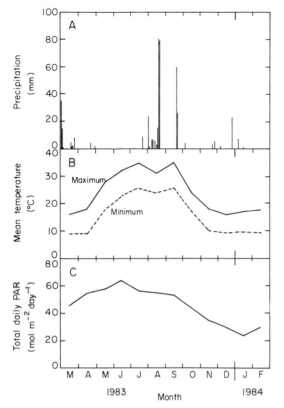

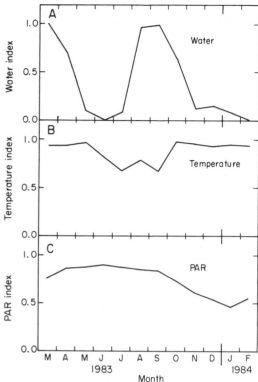

Figure 7.4. Environmental conditions at Agave Hill in the northwestern Sonoran Desert over a one-year period. (A) Precipitation events. (B) Daily maximum and minimum air temperatures averaged over the month. (C) Mean total daily PAR on a horizontal surface. Precipitation and temperatures were obtained from standard U.S. Weather Bureau instruments at the site. PAR was obtained by simulation for clear days and then corrected for cloudiness using the measured attenuation of shortwave irradiation by clouds. Modified from Nobel (1984a) and unpublished observations.

Figure 7.5 Component indices of the environmental productivity index, as determined monthly for *Agave deserti* under the rainfall, temperature, and PAR conditions indicated in Figure 7.4 (A)Water index. (B) Temperature index. (C) PAR index. Modified from Nobel (1984a).

take will eventually decline. For instance, for *A. deserti* a drought of 11 days decreases the net CO_2 uptake over a 24-h period by 50% (Fig. 3.17A), so the water index is then 0.50. After 33 days of drought, no net CO_2 uptake occurs over the 24-h period, and the water index is then 0.00 (Nobel, 1984a). By first considering the effect of rainfall on soil water potential (see Chapter 3) and then the effect of wet and dry conditions on net CO_2 uptake over 24-h periods, we can assign a water index.

As discussed in the previous subsection, the water index is averaged over a month, which we will now consider for *A. deserti* at a site in the northwestern Sonoran Desert (Fig. 7.5A). For the one-year period considered, the water index is 1.00 for March, reflecting the initially wet soil and the rainfall in that month (Fig. 7.4A), and again 1.00 for September, reflecting the heavy rainfall in August as well as in late September. On the other hand, the water index is 0.00 in June and February (Fig. 7.5A), reflecting the effect of extensive drought periods (Fig. 7.4A) on net CO_2 exchange by *A. deserti* (Fig. 3.17A).

Let us now turn to temperature, which has its highest values at Agave Hill in the summer and lowest values in the winter (Fig. 7.4B). In the laboratory, net CO_2 uptake over 24-h periods can be measured under various day/night air temperatures. Maximum net CO_2 uptake by *A. deserti* occurs at 25°C/15°C (Figs.

4.5A and 4.7), similar to the mean annual day/night temperatures at the Agave Hill site. At 25°C/15°C, the temperature index for *A. deserti* therefore is 1.00. When the day/night temperatures are lowered to 18°C/8°C, the net CO_2 uptake over the 24-h period is reduced 20%, corresponding to a temperature index of 0.80, and at day/night temperatures of 40°C/30°C the temperature index is 0.37 (Fig. 4.5A). Air temperatures prove not to be very limiting with respect to net CO_2 uptake by *A. deserti* in the field, because the temperature index averages 0.88 for the entire year (Fig. 7.5B). The temperature index drops below 0.80, indicating greater than 20% inhibition of net CO_2 uptake attributable to adverse temperatures, only during the summer with its relatively high temperatures (Fig. 7.4B).

As we indicated in Chapter 4, the nighttime temperature is more important than the daytime temperature for determining net CO_2 uptake over 24-h periods for agaves and cacti, so attention can be focused on the minimum daily temperatures averaged over a month. Depending primarily on leaf size, which can affect leaf temperatures for agaves (Fig. 4.21A), a nighttime temperature maintained during the night in the laboratory that is 0°C to 2°C above the minimum daily air temperature in the field gives approximately the same net CO_2 uptake over a 24-h period as does using the varying temperatures that occur in the field during the course of a night (Nobel, 1984a, 1985a; Nobel and Quero, 1986). Thus, laboratory measurements of net CO_2 exchange over 24-h periods required for determining the temperature index of some species for a particular month are generally done at about 1°C above the minimum daily air temperature averaged over that month at the field site.

The final component index in EPI is the PAR index, representing the fractional limitation of net CO_2 uptake over a 24-h period by suboptimal levels of total daily PAR. The average total daily PAR on a horizontal surface, which can be readily determined each month (Fig. 7.4C), must be converted to values incident on the leaf surfaces. Using the model discussed in Chapter 5, we see that the total daily PAR received by the leaves of *A. deserti* averages about 38% of the total daily PAR on a horizontal surface near the winter solstice

and about 32% near the summer solstice. Once the total daily PAR on the leaves is estimated, the PAR index can be determined from the response of net CO_2 uptake over 24-h periods to total daily PAR for *A. deserti* (Fig. 5.5A). For instance, the PAR index is 0.50 for a total daily PAR in the planes of the leaves of 10 mol m^{-2} day^{-1} (Nobel, 1984a). Because of seasonal changes in the sun's trajectory, the total daily PAR on a horizontal surface at 34°N is over twice as high near the summer solstice compared with the winter solstice (Fig. 7.4C). As a result, the PAR index for *A. deserti* is nearly twofold higher in the summer than in the winter (Fig. 7.5C).

Two matters with respect to PAR raised in Chapter 5 deserve reemphasis here. First, the response of net CO_2 uptake to total daily PAR is nonlinear (Fig. 5.2), so the PAR index should ideally be calculated separately for each leaf surface and then averaged over the plant. Second, PAR is always limiting net CO_2 uptake in the field, as shown here by the PAR index always being less than 1.00 for *A. deserti* (Fig. 7.5C).

Now that values of the component indices have been presented for a one-year period, using Equation 7.1 we can indicate monthly values of EPI for that period. The product of the water index (Fig. 7.5A), the temperature index (Fig. 7.5B), and the PAR index (Fig. 7.5C) indicates that EPI for *A. deserti* has a maximum value of 0.64 and decreases to 0.00 during times of prolonged drought (Fig. 7.6A). The average value of EPI for the year considered is 0.27. Because each of the component indices indicates the fractional limitation on net CO_2 uptake by that environmental factor, EPI indicates the overall limitation on net CO_2 uptake by the three factors combined, as we have already indicated. Thus, an EPI of 0.27 means that net CO_2 uptake would be nearly four times higher under wet conditions, optimal temperatures, and saturating PAR.

Let us next consider how leaf unfolding, which can be readily monitored nondestructively for agaves in the field, is related to EPI. Figure 7.6 indicates that the number of leaves unfolding monthly closely follows the pattern of monthly EPI values, both for the year considered heretofore (Fig. 7.6A) and for the suc-

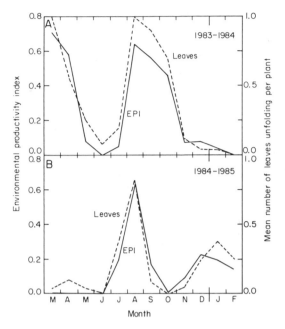

Figure 7.6 Monthly environmental productivity index and monthly leaf unfolding for *A. deserti* at Agave Hill for two sequential years. (**A**) 1983–1984, the period presented in Figures 7.4 and 7.5. (**B**) The subsequent one-year period. Leaf unfolding was averaged for the fifty plants examined each month. Modified from Nobel (1984a) and Nobel and Hartsock (1986c).

ceeding year (Fig. 7.6B). Because the fraction of the overall plant dry weight of *A. deserti* comprised by unfolded, living leaves is fairly constant at 0.60 (Nobel, 1984a), the increase in plant dry weight represented by the newly unfolded leaves indicates a net productivity for the plant as a whole. Thus, monthly productivity estimated morphologically by leaf unfolding is closely correlated with relative net CO_2 uptake represented by monthly values of EPI ($r^2 = 0.93$; Nobel, 1984a).

Annual net CO_2 uptake calculated using EPI can also be compared with productivity measured by dry-weight increases, a method that requires harvesting of plants. Annual EPI times the maximum net CO_2 uptake over a 24-h period (285 mmol m^{-2} day^{-1} for *A. deserti*; Nobel, 1984a) times the number of days per year indicates the annual net CO_2 uptake per unit leaf area. When this total net CO_2 uptake for the shoot is corrected for root respiration, the productivity estimate based on EPI is within 1% of that obtained from dry-weight measurements (Nobel, 1984a). Because *A. deserti* can occur in monospecific stands (Fig. 7.7A), net CO_2 uptake or dry-weight increases can be expressed per unit ground area occupied by a group of plants. To minimize uncer-

Figure 7.7. Monospecific stands suitable for measuring productivity per unit ground area. (**A**) *Agave deserti* at Agave Hill, where annual plants, especially grasses, occur but at a very low biomass per unit ground area. (**B**) *Opuntia streptacantha* near San Martín, Mexico, Mexico; this cultivated plant is also referred to as *O. amyclaea* and *O. ficus-indica* by various experts [photographed by Edmundo García Moya; in the background note the Pyramid of the Sun (64 m tall) on the left and the Pyramid of the Moon near the right, both at Teotihuacán].

tainties introduced by root distribution, only the innermost plants, whose roots are within the stands, are usually considered in such calculations. The dry-weight gain of *A. deserti* and other agaves can then be expressed per unit ground area, the conventional basis for productivity, especially for agronomic crops (Table 7.1).

Agave fourcroydes *and* A. lechuguilla

Let us next consider EPI for two species of agaves that are harvested for the hard fiber in their leaves, *Agave fourcroydes*, cultivated in Yucatán, Mexico (Fig. 1.7), and *A. lechuguilla* (Fig. 1.8), which is harvested from the wild in many states of Mexico. For these two species, we have already indicated the response of net CO_2 uptake over 24-h periods to water status (Fig. 3.17B,C), temperature (Fig. 4.5B,C), and PAR (Fig. 5.5B,C). We will begin by considering the component indices for *A. fourcroydes* (Fig. 7.8), just as we did for *A. deserti* (Fig. 7.5), followed by a briefer consideration for *A. lechuguilla*.

Because of higher rainfall in Yucatán, which is subtropical with up to about 1,000 mm of annual rainfall, than in the Sonoran Desert, the water index tends to be higher for *A. fourcroydes* (Fig. 7.8A) than for *A. deserti* (Fig. 7.5A). Indeed, the water index of *A. fourcroydes* is 1.00 for the six-month period from July through December, never falls below 0.30, and averages 0.81 for the one-year period (Fig. 7.8A). As is the case for *A. deserti*, temperature is not a major limiting factor for net CO_2 uptake by *A. fourcroydes* in the field. The late spring and summer are somewhat too warm for optimal net CO_2 uptake, but the monthly temperature index still averages 0.85 for the one-year study period (Fig. 7.8B). However, net CO_2 uptake is considerably limited by suboptimal levels of PAR. Under cultivation, *A. fourcroydes* is planted fairly close together, which leads to mutual shading of the plants but increases the productivity per unit ground area, as we will discuss later. Over the one-year study period, the PAR index averages 0.52 for plants initially one year old and 0.62 for initially six-year-old plants (Fig. 7.8C), which are much more widely spaced. Superimposed on the expected seasonal decline in the PAR index as winter approaches

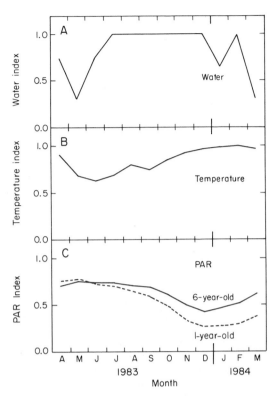

Figure 7.8. Monthly values of the component indices of EPI determined for *Agave fourcroydes* cultivated in Yucatán, Mexico. (**A**) Water index. (**B**) Temperature index. (**C**) PAR index, which was determined for plants of the two indicated initial ages. Modified from Nobel (1985a).

is a PAR decrease because of plant growth, which is particularly evident for the one-year-old plants (Fig. 7.8C). This decrease reflects an increased shading of older leaves due to the unfolding of new leaves. For instance, the *leaf area index* (area of both sides of the leaves per unit ground area occupied by the plants) increases from 4.1 to 8.4 over the one-year study period for the plants initially one year old, which substantially reduces the PAR per unit leaf area and hence reduces the PAR index (Nobel, 1985a).

As for *A. deserti* (Fig. 7.6), monthly leaf unfolding is highly correlated with monthly EPI for *A. fourcroydes* (Fig. 7.9). Maximal monthly leaf unfolding reaches nearly three leaves per plant for *A. fourcroydes*, compared with only one leaf per plant for *A. deserti*. EPI over the year averaged 0.34 for plants of *A. fourcroydes* initially one year old (Fig. 7.9A)

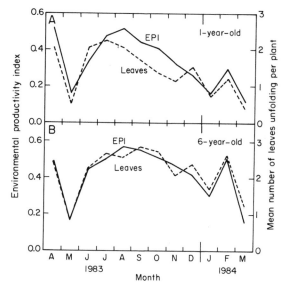

Figure 7.9. Monthly environmental productivity index and monthly leaf unfolding for initially one-year-old (**A**) and six-year-old (**B**) plants of *Agave fourcroydes* in Yucatán, Mexico. EPI was derived from the component indices indicated in Figure 7.8, including the different PAR indices for plants of the indicated initial ages. Leaf unfolding was averaged for the twenty plants examined each month. Modified from Nobel (1985a).

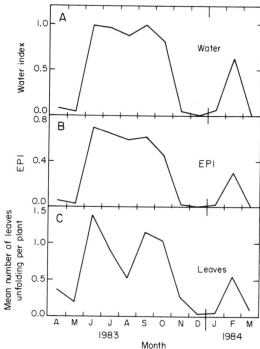

Figure 7.10. Monthly water index (**A**), monthly environmental productivity index (**B**), and monthly number of leaves unfolding (**C**) for *Agave lechuguilla* near Saltillo, Coahuila, Mexico. Leaf unfolding was averaged for the fifty-two plants examined each month. Modified from Nobel and Quero (1986).

and 0.42 for those initially six years old (Fig. 7.9B).

We can also use EPI and its component indices to see quantitatively how changes in various environmental conditions can affect productivity. For instance, if the water index were 1.00 for each month, as could be induced by irrigation in the winter and early spring (Fig. 7.8A), then the annual EPI would increase from 0.42 to 0.52 for the six-year-old plants (Nobel, 1985a). Thus, productivity of *A. fourcroydes* in Yucatán can be increased 24% by appropriate irrigation. Because temperatures in Yucatán are slightly too high for optimal net CO_2 uptake by *A. fourcroydes*, especially from May through September (Fig. 7.8B), let us see what would happen to EPI if the day/night temperatures are decreased by 5°C. The temperature index would then average 0.96, and EPI for the six-year-old plants would be 13% higher. Finally, if the plants are grown at the equator instead of at 21°N, the annual PAR index and annual EPI would both

increase about 10%. When all three environmental factors are close to optimal, such as under irrigation (increased water index) at moderate elevations (increased temperature index) near the equator (increased PAR index), EPI can be increased by 54%. The aboveground productivity of *A. fourcroydes* in Yucatán of 1.53 kg m^{-2} year^{-1} (Table 7.1) would then be increased to 2.36 kg m^{-2} year^{-1} under the more favorable conditions.

Component indices and EPI have also been determined monthly over a one-year period for *Agave lechuguilla* in the Chihuahuan Desert (Nobel and Quero, 1986). The monthly variations and the mean value of the PAR index are similar to those for the two agave species previously considered. The monthly temperature index ranges from 0.76 to 0.97 but is lowest during the winter for *A. lechuguilla*, when the mean daily minimum temperature for

Figure 7.11. *Agave salmiana* near Sahagun, Hidalgo, Mexico. (**A**) Two-year-old plants being planted in a field. (**B**) A similar plantation about one year later (photographed by Edmundo García Moya).

December and January is just above 0°C. As would be expected under desert conditions, monthly variations in net CO_2 uptake mainly reflect the variations in the availability of water. Indeed, monthly changes in the water index (Fig. 7.10A) are responsible for nearly all the month-to-month variation in EPI ($r^2 = 0.97$; Fig. 7.10B). Monthly EPI is again highly correlated with monthly unfolding of new leaves ($r^2 = 0.83$; Fig. 7.10C). The annual EPI is 0.28, which is actually not that low compared with other agaves under natural conditions in the field; the productivity expressed per unit ground area is rather low for *A. lechuguilla* (Table 7.1), because its leaf area index was only 1.23 (Nobel and Quero, 1986).

Agave salmiana *and* A. tequilana

The EPI approach has also been used to study the productivity of two species of agave that are commercially exploited for the manufacture of alcoholic beverages: *Agave salmiana* (Figs. 1.4 and 7.11), which is harvested from the wild for pulque and mescal; and *A. tequilana* (Fig. 1.5), which is cultivated for the manufacture of tequila. The influence of water

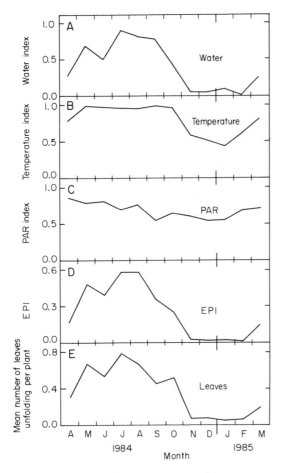

Figure 7.12. Monthly values of the component indices (**A–C**), EPI (**D**), and leaf unfolding (**E**) for *Agave salmiana* under natural conditions near Salinas de Hidalgo, San Luis Potosí, Mexico. The component indices are based on environmental effects on nocturnal acid accumulation. Leaf unfolding was averaged for the seventy-three plants examined each month. Modified from Nobel and Meyer (1985).

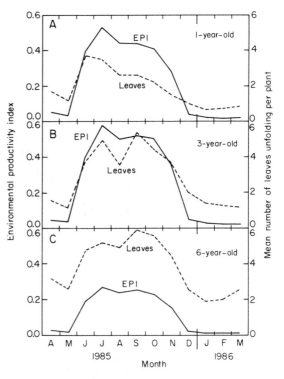

Figure 7.13. Monthly EPI and monthly leaf unfolding for initially one-year-old (**A**), three-year-old (**B**), and six-year-old (**C**) plants of *Agave tequilana* at Tequila, Jalisco, Mexico. Component indices are based on nocturnal acid accumulation. Leaf unfolding was averaged for the twenty plants of each of the indicated initial ages that were examined each month. Modified from Nobel and Valenzuela (1987).

status, temperature, and PAR on nocturnal acid accumulation, instead of on net CO_2 uptake over 24-h periods, will be examined for these two species. Nocturnal acid accumulation, which is more easily measured than is CO_2 exchange, can be related to net CO_2 uptake over 24-h periods for other agaves and for cacti at various PAR levels (Fig. 5.2 compared with Fig. 5.3), although such relationships have not been established for any environmental factors for *A. salmiana* and *A. tequilana*. We will thus see whether monthly leaf

unfolding is correlated with EPI based on nocturnal acid accumulation.

For *A. salmiana*, drought causes the water index to be low during the winter and early spring (Fig. 7.12A). Temperatures during the winter are too cold for optimal nocturnal acid accumulation (Fig. 7.12B), and the PAR index has its usual seasonal trend (Fig. 7.12C). Again, primarily because of the water index, EPI tends to be high during the summer and low during the winter (Fig. 7.12D). Of perhaps greater interest, monthly EPI, determined when the component indices are based on nocturnal acid accumulation, is also highly correlated with monthly leaf unfolding ($r^2 = 0.95$; Fig. 7.12E). Thus, the responses of nocturnal acid accumulation to individual environmental factors can apparently be used to predict the

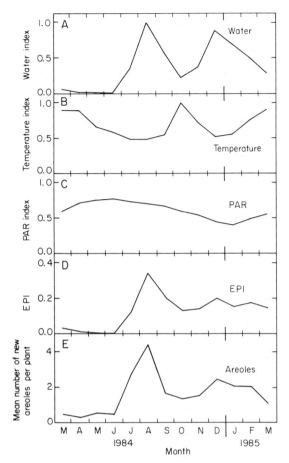

Figure 7.14. Monthly values of the component indices (**A–C**), EPI (**D**), and number of new areoles produced (**E**) for *Ferocactus acanthodes* at Agave Hill. The number of newly initiated areoles is averaged for the thirty-three plants examined each month. The position of each new areole was monitored for a few months by noting the position of its spines projecting through the apical pubescence; the time of initiation when an areole was first displaced from the apex was then determined by extrapolation. Modified from Nobel (1986a).

cumulation to various environmental factors. For plants that are initially one year old (Fig. 7.13A), three years old (Fig. 7.13B), or six years old (Fig. 7.13C), monthly EPI is highly correlated with monthly leaf unfolding (r^2 averages 0.89). EPI tends to be lower for the six-year-old plants because they have more leaf area per unit ground area and hence a lower PAR index compared with the two younger age classes (Nobel and Valenzuela, 1987).

Contrary to the results with the other four species of agave considered, appreciable monthly leaf unfolding for *A. tequilana* occurs even when EPI is low, especially for the older plants (Fig. 7.13). Such leaf unfolding in the absence of appreciable nocturnal acid accumulation may be supported by translocation of dry matter from the stem to the new leaves rather than by net CO_2 fixation by the existing leaves. The stem represents a relatively high fraction of the dry weight for *A. tequilana* (Table 7.2), and its fraction increases with plant age (Nobel and Valenzuela, 1987). Indeed, the stem of this species (Fig. 1.5B) is harvested for the production of tequila.

CACTI

In part because of the ease of correlating EPI, and therefore productivity, with leaf unfolding, the EPI approach has been applied more to agaves than to cacti. However, the responses of net CO_2 uptake to water status, temperature, and PAR have been determined for the two cactus species we have been considering in detail throughout this book, *Ferocactus acanthodes* and *Opuntia ficus-indica*. EPI has also been calculated under field conditions for these two species.

Ferocactus acanthodes

The influence of drought on the water index (Fig. 3.18), which generally accounts for most of the monthly variation in EPI for plants in desert regions, has been determined for *F. acanthodes* (Fig. 1.2A). After calculating the soil water potential in the root zone from individual precipitation events (Chapter 3), the water index for *F. acanthodes* at Agave Hill can be determined for the one-year period (Fig. 7.14A). The pattern for the water index is bimodal, approaching unity during wet pe-

productivity of *A. salmiana* in the field (Nobel and Meyer, 1985).

Unlike the case for *A. salmiana* occurring in natural populations, the ages of cultivated plants of *A. tequilana* are known, as are those of cultivated *A. fourcroydes*. Thus, possible influences of plant age on EPI can again be examined. As for *A. salmiana*, EPI for *A. tequilana* is based on component indices determined from responses of nocturnal acid ac-

riods in the summer (80 mm of rainfall) and again in the winter (90 mm of rainfall). The temperature index is seasonally out of synchrony with the water index. The summer is too warm and the winter is too cold for maximal CO_2 uptake by *F. acanthodes* (see Fig. 4.7 for temperature effects on CO_2 uptake), so the temperature index approaches unity only during relatively dry periods in the spring and the fall (Fig. 7.14B). Based on the response of net CO_2 uptake over 24-h periods to total daily PAR for *F. acanthodes* (Fig. 5.2) and the PAR available in the field, its PAR index exhibits the usual seasonal pattern, rising to 0.71 near the summer solstice and falling to 0.42 near the winter solstice (Fig. 7.14C). For the one-year study period, EPI (Eq. 7.2) is very low during the dry spring and early summer (March through June), reaches 0.34 in August, and averages 0.13 (Fig. 7.14D). Therefore, net CO_2 uptake by *F. acanthodes* in the field is here only 13% of the maximum possible under optimal conditions (Nobel, 1986a).

Productivity of *F. acanthodes* cannot be reliably monitored by monthly measurements of height and diameter, because these morphological parameters reflect stem volume, not stem dry weight. Indeed, both height and diameter can decrease during drought even when the dry weight remains unchanged. An alternative approach is to count the newly emerging areoles, a process that in some ways is analogous to monitoring the unfolding of agave leaves, because areoles are produced on modified leaf bases. The areoles produced by the apical meristem of *F. acanthodes* are hidden by apical pubescence and become evident only later when the newly initiated spines on each areole project through the pubescence. The time when a particular areole has just emerged from the apical meristem is estimated by monitoring the position of each new areole for two to four months and then extrapolating back in time to when that areole was within a few millimeters of the apex. The monthly production of new areoles by *F. acanthodes* (Fig. 7.14E) is highly correlated with monthly values of EPI ($r^2 = 0.81$; Fig. 7.14D).

Areole production can be monitored nondestructively and can indicate the monthly changes in the productivity of this cactus. However, besides the difficulty in determining the time of initiation of each areole, another problem is encountered. *Ferocactus acanthodes* tends to produce areoles in bunches over one- to two-month periods, the number of areoles in a bunch corresponding to the numbers in the Fibonacci series (Nobel, 1986a). In Chapter 1 we noted that the numbers of ribs of certain barrel cacti, including *F. acanthodes*, conform to the Fibonacci series (Fig. 1.30). Such plants generally increase in size by fairly synchronously adding a new areole to each of the existing ribs. Specifically, fourteen of the thirty-three plants monitored added 5, 8, 13, 21, or 34 new areoles during the year and an additional ten plants added two times a Fibonacci number (Nobel, 1986a). Because areoles are added in bunches, one areole for each of the already existing ribs, the monthly correlation between EPI and areole production for *F. acanthodes* does not hold for an individual plant. For agaves, the number of plants monitored must also be large enough so that the idiosyncratic nature of a particular plant does not dominate the average monthly change in the morphological parameter used to represent productivity.

Opuntia ficus-indica

Let us next consider the component indices and EPI for *O. ficus-indica*, which is cultivated worldwide (Fig. 1.22) for both fruit (Fig. 1.13B) and young cladodes (Fig. 7.15) for human consumption, as well as older cladodes for animal fodder. Under cultivation at Til Til in central Chile, the plants are irrigated three times during the dry season occurring in the late spring and the summer (Acevedo et al., 1983). The water index, based on known responses of *O. ficus-indica* to drought (Fig. 3.18), thus never drops below 0.70 and averages 0.84 for the twenty-month study period (Fig. 7.16A). The temperatures at Til Til are also near optimal for the growth of *O. ficus-indica*, the temperature index never falling below 0.80 and averaging 0.93 during the study period (Fig. 7.16B; see Fig. 4.7 for the responses of *O. ficus-indica* to temperature). Among other factors, the PAR index depends on the number of cladodes per plant and the plant spacing, which leads to considerable mutual shading at the site considered. The PAR index varies from 0.14 near the winter solstice

Figure 7.15. *Opuntia ficus-indica* at Milpa Alta in the District Federal, Mexico. (A) Mature cladodes recently planted in a field that received 200 tons of cattle manure ha^{-1}. (B) Small cladodes ("nopalitos"), whose harvest for human consumption can reach 300 tons fresh weight ha^{-1} year^{-1} (photographed by Ephríam Hernández Xolocotzi).

(in June in the southern hemisphere) to 0.56 near the summer solstice (Fig. 7.16C). Chiefly because of the seasonal variations in the PAR index, EPI is highest in early summer (0.51) and lowest in early winter (0.12; Fig. 7.16D).

A readily measured morphological feature, unambiguously related to productivity and amenable to nondestructive monitoring in the field, has not been identified for *O. ficus-indica*. Attempts have been made to use length, width, and thickness of cladodes, but thickness has been difficult to measure reliably. The number of cladodes can change rapidly, as dry matter from underlying cladodes is translocated to the new cladodes, whose dry-weight/volume ratios differ from those of

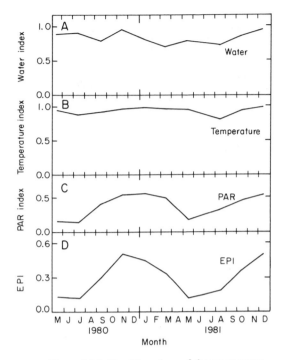

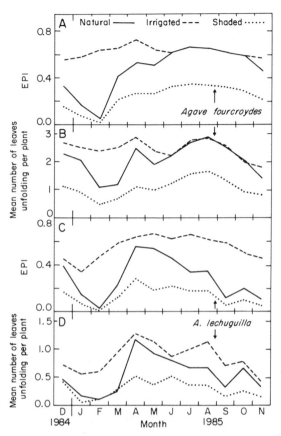

Figure 7.16. Monthly values of the component indices (**A–C**) and EPI (**D**) for *Opuntia ficus-indica* under cultivation at Til Til, Chile. Data are from plants initially one and a half years old, as for Figure 7.2. Modified from Acevedo et al. (1983), García de Cortázar et al. (1985), and unpublished observations.

Figure 7.17. Monthly variations in EPI (**A, C**) and leaf unfolding (**B, D**) for *Agave fourcroydes* (**A, B**) and *A. lechuguilla* (**C, D**). Twenty plants were maintained under natural conditions (————), watered weekly so that the soil in the root zone was continuously moist (– – – –), or shaded by screening such that the PAR level was reduced 60% (· · · ·). Modified from Nobel and García de Cortázar (1987).

the older cladodes. Many cladodes from adjacent plants have to be harvested to obtain reliable correlations between cladode area and cladode dry weight (Acevedo et al., 1983; García de Cortázar et al., 1985). Such periodic morphological measurements together with regression analysis relating size to dry weight for the harvested cladodes lead to an ongoing calculation of NAR (Eq. 7.1) to indicate productivity for *O. ficus-indica*, a procedure familiar to agronomists. Moreover, the seasonal variations in NAR for *O. ficus-indica* (Fig. 7.2) are highly correlated with seasonal variations in EPI (Fig. 7.16D). We therefore conclude that the EPI approach is also valid for this species.

FIELD TESTS

Before we examine refinements and applications of EPI, let us consider some field tests of the EPI approach. Although manipulating temperature in the field can be difficult, increasing water availability and decreasing PAR availability are relatively simple. Moreover, both of these manipulations can occur agriculturally, namely by irrigation and by close spacing of plants, respectively.

Weekly watering of *Agave fourcroydes* under cultivation and of *A. lechuguilla* in natural populations causes the water index to be unity. This increases the average monthly EPI for a one-year study period from 0.47 to 0.63 for *A. fourcroydes* (Fig. 7.17A) and from 0.29 to 0.55 for *A. lechuguilla* (Fig. 7.17C). Watering concomitantly increases the monthly rate of leaf unfolding by 21% for *A. fourcroydes* (Fig. 7.17B) and by 55% for *A. lechuguilla*

(Fig. 7.17D). An experimental shade treatment that decreases the total daily PAR by 60% approximately halves the PAR index and EPI for both species (Fig. 7.17A,C). This decrease reduces the rate of leaf unfolding by 48% for *A. fourcroydes* (Fig. 7.17B) and by 45% for *A. lechuguilla* (Fig. 7.17D). Thus both manipulations, watering and shading, affect leaf unfolding approximately as predicted by the changes in EPI (Nobel and García de Cortázar, 1987).

Refinements of EPI

Although considerable success has been achieved using the EPI approach to analyze the effects of three physical factors of the environment on the productivity of agaves and cacti, we should ask whether other factors should also be included. For instance, how about nutrients? Also, the increasing levels of CO_2 in the atmosphere should be considered. Air pollutants such as sulfur dioxide could affect productivity; 0.9 ppm SO_2 in the air (about 500 times higher than global background SO_2 levels) presented for three days does not decrease net CO_2 uptake or lead to visual symptoms of injury for *Agave deserti* and *Opuntia basilaris*, indicating a low sensitivity to SO_2 for these species, although 2 h at 3 ppm SO_2 does injure *O. basilaris* (Olszyk, Bytnerowicz, and Fox, 1987). We should also examine an underlying assumption of EPI, namely, that the three component indices do not depend on other than the indicated physical factor with respect to the effect on net CO_2 uptake (Eq. 7.2). Thus, we have assumed that the water index depends only on the water status, not on temperature or PAR, an assumption that we also will evaluate in this section.

NUTRIENTS

Perhaps the most obvious omission from EPI is a *nutrient index* indicating the effect of soil mineral elements on net CO_2 uptake over 24-h periods. Part of the reason for this omission is that we still have incomplete information on the responses of these desert succulents to nutrients. Moreover, nutrients interact in a complex manner. For instance, iron and zinc are less available to plants in a soil of high pH, and a deficiency in potassium can often be overcome by an excess of sodium. In ad-

dition, high levels of iron can lead to manganese deficiency in plants, and vice versa, just as high levels of calcium can lead to magnesium deficiency (Chapman, 1966; Berry, 1986).

We can circumvent the lack of a nutrient index by determining net CO_2 uptake under optimal conditions (wet, optimal nighttime temperatures, saturating PAR) for the soil condition at each site being considered. This is effectively the procedure adopted for the species of agaves and cacti considered so far (Table 7.4). For *Ferocactus acanthodes*, the net CO_2 uptake over a 24-h period at Agave Hill under wet conditions, saturating PAR, and corrected for optimal temperatures is within 10% of that measured under optimal conditions in the laboratory (Table 7.4). Some differences might be expected because roots in the field would explore a greater volume of soil, although possible nutrient exhaustion in the laboratory is avoided by application of a dilute nutrient solution (0.05- or 0.10-strength Hoagland solution Number 1; Hoagland and Arnon, 1950). Net CO_2 uptake by *Agave lechuguilla* over a 24-h period under optimal conditions is 29% lower in the field than when irrigated by a dilute nutrient solution in the laboratory (Table 7.4), possibly because of the low boron levels that occur for soils in Coahuila (Nobel et al., 1987). Net CO_2 uptake over 24 h is 50% higher for *Opuntia ficus-indica* grown in a commercial soil mixture, which among other differences contains substantial levels of nitrogen, compared with when the plants are grown in soil from Agave Hill. Interspecific variations in net CO_2 uptake over 24 h occur even when the same soil is used; CO_2 uptake is over twice as high for *Agave tequilana* as for *F. acanthodes* in soil from Agave Hill (Table 7.4). When a new species is considered or a previously measured species is considered under a different soil regime, the net CO_2 uptake under optimal conditions can again be determined.

Let us next consider an approach that could be used to develop a nutrient index. Based on the responses of agaves and cacti to various elements (Chapter 6), we will begin by considering the nutrient proved to be most influential on growth, nitrogen. For seedlings of four different species, dry weight increases ap-

Table 7.4. *Net CO_2 uptake for various species of agaves and cacti under wet conditions, optimal temperatures, and saturating PAR*

Species	Soil	Net CO_2 uptake for 24-h period (mmol m^{-2} day^{-1})	Reference
Agave deserti	From Agave Hill	285	Nobel (1984a)
A. fourcroydes	From Mérida, Yucatán	380	Nobel (1985a)
A. lechuguilla	From Agave Hill	223	Nobel and Quero (1986)
	In field in Saltillo, Coahuila	158	Quero and Nobel (1987)
A. salmiana	From Agave Hill	325	P. S. Nobel (unpublished observations)
A. tequilana	From Agave Hill	350	Nobel and Valenzuela (1987)
Carnegiea gigantea	Commercial mixture	250	P. S. Nobel (unpublished observations)
Ferocactus acanthodes	In field at Agave Hill	148	Nobel (1977b)
	From Agave Hill	165	Nobel and Hartsock (1986d)
Opuntia ficus-indica	Commercial loamy sand	344	Nobel and Hartsock (1983 and 1984)
	From Agave Hill	231	Nobel and Hartsock (1986d)

Note: Plants were maintained in the laboratory (except where indicated) in the indicated soil.

proximately linearly with the logarithm of N concentration up to the rather high concentration of 15 mM nitrate in full-strength Hoagland solution (Fig. 6.2). The growth of seedlings of *Agave deserti* increases approximately logarithmically with levels of N in the soil up to 0.1% by dry weight (Fig. 6.3A). Also, the growth of *A. deserti* measured by changes in dry weight at Agave Hill varies logarithmically with the level of applied N from a low level of 1 g m^{-2} of ground area to a relatively high applied level of 10 g m^{-2} (100 kg ha^{-1}; Fig. 6.3B). We thus surmise that our proposed index should contain the logarithm of the N level in the soil or in the hydroponic solution; that is, net CO_2 uptake can be proportional to the logarithm of the N level plus an additive constant.

Within the limits of low and high N just enumerated, the net CO_2 uptake over 24 h under a new nitrogen condition could be predicted based on that measured under some original condition, such as at a particular site or for a particular nutrient solution, as follows:

Nitrogen index

$$= \frac{\text{Net } CO_2 \text{ uptake under new N condition}}{\text{Net } CO_2 \text{ uptake under original N condition}}$$

$$= \frac{m \ln(N_{new}) + b}{m \ln(N_{orig}) + b} = \frac{\frac{m}{b}\ln(N_{new}) + 1}{\frac{m}{b}\ln(N_{orig}) + 1}$$

$$= \frac{1 + n_N \ln(N_{new})}{1 + n_N \ln(N_{orig})} \tag{7.3}$$

where N_{new} and N_{orig} represent the nitrogen levels under the new and the original conditions, respectively; ln represents the natural logarithm; and n_N is an empirically derived, dimensionless coefficient that we will call the *nutrient coefficient*, in this case for nitrogen. The units for N to be used in Equation 7.3 are the relative strength of Hoagland solution for plants grown hydroponically or the % N level by dry weight in the root zone for plants grown in soil (other units would necessitate changes in the value of n_N). The N level under the new condition can include the amount of nitrogen applied as fertilizer, so Equation 7.3 can be

used in an agronomic context. Such an empirically derived nutrient index, which would enter into EPI as a multiplicative factor just as for the other component indices (Eq. 7.2), is meant only as a first approximation toward handling the effects of N on net CO_2 uptake by agaves and cacti.

To be specific, let us consider what would happen if the available soil N were to increase. The key to any prediction is the value of the nutrient coefficient n_N in Equation 7.3. For seedlings of the four species grown hydroponically, n_N averages 0.20 (Fig. 6.2). In the field, n_N is 0.24 for *A. deserti* (Fig. 6.3A) and 0.25 for *Opuntia phaeacantha* (Table 6.3), so we will here assume a value of 0.23 for n_N. Based on Equation 7.3, increasing soil N by 50% increases growth by 29% over that in native soils, and doubling soil N increases growth by 51%. However, above fairly high levels of N (for instance, 0.1% N by dry weight in the soil), additional N can decrease growth (Figs. 6.2 and 6.3), a response not considered by Equation 7.3.

Effects of nutrients like K and P could be incorporated into an overall nutrient index by relations such as Equation 7.3, because again increases in growth for agaves and cacti tend to be logarithmic with nutrient level over the growth-limiting levels of these nutrients. Based on responses of seedlings to K and P (Fig. 6.5), and responses of adult plants to P (Table 6.3), the nutrient coefficients n_K and n_P for such nutrient responses are less than n_N (Eq. 7.3) expressed on an equivalent basis, meaning that changes in K or P have less effect on growth than do similar changes in N. Over the ranges expected for soils, other elements such as Ca (Fig. 6.6) or Cu and Zn (Fig. 6.8) generally have little effect on growth of agaves and cacti; B can again have a positive effect (Fig. 6.9) and Na can have a negative effect (Figs. 6.11 and 6.12B), which would result in a negative n_{Na}. A comprehensive, analytic nutrient index must incorporate the effects of a multitude of elements – a worthwhile goal that requires further research.

CARBON DIOXIDE

Changes in the atmospheric CO_2 level will affect the rate of CO_2 diffusion into the shoots of agaves and cacti and so should affect productivity. The CO_2 level in the atmosphere is steadily rising, primarily reflecting the continuing and increasing combustion of fossil fuels, although deforestation in the tropics and CO_2 absorption into oceanic waters also influence global CO_2 levels (Gates, Strain, and Weber, 1983; Lemon, 1983). CO_2 is presently increasing at about 2 ppm by volume per year; that is, $N_{CO_2}^{ta}$ is increasing at about 0.000002 per year from a mean level in the mid-1980s of about 0.000350 (350 ppm). Based on Equation 2.8 $[J_{CO_2} = g_{CO_2}^{gas}(N_{CO_2}^{ta} - N_{CO_2}^{ias})]$, an increasing CO_2 level in the atmosphere would be expected to increase net CO_2 uptake, but CO_2 levels can also affect the stomatal opening and hence the CO_2 conductance. For instance, an increased CO_2 level in the intracellular air spaces can lead to immediate partial stomatal closure for agaves and cacti (Table 3.3).

Growth of *Agave vilmoriniana* for six months is 0% to 300% greater at 700 ppm than at 350 ppm CO_2, depending on watering protocol and initial plant size (Idso et al., 1986). Under well-watered conditions, net CO_2 uptake over 24-h periods by *A. vilmoriniana* is similar under the two atmospheric CO_2 levels, suggesting that drought and elevated CO_2 effects may interact (Szarek, Holthe, and Ting, 1987), possibly because of the relatively large daytime net CO_2 uptake by this species (Fig. 4.6). Studies of net CO_2 uptake over 24-h periods for well-watered *Agave deserti* and *Ferocactus acanthodes* show that when the ambient CO_2 level is suddenly raised from 350 to 650 ppm, net CO_2 uptake is greatly increased in the late afternoon and early evening (Nobel and Hartsock, 1986e). However, net CO_2 uptake toward the end of the night is lower for plants at 650 ppm CO_2 than at 350 ppm, presumably because the glucan leading to the three-carbon acceptor for CO_2 (Fig. 2.13) becomes depleted earlier in the night in response to the higher external CO_2 level. As a result, net CO_2 uptake over a 24-h period is only about 2% higher immediately upon shifting the plants from 350 to 650 ppm CO_2 (Nobel and Hartsock, 1986e).

In contrast to the small immediate effect of increased CO_2 level on net CO_2 uptake over a 24-h period, major increases are observed in a few days for *A. deserti* and *F. acanthodes* maintained at 650 ppm compared with at 350

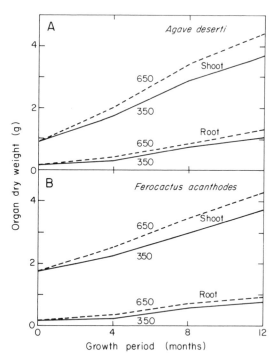

Figure 7.18. Influence of atmospheric CO_2 concentration on the growth of *Agave deserti* (**A**) and *Ferocactus acanthodes* (**B**) over a one-year period, as measured by changes in organ dry weight. Dry weights are averaged for six well-watered plants harvested at four-month intervals for *A. deserti* (initially one year old) and *F. acanthodes* (initially one and a half years old) that were growing under the indicated CO_2 level in ppm (1 ppm = 1 μL L^{-1} = 1 Pa MPa^{-1}). Modified from Nobel and Hartsock (1986e).

ppm CO_2 (Nobel and Hartsock, 1986e). After ten days, net CO_2 uptake over 24-h periods increases about 30% for both species at the higher external CO_2 level. What happens over an even longer period? Raising the ambient CO_2 level from 350 ppm to 650 ppm, an increase which many predict will occur in the earth's atmosphere before the year 2100 (Gates et al., 1983; Lemon, 1983), increases the growth rate of *A. deserti* by 28% and that of *F. acanthodes* by 30% (Fig. 7.18). At an intermediate external CO_2 level of 500 ppm, the enhancement of growth by elevated CO_2 over the one-year period is about half as great (Nobel and Hartsock, 1986e). Thus, the daily net CO_2 uptake and productivity of agaves and cacti may increase about 1% for each 10 ppm rise in atmospheric CO_2 up to about 650 ppm.

This fractional increase can be used to adjust the net CO_2 uptake over 24-h periods measured for agaves and cacti under optimal conditions but at ambient CO_2 levels of 340–350 ppm (Table 7.4) to the CO_2 uptake expected at the atmospheric levels of CO_2 that will occur in the future.

PHYSICAL FACTOR DEPENDENCE OF
COMPONENT INDICES

In devising an environmental productivity index, emphasis is on a simple relationship that closely predicts the primary effects of the climatic factors. An important assumption is that each component index depends on only one physical factor. The product of component indices constituting EPI (Eq. 7.2) thus recognizes that the various physical factors can simultaneously be limiting net CO_2 uptake to different extents, but each component index is obviously oversimplified. The real questions concern the magnitude of the dependence of a particular component index on multiple physical factors and whether such interdependencies significantly change the predictions based on EPI. Because we have identified three physical factors in EPI, only three different pairs of interdependencies of physical factors can occur: water–temperature, water–PAR, and temperature–PAR. We will evaluate each of these possible effects in turn.

We will begin by asking does the temperature dependence of net CO_2 uptake by agaves and cacti over 24-h periods depend on the water status of the plants? Let us first consider the temperature response of net CO_2 uptake for desert perennials other than agaves and cacti during drought. As water stress increases, the optimal temperature for net CO_2 uptake by the desert brittle bush *Encelia farinosa* and the desert fern *Notholaena parryi* shifts downward by 7–10°C (Nobel, Longstreth, and Hartsock, 1978). This downward shift reflects stomatal closure induced by higher temperatures, just as for agaves and cacti (Fig. 4.1), which becomes more important during drought, and the favoring of respiration over net CO_2 uptake as temperature increases, again just as for agaves and cacti (Nobel, 1984a). The limited studies on agaves and cacti also suggest that the optimal temperature for net CO_2 uptake decreases as

drought increases. For instance, for *Agave deserti* grown at day/night temperatures of 25°C/15°C, the optimal nocturnal temperature shifts from 15°C under well-watered conditions to 11°C after 20 days of drought (P. S. Nobel, unpublished observations). This shift decreases the temperature index at 25°C/15°C by 7%, but the overriding effect is the direct effect of drought on the water index, which in this case goes from 1.00 to 0.15 at the optimal temperature for net CO_2 uptake.

We might next ask what happens to the water index during drought as the day/night temperatures are changed? Actually, higher ambient temperatures reduce stomatal opening for many agaves and cacti (see Fig. 4.1). Even though the driving force on water loss is greater at higher temperatures (Eq. 2.6), the actual water loss is not affected much by temperature, as we discussed in Chapter 4. Therefore, the decrease in the water index during drought is little affected by ambient temperature. We conclude that the water–temperature interaction is rather minor with respect to net CO_2 uptake by *A. deserti* and other desert succulents.

The influence of the water index on EPI also dominates the water–PAR interaction. As the water index decreases, indicating limitation on net CO_2 uptake because of the water status, the total daily PAR required for 90% saturation of net CO_2 uptake decreases. For *A. deserti*, 90% saturation under well-watered conditions occurs at about 20 mol PAR m^{-2} day^{-1}, which decreases to 16 mol m^{-2} day^{-1} at a water index of 0.30 (P. S. Nobel, unpublished observations).

The influence of temperature on the PAR response has been studied in the most detail (Fig. 7.19). Nocturnal acid accumulation for *Opuntia ficus-indica* is close to maximal at a minimum nighttime air temperature of 9°C, at which temperature 90% PAR saturation occurs for a total daily PAR of 23 mol m^{-2} day^{-1} (Fig. 7.19A). At minimum nighttime temperatures of 13°C and 17°C, nocturnal acid accumulation is lower and 90% of PAR saturation occurs at 20 and 18 mol m^{-2} day^{-1}, respectively (Fig. 7.19A). For *Stenocereus gummosus*, maximal nocturnal acid accumulation occurs at a minimum nighttime temperature near 14°C, at which

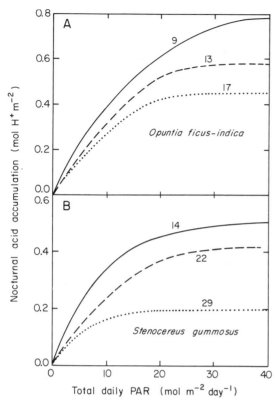

Figure 7.19. Effect of total daily PAR on nocturnal acid accumulation at different minimum nighttime temperatures (indicated next to the curves in °C) for (**A**) *Opuntia ficus-indica* and (**B**) *Stenocereus gummosus*. Data were obtained under wet conditions and are adapted from Nobel (1980a) and Acevedo et al. (1983).

temperature 90% of PAR saturation is attained at a total daily PAR of 21 mol m^{-2} day^{-1} (Fig. 7.19B). At higher minimum nighttime temperatures of 22°C and 29°C, the nocturnal acid accumulation at PAR saturation decreases and the total daily PAR for 90% saturation is lower, 19 and 14 mol m^{-2} day^{-1}, respectively (Fig. 7.19B). Thus, as the nighttime temperature deviates from the optimal, less CO_2 uptake occurs and the PAR required for 90% saturation decreases, just as for other plants. Because a change of about 0.5 in the temperature index, which is about as large as normally occurs during the course of a year under field conditions (Figs. 7.5B, 7.8B, 7.12B, 7.14B, and 7.16B), causes only about a 10% change in the PAR index at a total daily PAR of about 15 mol m^{-2} day^{-1} (Fig. 7.19), the temperature–

PAR interaction is relatively minor with respect to net CO_2 uptake over 24-h periods.

EPI, which is based on the product of the limitations of net CO_2 uptake by water status, temperature, and PAR (Eq. 7.2), represents a simple means of quantitatively predicting the major effects of environmental factors on net CO_2 uptake and, by extension, on productivity. It ignores potential dependencies of the component indices on multiple physical factors, which for agaves and cacti prove to be only secondary effects anyhow. Moreover, the small discrepancies between monthly EPI and monthly morphological changes cannot be readily explained by interdependencies in the component indices. We thus conclude that, although the effect of interdependencies can be included, EPI as originally proposed (Nobel, 1984a) is useful for predicting the primary effects of environmental factors on net CO_2 uptake and productivity of agaves and cacti and, in principle, can be extended to other plant groups.

Ecological applications of EPI

Now that we have introduced the concept of EPI and examined its limitations, we will illustrate its use in ecology. We will first consider the productivity of an agave at various elevations. Plant productivity has been difficult to predict as a function of elevation, as changes in elevation lead to simultaneous changes in more than one environmental variable. We will next consider ages and life spans of plants under natural conditions. Finally, we will illustrate how various aspects of competition can be quantitatively analyzed using the EPI approach.

ELEVATION

Although temperature and rainfall are known to vary substantially with elevation, rarely has productivity of a particular species been determined along an elevational transect. We have already indicated that EPI has been calculated over a two-year period for *Agave deserti* at Agave Hill (Fig. 7.6). Moreover, EPI has been successfully correlated with leaf unfolding and dry-weight productivity for this species (Nobel, 1984a). Thus, *A. deserti* was selected to determine the influence of eleva-

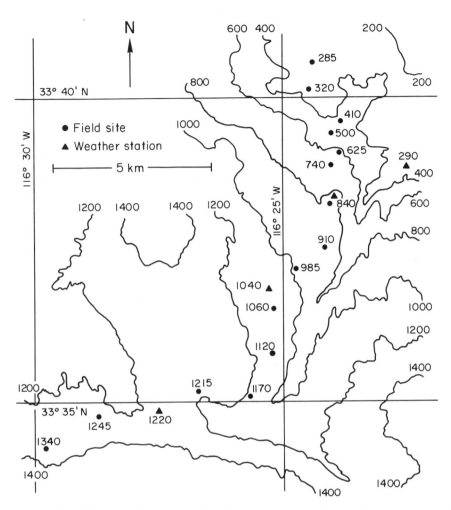

Figure 7.20. Field sites and weather stations in the northwestern Sonoran Desert near Agave Hill (850 m), where the productivity of *Agave deserti* was studied. Elevations in meters are indicated next to the field sites, weather stations, and contour lines. *Agave deserti* did not occur at the lowest and highest elevation sites (285 and 1,340 m, respectively). Used by permission from Nobel and Hartsock (1986c). © 1986 by Springer-Verlag/Publishers.

tion on productivity, as estimated using the EPI approach.

Agave deserti occurs over a considerable range in elevations – about 500 m above and 500 m below the Agave Hill field site at 840 m (Fig. 7.20). Because of the steepness of the terrain, the local 1,000-m elevational range occurs over a horizontal distance of only about 7 km (Fig. 7.20). Therefore, *A. deserti* may be genetically quite similar over the elevational transect.

The monthly values of air temperature and the amount of rainfall in each precipitation event can be estimated for each field site by linearly interpolating between the four weather stations. Based on the known responses of *A. deserti* to water status, temperature, and PAR, EPI can then be calculated for each field site. Instead of presenting monthly values of EPI, we will average EPI for the three-month wet periods occurring in the summer and in the winter. Such seasonal values of EPI will then be used to predict leaf unfolding (Fig. 7.21) based on the previously established correlation between EPI and leaf unfolding (Fig. 7.6A). The predicted leaf un-

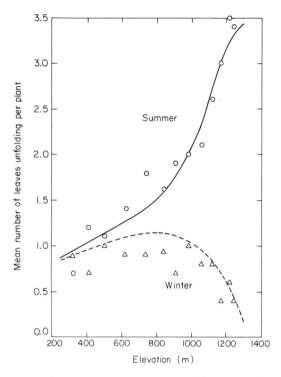

Figure 7.21. Relation between field site elevation and predicted (lines) or measured (symbols) number of leaves unfolding for *Agave deserti* during a summer period (July–September 1984) or a winter period (December 1984–February 1985). Predictions of the number of leaves unfolding were based on previously determined correlations between leaf unfolding and EPI (Nobel, 1984a), and measurements of leaf unfolding were means for ten plants at each field site (Fig. 7.20). Used by permission from Nobel and Hartsock (1986c). © 1986 by Springer-Verlag/ Publishers.

folding is similar to the measured leaf unfolding for the thirteen field sites involved (Fig. 7.21). Thus, variations in productivity with elevation in response to changes in the water availability, temperature, and PAR can presumably be correctly interpreted for *A. deserti* using the EPI approach.

Let us next consider the ecological implications of the variations in seasonal EPI and seasonal leaf unfolding for *A. deserti* (Fig. 7.21). Rainfall in the Sonoran Desert tends to be annually bimodal (see Fig. 3.3), so growth and net productivity of *A. deserti* tend to occur in the summer and in the winter, as is the case for the year considered. Growth in the summer increases monotonically with elevation (Fig.

7.21). This increase in growth in part reflects the increase in rainfall with elevation – summer rainfall increases from 79 mm at 290 m to 281 mm at 1,220 m (Nobel and Hartsock, 1986c). The increase in growth with elevation also reflects a decrease in temperature – the daily minimum temperature in the summer averages 26°C at 290 m and 16°C at 1,220 m. On the other hand, growth and net productivity are maximal at intermediate elevations in the winter (Fig. 7.21). The increase in rainfall with elevation, which raises the water index, is offset by the decrease in temperature with elevation, which causes a substantial decrease in the temperature index in the winter for *A. deserti*. For the one-year period considered, the highest annual productivity occurs at the upper elevational limit. Thus, the upper elevational distributional boundary for *A. deserti* is apparently not a result of reduced potential productivity. Rather, the upper elevational limit apparently results from a reduction in PAR by competition from larger, overtopping perennials, as well as possible episodic freezes; the lower elevational limit of *A. deserti* appears to result from high-temperature limitations on seedling establishment (Nobel, 1984c; Nobel and Hartsock, 1986c). We also note that predicted productivity for the year considered is highest in the summer at all elevations (Fig. 7.21). This is not always the case, because the summer rainfall in the northwestern Sonoran Desert is quite variable from year to year (Fig. 3.3).

Most of the measurements of leaf unfolding during the wintertime over the elevational transect for *A. deserti* are somewhat below those predicted (Fig. 7.21). This discrepancy relates to using average temperatures for the prediction when dealing with the nonlinear dependency of net CO_2 uptake on temperature (Fig. 4.7). A typical minimum daily air temperature averaged over a month during the winter is 5°C, which leads to a temperature index of 0.64 for *A. deserti*. Such an average temperature could be caused by half a month with minimum nighttime temperatures of 0°C, which leads to a temperature index of 0.07, and half a month with minimum temperatures of 10°C, which leads to a temperature index of 0.92, yielding an average monthly temperature index of 0.50. The temperature index based on actual temperatures (0.50), which is what the plants would be responding to, is then 22% lower than that based on average temperatures (0.64). For some applications of the EPI approach, use of daily temperatures, as well as daily PAR, may be necessary.

PLANT AGES

Field studies of productivity of agaves and cacti in natural populations involve plants of many ages, unlike most studies on cultivated fields. Moreover, the ages of plants in natural populations are usually unknown. The ages of agaves and cacti are generally difficult to determine, because of the absence of annual growth rings. By calculating the relative growth expected each year, we can use EPI to help estimate ages of plants in natural populations.

Based on net CO_2 uptake during a particular one-year period, which turned out to have more wet days than the average year, a 90-cm tall *Ferocactus acanthodes* was estimated to be 54 years old (Nobel, 1977b). Refinements introduced by the equivalent of a water index, considering the growth expected for wet conditions of each year, lengthen this estimate by about 30 years. Also, subsequent studies have shown that an *F. acanthodes* 8 cm in diameter would be 7 or 8 years old at the study site involved (Nobel, 1980e; Jordan and Nobel, 1982), not 4 years old as originally proposed (Nobel, 1977b). Thus, a 90-cm tall *F. acanthodes* growing at Agave Hill is probably about 90 years old.

Ages of individuals of *Agave deserti* at Agave Hill have also been estimated, this time directly using the EPI approach. First we will introduce a new term that is commonly used in growth studies, the *relative growth rate* (RGR):

$$\text{RGR} = \frac{\ln(\text{DW}_2) - \ln(\text{DW}_1)}{t_2 - t_1} \qquad (7.4)$$

where DW_1 and DW_2 are the plant dry weights at times t_1 and t_2, respectively (Hunt, 1982). We can use EPI to calculate dry-weight changes, which in turn can be used to calculate RGR.

For *A. deserti*, RGR is about 0.8 month^{-1} for ten-day-old seedlings under wet conditions, decreasing to 0.3 month^{-1} for six-month-old seedlings (Jordan and Nobel, 1979; Raphael and Nobel, 1986). Decreases in RGR continue to occur as plant age increases, becoming 0.046 month^{-1} for seedlings and 0.071 month^{-1} for ramets at about five years of age under wet conditions; RGR averages 0.000 month^{-1} under drought (Raphael and Nobel, 1986). Using RGR, we estimate a ramet of 300 g dry weight to be 14 years old at Agave Hill. At this stage, the ramet becomes independent of the mother plant, because severing the rhizome between ramet and parent for *A. deserti* has no effect on the rate of ramet growth after about 300 g (Raphael and Nobel, 1986).

Like most agaves, *A. deserti* is monocarpic. Its inflorescence, whose production is accompanied by a dry-weight loss of 1.8 kg from the leaves, has a dry weight of 1.3 kg at maturity, which is about the same as the annual productivity of an entire large plant (Nobel, 1977a). The average dry weight of *A. deserti* at flowering is about 5.7 kg (Nobel, 1987a), and at least 95% of its rosettes with inflorescences at Agave Hill begin as ramets, not as seedlings (Nobel, 1977a). Using EPI to evaluate the year-to-year environmental conditions, we estimate the length of time required for an independent ramet (300 g dry weight) to grow to the size of a mature flowering plant to be 36 years at Agave Hill (Nobel, 1987a). When the 14 years required for the ramet to become independent of the parent is added to this, the age at reproduction based on EPI calculations is 50 years. On the other hand, an average of 1.82% of the rosettes of *A. deserti* at Agave Hill produce an inflorescence each year (Nobel, 1987a), so the average age at flowering is 1/(0.0182 per year), or 55 years. This good agreement between two entirely different methods of estimating age gives us further confidence in using an approach based on CO_2 exchange measured in the laboratory and environmental conditions in the field to estimate the ages of agaves and cacti.

COMPETITION – NURSE PLANTS

We have described specific ecological applications of the EPI approach with regard to elevational effects on productivity and estimates of plant age in the field. Interactions, such as plant–plant competition, may provide

Table 7.5. *Proposed influence of physical factors on the growth or productivity of agaves and cacti in potentially competitive situations*

Species	Physical factor invoked to explain variations	Reference
Agave salmiana	PAR, water	Yeaton and Manzanares (1986) (**A**)
Carnegiea gigantea	Water	Yeaton, Travis, and Gilinsky (1977) (**B**)
	Water	Vandermeer (1980)
	Water	McAuliffe (1984)
	Water	McAuliffe and Janzen (1986)
Echinocereus triglochidiatus	PAR, water	Lozano and Reid (1982)
Eulychnia iquiquensis	Water	Rundel and Mahu (1976)
Opuntia acanthocarpa	PAR, water	Yeaton and Cody (1976) (**C**)
O. fulgida	Water	**B**
O. imbricata	Water	Fraser and Peiper (1972)
O. littoralis	PAR, temperature	Yeaton et al. (1983) (**D**)
O. parryi	PAR, temperature	**D**
O. ramosissima	PAR, water	**C**
O. streptacantha	PAR, water	**A**
Stenocereus thurberi	PAR, temperature	Parker (1987)
Eleven species of cacti	PAR, water	Yeaton and Cody (1979)

Note: Studies cited more than once are assigned a boldface letter that is used the next time they are cited.

an even greater area for application of the EPI approach.

Physical factors of the environment have often been invoked to explain productivity variations for agaves and cacti in potentially competitive situations, although such speculations are rarely substantiated by measurements of net CO_2 uptake or changes in dry weight for the species involved (Table 7.5). When the ecological effect of slope is discussed with respect to productivity and competition, both the influence on PAR interception and soil water potential are generally mentioned. Physical factors are also invoked to explain patterns of plant distribution based on competition or microhabitat preference by certain species of agaves and cacti (Fowler, 1986). Similar root depths and hence intraspecific competition among individuals of *Opuntia ramosissima* may cause its nearest neighbors to be shrubs with different rooting depths (Cody, 1986c). The removal of grasses by heavy grazing can lead to an increase in the productivity of platyopuntias such as *Opuntia polyacantha* in rangelands (Houston, 1963)

and natural ecosystems (Janzen, 1986), although not in all cases (Bement, 1968). These situations could be analyzed using the EPI approach to quantify the effects of neighboring plants on soil water potential, air temperature, and available PAR.

Another case of competition between plants involves nurse plants and their associated seedlings. The shading by the nurse plant moderates the temperature extremes but also reduces the PAR available to the seedling, as we discussed in Chapters 4 and 5. Let us now return to the *Agave deserti* pictured in Figure 1.1 that is surrounded by a perennial bunchgrass, *Hilaria rigida*. Because we know the response of net CO_2 exchange by *A. deserti* to various environmental factors (Figs. 3.17A, 4.5A, 4.7, and 5.5A), the EPI approach can be used to predict the effects of the changes in microclimate provided by the nurse plant on the growth and productivity of the agave.

We will begin with the effects of the nurse plant on temperature. To be specific, we will portray measurements for a day near the spring equinox; the calculations need to be

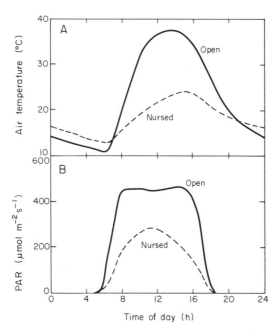

Figure 7.22. Influence of a nurse plant on the air temperature (**A**) and PAR in the planes of the leaves (**B**) at midheight on a seedling of *Agave deserti* compared with a plant in an open, unprotected site. Data were obtained on 16 March 1986 for the nurse plant and agave seedling pictured in Figure 1.1 or for a seedling of similar size in an exposed location (P. S. Nobel, unpublished observations).

done for various times of the year to calculate the influence of *H. rigida* on the annual productivity of *A. deserti*. The moderation of air temperature at midheight on the seedling (2 cm) increases the minimum nocturnal temperature by about 2°C (Fig. 7.22A), thereby raising the temperature index from 0.92 to 0.97 for the day involved. Such moderation of air temperature near the photosynthetic surfaces tends to raise the temperature index from approximately October through April. However, a quantitatively much more important effect with respect to productivity occurs for the PAR index. Shading reduces the total daily PAR for the *A. deserti* considered by approximately 50% (Fig. 7.22B), which, given the PAR levels involved and the nonlinear response of net CO_2 uptake to total daily PAR (Fig. 5.2), reduces the PAR index from 0.72 to 0.24, a threefold reduction. For the specific plants (Fig. 1.1) and day considered, the water index is unity. However, the effect of the nurse

plant on the water index of the agave is fairly complicated and depends on the relative sizes of the two plants; for the plant sizes considered and a year with an average amount of rainfall, water uptake by *H. rigida* reduces the water available to *A. deserti*, lowering its mean water index by approximately 40% (Franco and Nobel, 1988).

When the effects of the three environmental factors involved (Eq. 7.2) are combined under wet conditions, EPI for the day in question is 0.66 for an *A. deserti* in an exposed site and 0.23 under *H. rigida*. The nurse plant thus considerably reduces the net CO_2 uptake over the 24-h period considered, primarily by reducing the available PAR. Nurse plants can also affect the nutrient relations for the associated seedling. For instance, the soil nitrogen level in the root zone of *A. deserti* is about 0.029% by dry weight for an exposed site but 0.046% near *H. rigida* (Franco and Nobel, 1988), which would lead to a nitrogen index (Eq. 7.3) under the nurse plant compared with an exposed site of 1.57. Considering the water, temperature, PAR, and nitrogen indices, the annual growth of *A. deserti* would be predicted to be nearly threefold lower under *H. rigida* than at an exposed site. Such nurse plants clearly do not enhance the growth or productivity of the associated seedling. Rather, the advantage of nurse plants for *A. deserti* must be ascribed to an amelioration of the thermal environment by preventing high temperatures (Chapter 4) and thus allowing seedling establishment to occur (Nobel, 1984c).

Predicting agronomic productivities

Let us next use EPI to predict productivities over wide geographical regions, an endeavor that has both ecological and agronomic ramifications. Cultivated plants are generally grown so close together that considerable shading occurs. Indeed, the PAR index is often the limiting component of EPI for agaves and cacti in commercial plantations. We will therefore outline a ray-tracing method that facilitates the calculation of PAR interception by complex but repeating entities, such as plants along a row. We will conclude with some comments on the future agronomic potential of agaves and cacti.

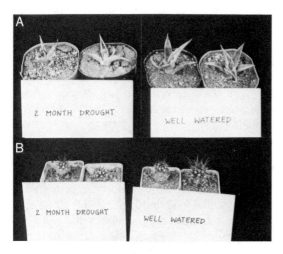

Figure 7.23. Influence of watering protocol on shoot volume of approximately one-year-old seedlings of (**A**) *Agave deserti* and (**B**) *Ferocactus acanthodes*. Experimental conditions are described in Nobel and Hartsock (1986d).

GEOGRAPHICAL VARIATION

An objective of modern experimental plant science is to gain sufficient understanding of the responses of plants to enable meaningful predictions. This is one of the key reasons for introducing the EPI approach. Using EPI we can predict the relative productivity of the widely cultivated *Opuntia ficus-indica* in various regions. For comparison we will calculate the annual EPI for *Agave deserti* and *Ferocactus acanthodes* over a wide region in the southwestern United States. We will also predict EPI for *Agave lechuguilla*, which is harvested for its leaf fiber in many states of Mexico, over a wide region extending beyond the boundaries of the Chihuahuan Desert.

Three succulents in the southwestern United States

Most of the southwestern United States is arid or semiarid, a region well suited to agaves and cacti, as exemplified by the many species that occur natively or are cultivated, especially horticulturally. Yet we should ask how close do agaves and cacti come to attaining their maximum productivity in this region. More specifically, what is their EPI at various locations in the southwestern United States? We will discuss the three species that have been considered in detail throughout this book

– *A. deserti, F. acanthodes*, and *O. ficus-indica* – because we already know the effect of environmental factors on EPI for these species. Moreover, all three species occur natively or are cultivated in the region.

To check the influence of water status, temperature, and PAR on growth, all three species were maintained in environmental chambers with day/night temperatures and PAR levels changed monthly to the average values for Agave Hill (Nobel and Hartsock, 1986d). Various watering protocols were used to vary the water index monthly from that of a dry year to a normal year to a continuously wet year, which can have a substantial effect on plant size (Fig. 7.23). Shoot volume was measured nondestructively by carefully inverting the plants and inserting them up to the soil level in water so that the volume of displaced water could be determined, an application of Archimedes' principle of buoyancy. Shoot volume was converted to dry weight using factors obtained by harvesting and drying plants that had been treated similarly to those whose volume was monitored monthly. Such dry-weight estimates were in agreement with predictions based on EPI for all three species (Nobel and Hartsock, 1986d). Therefore, EPI should also lead to accurate prediction of growth at various locations in the field for these species.

Let us now consider the EPI predictions for these three species over an area exceeding 600,000 km^2 in the southwestern United States (Fig. 7.24). Annual EPI averages 0.27 for *A. deserti*, 0.09 for *F. acanthodes*, and 0.26 for *O. ficus-indica* for the twenty-one specific sites. The lower EPI for the barrel cactus results mainly from a lower PAR index because of its spines and the convolutions of its stem by ribbing, both of which reduce the PAR per unit stem surface area (Nobel and Hartsock, 1986d). Low rainfall causes the water index to be quite low at the three sites in the Colorado River valley along the Arizona–California border (Fig. 7.24). The low elevation of these sites and the site in southeastern California results in hot summers, leading to a lowered temperature index. As a consequence, EPI averages only 0.04 for the three species at these four sites. For the four sites in the northcentral part of the region considered, the rainfall tends to

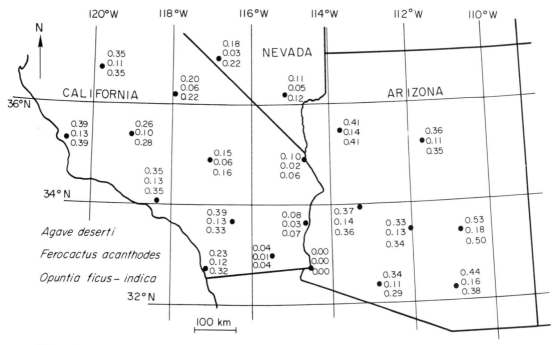

Figure 7.24. Annual average values of the monthly EPI for *Agave deserti* (upper number at each site), *Ferocactus acanthodes* (middle number), and *Opuntia ficus-indica* (lower number). Extensive weather records were used to determine the monthly precipitation, temperatures, and PAR at each of the twenty-one sites indicated. Based on the known responses of net CO_2 uptake to the three environmental factors, EPI was then calculated for each month for each species. Used by permission from Nobel and Hartsock (1986d). © 1986 by Blackwell Scientific Publications Ltd.

occur in the winter when nighttime temperatures are low and the sun has a low trajectory in the sky; thus, for the part of the year with favorable water indices, the temperature and PAR indices are low, leading to an annual average EPI of 0.13. The remaining seven sites in Arizona and six sites in California all have higher EPIs, averaging 0.28 for the three species (Fig. 7.24).

We conclude that annual productivity of *A. deserti* and *F. acanthodes* is severely limited by water status, temperature, or PAR at some time during the year at each of the twenty-one sites considered in the southwestern United States. Indeed, EPI rarely approaches 1.00 for any of the months. Just as we concluded when discussing the relationship between productivity and elevation for *A. deserti* (see Fig. 7.21), species are not always native in regions where their annual EPI is high. For instance, neither *A. deserti* nor *F. acanthodes* occurs natively at the three sites in coastal southwestern California, although

their annual EPIs are considerably above average there (Fig. 7.24). Factors such as seedling establishment and competition with other species for PAR can prove to be far more important than productivity in monospecific stands in determining distributional boundaries, although a very low annual EPI would be a severe handicap.

Even though EPI cannot be used to predict the distributions of native desert succulents, it does indicate the productivity expected in various locations, which can be quite important for a cultivated species such as *O. ficus-indica*. For instance, productivity of this platyopuntia is expected to be substantial in southcentral Arizona and southwestern California (Fig. 7.24). However, we must be careful in extrapolating from calculated EPI to realized productivity, such as when deciding whether to plant expensive specimens out of doors or investing money in commercial plantations. For instance, although annual EPI is relatively high for *O. ficus-indica* in northern

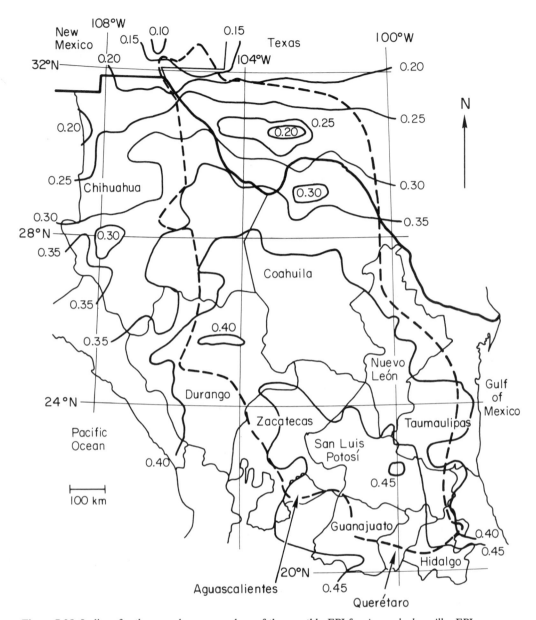

Figure 7.25. Isolines for the annual average values of the monthly EPI for *Agave lechuguilla*. EPI was calculated at 0.5° intervals of latitude and longitude, and the indicated isolines were interpolated between these locations for all the states in which *A. lechuguilla* occurs naturally in Mexico and the United States. Rainfall and temperature data were interpolated from 300 weather stations having at least ten years of records, and PAR was obtained by simulation. The approximate range of *A. lechuguilla* is indicated by the dashed line. Used by permission from Quero and Nobel (1987). © 1987 by the British Ecological Society.

Arizona (Fig. 7.24), plants would most likely succumb to the low wintertime temperatures there. Breeding of platyopuntias, especially spineless forms, for tolerance of low temperature has long been recognized as an important objective for increasing their agronomic potential (Griffiths, 1915b).

Agave lechuguilla – *Chihuahuan Desert*

Agave lechuguilla is harvested from the wild over most of its geographical distribution, which extends from Texas and New Mexico about 1,300 km southward to the states of Guanajuato, Hidalgo, and Querétaro (Fig. 7.25).

Annual EPI averages 0.14 or less in the north-western part of its distribution compared with over 0.48 in the southern part. Even though the southern part has slightly more advantageous temperatures and PAR, the chief cause of the variations in annual EPI is annual precipitation, which increases from less than 200 mm in the north to about 800 mm in the south. The importance of water for the productivity of desert vegetation has long been recognized (Cannon, 1916; Evenari, Shanan, and Tadmor, 1971; Noy-Meir, 1973), but use of EPI allows a quantitative presentation of the individual effects of water, temperature, and PAR.

Even where environmental conditions are most favorable for net CO_2 uptake by *A. lechuguilla*, EPI is less than 0.5 (Fig. 7.25). Thus, productivity is never greater than half of that expected under wet conditions, optimal temperatures, and saturating PAR. Also, except to the north, EPI does not decrease as the distributional boundary of *A. lechuguilla* is crossed (Fig. 7.25). Again, seedling establishment, survival of episodic freezes or hot spells, competition with other plants, or interactions with animals are apparently more important for determining distributional boundaries than is the potential productivity. Nevertheless, annual EPI can be used to assess the productivity of *A. lechuguilla* in regions where it is currently harvested, and examining the geographical variations in EPI may suggest regions where it could be more successfully exploited.

RAY-TRACING TECHNIQUES – PLANT SPACING

Productivity is of primary agronomic concern and depends among other factors on plant spacing and canopy architecture, both of which affect PAR interception. Models predicting direct and diffuse PAR can therefore be very important to the future agronomic use of agaves and cacti, suggesting optimal row spacing, optimal row orientation, and optimal plant design. Using models, we can determine what leaf area index (LAI) or *stem area index* (SAI, total stem surface area per unit ground area), a more appropriate index for cacti, leads to the highest productivity. Also, models can determine the productivity at different positions in a canopy, such as for different leaves

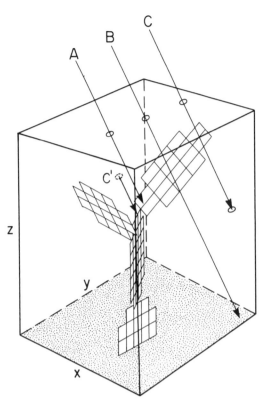

Figure 7.26. Three-dimensional "cell" containing a platyopuntia with four cladodes. Cladode surfaces are divided into squares, only a few of which are illustrated here. Direct PAR is indicated by rays that are incident on a cladode (ray A), incident on the ground (ray B), or entering and then exiting the cell without being intercepted (ray C). Ray C can enter an adjacent identical cell, which is geometrically equivalent to reentering at the opposite side of the same cell (ray C'); ray C' is intercepted by the lower portion of the plant. Used by permission from García de Cortázar et al. (1985). © 1985 by Elsevier Science Publishers B.V.

of agaves or different cladodes of platyopuntias. We will begin by considering a method for tracing the sun's rays and thus analyzing where PAR is intercepted on an agave or a cactus. The method described is especially suited to structures repeating at given intervals in space, and therefore is readily adapted to plants under cultivation.

Model

The model involves dividing each photosynthetic surface into squares (Fig. 7.26), the number of which depends on the precision required. The number of squares used has var-

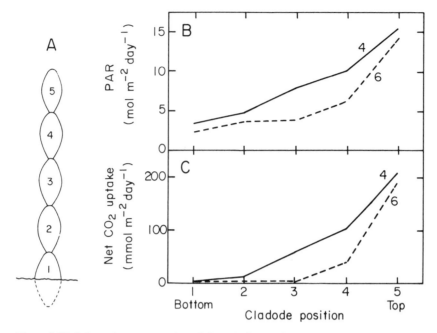

Figure 7.27. Schematic representation of *Opuntia ficus-indica* (**A**) used to calculate the total daily PAR (**B**) and net CO_2 uptake per unit cladode area (**C**) at various positions in the canopy for the indicated stem area indices of 4 or 6. The ray-tracing technique was used to calculate the incident PAR for clear days on an equinox at 33° latitude. Modified from García de Cortázar and Nobel (1986).

ied from about 300 for the two sides of a cladode (García de Cortázar et al., 1985) to over 5,000 squares for an agave leaf (Nobel and García de Cortázar, 1987). The direct solar radiation is divided into 5,000 to 200,000 parallel rays that enter the top of a "cell" containing the plant of interest (Fig. 7.26). The computer program then decides which squares are directly struck by rays and at what angle, and which squares are shaded by overlying squares. By adding up the direct PAR and the diffuse PAR, the latter being calculated using the actual exposure of a particular square to the sky and to the ground, at a series of times during the day, we can determine the total daily PAR for each square. Such a total daily PAR can be converted into a PAR index (using Fig. 5.2), which indicates the relative net CO_2 uptake by that square, and then integrated over an entire leaf or stem.

One of the most important features of the model is how it processes a ray that is not intercepted by any photosynthetic surface. Such a ray (ray C in Fig. 7.26) leaves the cell surrounding the modeled plant at a specific angle that could cause it to enter the cell of a neigh-

boring plant. When the cell of this neighboring identical plant is adjacent to the original plant, as would occur when the plants are closely spaced, the ray enters this new cell at the same height above the ground and at the same angle as it exited from the original cell. In terms of PAR interception, this is equivalent to reentering the original cell at the opposite side (ray C′ in Fig. 7.26). This continuing ray then has another chance of striking a photosynthetic surface of the plant of interest.

Influence of canopy position

Let us use the model to examine PAR distribution and net CO_2 uptake down through an idealized canopy of *Opuntia ficus-indica* at two different spacings. For simplicity, each plant is represented as a basal cladode, half of which is buried in the ground, with three intermediate-sized cladodes extending vertically from it, topped by a slightly smaller terminal cladode (Fig. 7.27A). Total daily PAR is highest for the terminal cladode and progressively decreases down through the canopy (Fig. 7.27B). Also, the total daily PAR for each clad-

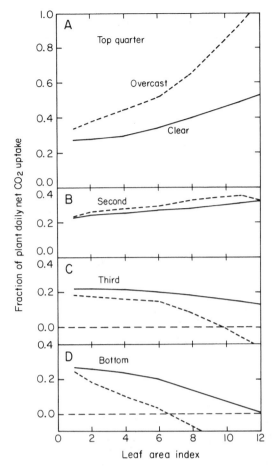

Figure 7.28. Partitioning of net CO_2 uptake by the whole plant into the four parts of the canopy for large agaves with 160 leaves at plant spacings leading to various leaf area indices (total area of both sides of all leaves per unit ground area). The ray-tracing technique was used to determine the intercepted PAR by the forty leaves pointing into each canopy quarter (the "top quarter," panel **A**, thus contains the youngest forty leaves) on an equinox at 21° latitude, the mean latitude of the field sites for *Agave fourcroydes* and *A. tequilana* (the species used to define the morphology of the modeled plant). Direct PAR and diffuse PAR were determined for over 2,000 subsurfaces on each leaf at approximately 0.8-h intervals and then summed to obtain the total daily PAR for each subsurface; from the total daily PAR, the relative net CO_2 uptake by each subsurface was calculated using Figure 5.2 and then summed to obtain the CO_2 uptake by each whole leaf. The indicated atmospheric conditions were either a clear sky or a uniformly overcast sky, for which the total daily PAR was one-fourth of the clear sky condition. Modified from Nobel and Garciá de Cortázar (1987).

ode becomes less when the plants are placed closer together so that the stem area index is raised from 4 to 6. Net CO_2 uptake per unit cladode area decreases even more rapidly down through the canopy than does PAR (Fig. 7.27C), a consequence of the nonlinear relationship between net CO_2 uptake and total daily PAR (Fig. 5.2). The lower cladodes contribute very little to the net CO_2 uptake of the plant as a whole, especially at the closer spacing.

Similar conclusions about changes in net CO_2 uptake with canopy position are reached for large agaves with 160 leaves, computations that are realistically possible only recently as computers became more powerful and could handle the approximately 100 billion (10^{11}) calculations involved under each condition (if each calculation were made using a hand calculator in 10 s, then it would take all the scientists in the world over one year to make the computation for one value of LAI). The morphology of these modeled plants, whose average leaf length is 130 cm, is based on ten-year-old plants of *Agave fourcroydes*, for which maximal harvests of leaves occur, and on seven-year-old plants of *A. tequilana*, an age at which harvesting of the stem typically begins (Nobel and García de Cortázar, 1987).

At a low LAI, each agave leaf receives about the same total daily PAR, so the fraction of total plant net CO_2 uptake is similar for the forty leaves in each of the four canopy positions considered (Fig. 7.28). When PAR becomes more limiting, such as for overcast days or higher LAIs, the fractional contribution of lower leaves to the net CO_2 uptake by the plant decreases. At a leaf area index of 11 on clear days and at a lower LAI of 6 on overcast days, half of the plant's net CO_2 uptake is contributed by the upper one-quarter of the leaves (Fig. 7.28A). On overcast days, spacings that lead to high LAIs result in the lower leaves becoming net consumers of carbon. When the mixture of both clear and overcast days is considered for an entire year, the lowermost (oldest) quarter of the leaves contribute less than 10% to the net CO_2 uptake by the entire plant at LAIs exceeding 7. Thus, the harvest of the older leaves for the production of fiber by *A. fourcroydes* in Yucatán, Mexico (Figs. 1.7 and 7.1A), and by *A. sisalana* in Kenya (Figs. 1.6

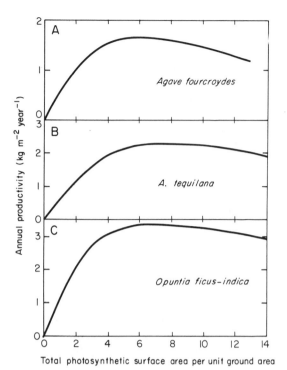

Figure 7.29. Annual dry-weight productivity per unit ground area predicted for shoots of *Agave fourcroydes* (**A**), *A. tequilana* (**B**), and *Opuntia ficus-indica* (**C**) at various leaf area indices or stem area indices. LAI (**A, B**) and SAI (**C**) were varied by changing the spacing of field plants. PAR was calculated based upon field measurements of PAR distribution (**A**) or using the ray-tracing technique (**B, C**). Modified from García de Cortázar et al. (1985), Nobel (1985a), Nobel and García de Cortázar (1987), and Nobel and Valenzuela (1987), which describe the water and temperature conditions in the field for each species.

and 7.1B) does not greatly reduce the overall productivity of the plants.

Influence of leaf and stem area indices

We have already considered the actual productivity of various species (Table 7.1), usually under favorable conditions; we now turn to the influence of spacing on productivity, an influence that has generally been determined empirically. Specifically, we will consider annual productivity as a function of LAI and SAI, a consideration vital to the economic success of agaves and cacti grown for food, fodder, and fiber. The highest annual productivity occurs for an LAI or SAI of 6–8

for *Agave fourcroydes* (Fig. 7.29A), *A. tequilana* (Fig. 7.29B), and *Opuntia ficus-indica* (Fig. 7.29C). We again see that agaves and cacti can be very similar in their environmental responses, in this case the dependency of annual productivity on the total photosynthetic surface area per unit ground area.

Even though the architecture of agaves and of cacti is quite different, their responses to total daily PAR are similar (Figs. 5.2 and 5.3). Moreover, the predicted annual productivities can also be similar. Specifically, the maximal annual productivity of *A. fourcroydes* is just under 2 kg m^{-2} year^{-1}, that of *A. tequilana* is just over 2 kg m^{-2} year^{-1}, and that of irrigated *O. ficus-indica* is just over 3 kg m^{-2} year^{-1} (Fig. 7.29). The difference between the latter two species can be explained by differences in the water index, which is 0.52 for *A. tequilana* and 0.84 for *O. ficus-indica* (García de Cortázar et al., 1985; Nobel and Valenzuela, 1987). If rainfall and irrigation cause the water index to be unity for these two species, which is certainly within the realm of possibility, then under optimal spacing the maximal annual productivity would be 4.4 kg m^{-2} year^{-1} for *A. tequilana* and 4.0 kg m^{-2} year^{-1} for *O. ficus-indica*. In practice, such potential productivities are only exceeded by those of sugar cane (Table 7.1).

Because the pertinent parameters, LAI and SAI, are rarely measured agronomically, we will next convert such measures of photosynthetic surface area per unit ground area into another framework. In particular, an LAI of 7 for large agaves corresponds to just over 3,000 plants ha^{-1} (0.3 plants m^{-2}). For *O. ficus-indica* of the size modeled, the number of plants per hectare at optimal productivity is about 20,000; for the larger plants that are commercially harvested, the same number of cladodes per unit ground area corresponds to about 6,000 plants ha^{-1}.

How do predicted optimal spacings compare with those developed by farming experience? In the commercial plantations of Mexico, *A. fourcroydes* is usually planted at 2,500–3,000 plants ha^{-1} (1,000–1,200 plants acre^{-1}), although as many as 4,000 plants ha^{-1} can be found. *Agave sisalana*, which is somewhat smaller than *A. fourcroydes* at maturity, is usually planted at 3,500–5,000 plants ha^{-1} in

East Africa (Weindling, 1947) and 6,000 plants ha^{-1} can occur (Kenya Sisal Board, 1984). For *A. sisalana*, closer spacing requires a longer growth period before the leaves can initially be profitably harvested. Thus, as the number of plants increases up to 10,000 plants ha^{-1}, eventual productivity is high, but the average productivities are greatest for 4,000–5,000 plants ha^{-1}, 6,000 plants ha^{-1} being favorable only under the best soil conditions (Lock, 1962, 1985). For *O. ficus-indica*, the number of plants in commercial plantations is often 7,000 ha^{-1}. At such close spacings, nearly all of the incident PAR is absorbed and so row orientation loses its importance. In particular, at an SAI of 6, variations in row and cladode orientation change annual productivity only 9% for *O. ficus-indica*, the best productivity occurring for rows oriented northeast-southwest (or equivalently northwest-southeast; García de Cortázar and Nobel, 1986). Of particular interest is the spacing developed by the Mayas over millenia for *A. fourcroydes* in Yucatán of nearly 4,000 plants ha^{-1}, which leads to a productivity within 2% of the maximum predicted by modern computer models (Nobel, 1985a).

FUTURE PROMISE

We have gained much knowledge since the early research on agaves and cacti described in Chapter 1. Many questions perceived a century ago can now be answered. We can also presume that the horticultural and agronomic interest in these plants will continue the ascension conspicuous since the fifteenth century.

Based particularly on the integration of information presented in this chapter, we realize that agaves and cacti are capable of high productivities. Moreover, because of CAM, such high productivities can be achieved without the high rainfall required by most other plants, an important consideration in the continuing development of arid and semiarid regions (Hoffmann, 1983). Underutilization of certain desert areas for agriculture has long been recognized (Duisberg, 1952), and expanding the cultivation of platyopuntias for cattle and sheep has been highly recommended for North Africa (Le Houerou, 1970). Where some return from the land is needed

each year, intercropping of species like *Opuntia ficus-indica*, which requires about six years to achieve high productivity, with annual crops like beans or corn can be a viable proposition. Also, intercropping with legumes can increase the soil nitrogen, which can enhance the productivity of *Agave sisalana* (Hopkinson, 1969), a species that has even been considered for cleared transects through forests (Banerjee, 1972). With regard to the commercialization of new crops, we should consider the many uses for agaves and cacti devised over the centuries by the Indians of Mexico, the United States, and other areas (Felger, 1979; Felger and Moser, 1985; Ebeling, 1986). The efforts of the Mexican government to promote the use of desert succulents must be commended (Fig. 7.30; Hernández Xolocotzi, 1970). Cultivation of various agaves and cacti is also being considered for the United States (McDaniel, 1985; Russell, 1985).

We are entering a phase favoring breeding and selection of agaves and cacti with particular traits, such as better fibers (Kirby, 1963; Wienk, 1970; Boulanger, 1985). Because of their relatively late sexual maturity, agaves and cacti have not been subjected to much experimental breeding. However, an agave hybrid has been developed in Tanzania that has a much higher fiber productivity than *A. sisalana* and lacks marginal spines on the leaves (No. 11648; Lock, 1985). Tissue culture has also been exploited to produce many genetically identical agave plants and in the future should lead to lines selected for agronomically favorable traits (Cruz-Ramos et al., 1985; Robert and Garcia, 1985). For instance, using stem and rhizome tissue, 4,000 plants can be propagated from a single plant of *Agave fourcroydes* (Robert et al., 1987). Cacti have been successfully propagated from the axillary buds that develop into areoles (Mauseth, 1979). Tissue culture has been successfully used to propagate *Ferocactus acanthodes* (Ault and Blackmon, 1987), *Leuchtenbergia principis* (Starling, 1985), *Mammillaria elongata* (Johnson and Emino, 1979), and *M. woodsii* (Kolar, Bartek, and Vyskot, 1976). For *Opuntia amyclaea*, such micropropagation techniques lead to 25,000 rooted plantlets from a single cladode in 100 days (Escobar A., Villalobos A., and Villegas M., 1986).

Figure 7.30. Agaves and cacti that are bred or selected for high productivity. (**A**) *Agave atrovirens* cultivated near Mexico City. (**B**) Various platyopuntias growing near San Luis Potosí, San Luis Potosí, Mexico.

Various practices can improve the utilization of mature plants. For instance, the level of nonstructural carbohydrates such as sugars varies seasonally, being lowest during drought in late summer for *Opuntia basilaris* in California (Sutton, Ting, and Sutton, 1981) and for *O. lindheimeri* in Texas (Potter, Petersen, and Ueckert, 1986); this should be taken into consideration when planning harvest times of desert succulents used for fermented beverages, food, and cattle fodder. The singeing of spines off both native and cultivated platyopuntias can enhance their exploitation as fodder, which can be especially crucial for maintaining animals through drought periods (Shoop et al., 1977; Russell, 1985). Agave leaves can also be used as cattle fodder, a by-product that, in addition to extraction of steroids, could dramatically enhance the economic value of agaves presently used for leaf fiber or stem carbohydrates. When new or expanded uses for agaves and cacti are contemplated, the impact of insect pests must be carefully evaluated, as well as the secondary chemical compounds produced by the plants that can act as a defense against herbivory (Houston, 1963; Hernández Xolocotzi, 1970; Curtis, 1977).

Where do we go from here? This question becomes especially pertinent when considering the economic potential of agaves and cacti. For instance, 800,000 metric tons of hard fiber were produced from *A. fourcroydes* and *A. sisalana* in 1965 (Wienk, 1970). Although most of the worldwide harvesting of agaves is for hard fibers, the demand for alcoholic beverages from *Agave salmiana* and *A. tequilana* is increasing rapidly. Also, the use of *Opuntia ficus-indica* and other platyopuntias for cattle fodder in arid regions of northern Africa, northern Mexico, and other areas is increasing (Le Houerou, 1970; Medina et al., 1988). Deserts are generally believed to have low average annual productivities (Noy-Meir, 1973; Lieth and Whittaker, 1975), but because of CAM this generalization is naive when applied to agaves and cacti. Agaves have been proposed to be economically viable in Australia (Wood and Angus, 1974), where cacti, without much encouragement, previously created a stranglehold on the land (Osmond and Monro, 1981). Armed with knowledge of the environmental responses that we have considered throughout this book, enterprising individuals can develop agronomic practices that will increase yields, and collectors and gardeners can also make better-informed decisions on the selection and care of their succulent plants. The future will bring not only an expanded role for agaves and cacti for pleasure and profit but also a deeper understanding of the environmental biology of these remarkable plants.

References

Abd El Rahman, A. A., A. M. El Gamassy, and M. S. Mandour. 1968. Water economy of *Agave sisalana* under desert conditions. *Flora, Section B*, 157: 355–378.

Abd El-Rahman, A. A., and M. Hassib. 1972. Structural modifications in relation to some physiological characteristics of *Agave sisalana* under various environmental conditions. *Gartenbauwissenschaft* 37: 361–369.

Abrahamson, W. G., and J. Rubinstein. 1976. Growth forms of *Opuntia compressa* (Cactaceae) in Florida sandridge habitats. *Bulletin of the Torrey Botanical Club* 103: 77–79.

Acevedo, E., I. Badilla, and P. S. Nobel. 1983. Water relations, diurnal acidity changes, and productivity of a cultivated cactus, *Opuntia ficus-indica*. *Plant Physiology* 72: 775–780.

Adams, W. W., III, K. Nishida, and C. B. Osmond. 1986. Quantum yields of CAM plants measured by photosynthetic O_2 exchange. *Plant Physiology* 81: 297–300.

Adams, W. W., III, S. D. Smith, and C. B. Osmond. 1987. Photoinhibition of the CAM succulent *Opuntia basilaris* growing in Death Valley: evidence from 77K fluorescence and quantum yield. *Oecologia* 71: 221–228.

Alcorn, S. M., and E. B. Kurtz, Jr. 1959. Some factors affecting the germination of seed of the saguaro cactus (*Carnegiea gigantea*). *American Journal of Botany* 46: 526–529.

Ananthan, J., A. L. Goldberg, and R. Voellmy. 1986. Abnormal proteins serve as eukaryotic stress signals and trigger the activation of heat shock genes. *Science* 232: 522–524.

Annecke, D. P., and V. C. Moran. 1978. Critical reviews of biological pest control in South Africa. 2. The prickly pear, *Opuntia ficus-indica* (L.) Miller. *Journal of the Entomological Society of Southern Africa* 41: 161–188.

Annecke, D. P., and S. Neser. 1977. On the biological control of some Cape pest plants. *National Weeds Conference of South Africa Stellenbosch, Proceedings, 2nd*: 303–319.

Anonymous. 1883. Contraction of vegetable tissues under frost. *Bulletin of the Torrey Botanical Club* 10: 83–84.

Arias, I., and L. Lemus. 1984. Interaction of light, temperature and plant hormones in the germination of seeds of *Melocactus caesius* Went. (Cactaceae). *Acta Científica Venezolana* 35: 151–155.

Arnold, L. A., Jr., and D. L. Drawe. 1979. Seasonal food habits of white-tailed deer in the South Texas Plains. *Journal of Range Management* 32: 175–178.

Arp, G. K., and D. E. Phinney. 1980. Ecological variations in thermal infrared emissivity of vegetation. *Environmental and Experimental Botany* 20: 135–148.

Ault, J. R., and W. J. Blackmon. 1987. In vitro propagation of *Ferocactus acanthodes* (Cactaceae). *HortScience* 22: 126–127.

Backeberg, C. 1958–1962. *Die Cactaceae. Handbuch der Kakteenkunde. Volume I* (*Einleitung und Beschreibung der Peireskioideae und Opuntioideae* 1958, 638 pp.); *Volume II* (*Cereoideae*, 1959, pp. 639–1360); *Volume III* (*Cereoideae*, 1959, pp. 1361–1926); *Volume IV* (*Cereoideae*, 1960, pp. 1927–2629); *Volume V* (*Cereoideae*, 1961, 2631–3543); and *Volume VI* (*Nachtrage und Index*, 1962, pp. 3645–4041). Gustav Fischer Verlag, Jena.

Bahre, C. J., and D. E. Bradbury. 1980. Manufacture of mescal in Sonora, Mexico. *Economic Botany* 34: 391–400.

Bailey, I. W. 1968. Comparative anatomy of the leaf-bearing Cactaceae, XVII. Preliminary observations on the problem of transitions from broad to terete leaves. *Journal of the Arnold Arboretum* 49: 370–376.

Baker, D. M. 1985. Alternative uses for sisal fiber. *In* C. Cruz, L. del Castillo, M. Robert, and R. N. Ondarza (eds.) *Biología y Aprovechamiento Integral del Henquén y Otros Agaves*. Centro de Investigación Científica de Yucatán, A.C., Mérida, Yucatán, pp. 177–186.

Banerjee, A. K. 1972. Trial of agave species in lateritic areas of West Bengal. *Indian Forester* 98: 432–436.

Barbera, G. 1984. Ricerche sull'irrigazione del ficodindia. *Frutticoltura*, Anno 46, Number 8: 49–55.

Barcikowski, W., and P. S. Nobel. 1984. Water

relations of cacti during desiccation: distribution of water in tissues. *Botanical Gazette* 145: 110–115.

Barrientos Pérez, F. 1983. Nopal y agaves como recursos de las zonas aridas y semiaridas de Mexico. *In* J. D. Molina Galan (ed.) *Recursos Agrícolas de Zonas Aridas y Semiáridas de México*, Colegio de Postgraduados, Chapingo, Mexico, pp. 133–143.

Barthlott, W., and I. Capesius. 1974. Wasserabsorption durch Blatt- und Sprossorgane einiger Xerophyten. *Zeitschrift für Pflanzenphysiologie* 72: 443–455.

Bartholomew, D. P. 1982. Environmental control of carbon assimilation and dry matter production by pineapple. *In* I. P. Ting and M. Gibbs (eds.) *Crassulacean Acid Metabolism*. American Society of Plant Physiologists, Rockville (MD), pp. 278–294.

Bartholomew, D. P., and S. B. Kadzimin. 1977. Pineapple. *In* P. de T. Alvim and T. T. Kozlowski (eds.) *Ecophysiology of Tropical Crops*. Academic Press, New York, pp. 113–156.

Baskin, J. M., and C. C. Baskin. 1973. Pad temperature of *Opuntia compressa* during daytime in the summer. *Bulletin of the Torrey Botanical Club* 100: 56–59.

Baskin, J. M., and C. C. Baskin. 1977. Seed and seedling ecology of *Opuntia compressa* in Tennessee cedar glades. *Journal of the Tennessee Academy of Science* 52: 118–122.

Batschelet, E. 1981. *Circular Statistics in Biology*. Academic Press. New York, 371 pp.

Beatley, J. C. 1976. *Vascular Plants of the Nevada Test Site and Central-Southern Nevada: Ecologic and Geographic Distribution*. Energy Research and Development Administration, Washington, D.C., 308 pp.

Becerra Rodríguez, S., F. Barrientos Pérez, and D. Diaz Montenegro. 1976. Eficiencia fotosintetica del nopal (*Opuntia* spp.) en relacion con la orientacion de sus cladodios. *Agrociencia* 24: 67–77.

Belmares, H., J. E. Castillo, and A. Barrera. 1979. Natural hard fibers of the North American continent. Statistical correlations of physical and mechanical properties of lechuguilla (*Agave lechuguilla*) fibers. *Textile Research Journal* 49: 619–622.

Bement, R. E. 1968. Plains pricklypear: relation to grazing intensity and blue grama yield on central Great Plains. *Journal of Range Management* 21: 83–86.

Bender, M. M., I. Rouhani, H. M. Vines, and C. C. Black, Jr. 1973. $^{13}C/^{12}C$ ratio changes in Crassulacean acid metabolism plants. *Plant Physiology* 52: 427–430.

Benson, L. 1982. *The Cacti of United States and Canada*. Stanford University Press, Stanford. 1044 pp.

Benson, L., and D. L. Walkington. 1965. The southern California prickly-pears – invasion, adulteration, and trial by fire. *Annals of the Missouri Botanical Garden* 52: 262–273.

Berry, J. A., and J. K. Raison. 1981. Responses of macrophytes to temperature. *In* O. L. Lange, P. S. Nobel, C. B. Osmond, and H. Ziegler (eds.) *Physiological Plant Ecology I Responses to the Physical Environment. Encyclopedia of Plant*

Physiology, New Series, Volume 12A. Springer-Verlag, Berlin, pp. 277–338.

Berry, W. L. 1986. Plant factors influencing the use of plant analysis as a tool for biogeochemical prospecting. *In* D. Carlisle, W. L. Berry, I. R. Kaplan, and J. R. Watterson (eds.) *Mineral Exploration: Biological Systems and Organic Matter*, Rubey Volume V. Prentice-Hall, Englewood Cliffs (NJ), pp. 13–32.

Berry, W. L., and P. S. Nobel. 1985. Influence of soil and mineral stresses on cacti. *Journal of Plant Nutrition* 8: 679–696.

Bethlenfalvay, G. J., S. Dakessian, and R. S. Pacovsky. 1984. Mycorrhizae in a southern California desert: ecological implications. *Canadian Journal of Botany* 62: 519–524.

Biebl, R. 1962. *Protoplasmatische Ökologie der Pflanzen: Wasser und Temperature. Protoplasmatologia, Handbuch der Protoplasmaforschung, Band XII*. Springer-Verlag, Vienna. 344 pp.

Björkman, O. 1981. Responses to different quantum flux densities. *In* O. L. Lange, P. S. Nobel, C. B. Osmond, and H. Ziegler (eds.) *Physiological Plant Ecology I Responses to the Physical Environment. Encyclopedia of Plant Physiology, New Series, Volume 12A*. Springer-Verlag, Berlin, pp. 57–107.

Black, C. C., N. C. Carnal, and W. H. Kenyon. 1982. Compartmentation and the regulation of CAM. *In* I. P. Ting and M. Gibbs (eds.) *Crassulacean Acid Metabolism*. American Society of Plant Physiologists, Rockville (MD), pp. 51–68.

Black, C. C., and S. Williams. 1976. Plants exhibiting characteristics common to Crassulacean acid metabolism. *In* R. H. Burris and C. C. Black (eds.) *CO_2 Metabolism and Plant Productivity*. University Park Press, Baltimore, pp. 407–424.

Blancas, A., L. Alpizar, G. Larios, S. Saval, and C. Huitron. 1982. Conversion of henequen pulp to microbial biomass by submerged fermentation. *In* C. D. Scott (ed.) *Fourth Symposium on Biotechnology in Energy Production and Conservation*. John Wiley & Sons, New York, pp. 171–175.

Blunden, G., C. Culling, and K. Jewers. 1975. Steroidal sapogenins: a review of actual and potential plant sources. *Tropical Science* 17: 139–154.

Boke, N. H. 1953. Tubercle development in *Mammillaria heyderi*. *American Journal of Botany* 40: 239–247.

Boke, N. H. 1980. Developmental morphology and anatomy in Cactaceae. *BioScience* 30: 605–610.

Bolio, J. A. 1914. *Manual practico del Henequen su cultivo y Explotacion*. Empressa Editorial Católica, Mérida, Yucatán.

Boulanger, J. 1985. Mejoramiento de la variabilidad genetica del sisal (*A. sisalana* Perrine) y del henequén (*A. fourcroydes* Lemaire). *In* C. Cruz, L. del Castillo, M. Robert, and R. N. Ondarza (eds.) *Biología y Aprovechamiento Integral del Henequén y Otros Agaves*. Centro de Investigatión Científica de Yucatán, A.C., Mérida, Yucatán, pp. 75–81.

Boyer, J. S. 1985. Water transport. *Annual Review of Plant Physiology* 36: 473–516.

Braun W., R. H., A. Cordero, and J. Ramacciotti. 1979. Productividad, ecologia y valor forrajero de tunales (*Opuntia ficus-indica*) de los Llanos, provincia de la Rioja. *Cuaderno Técnico, Iadiza* 1–79: 29–37.

Braun W., R. H., and J. J. López G. 1978. Biomasa y produccion ecologica del nopal cardon (*Opuntia streptacantha* Lemaire). *Deserta* 5: 73–81.

Bruhn, J. G. 1971. *Carnegiea gigantea*: The saguaro and its uses. *Economic Botany* 25: 320–329.

Brulfert, J., D. Guerrier, and O. Queiroz. 1984. Rôle de la photopériode dans l'adaptation à la sécheresse: cas d'une plante à métabolisme crassulacéen, l'*Opuntia ficus-indica* Mill. *Bulletin de la Société Botanique de France, Actualités Botaniques* 131: 69–77.

Brulfert, J., M. Kluge, D. Guerrier, and O. Queiroz. 1987. Characterization of carbon metabolism in *Opuntia ficus-indica* Mill. exhibiting the idling mode of Crassulacean acid metabolism. *Planta* 170: 92–98.

Brum, G. D. 1973. Ecology of the saguaro (*Carnegiea gigantea*): phenology and establishment in marginal populations. *Madroño* 22: 195–204.

Burgess, T. L. 1985. Agave adaptations to aridity. *Desert Plants* 7: 39–50.

Bustos, O. E. 1981. Alcoholic beverages from Chilean *Opuntia ficus-indica*. *American Journal of Viticulture* 32: 228–229.

Buxton, P. A. 1925. The temperature of the surface of deserts. *Journal of Ecology* 12: 127–134.

Calkin, H. W., and P. S. Nobel. 1986. Nonsteady-state analysis of water flow and capacitance for *Agave deserti*. *Canadian Journal of Botany* 64: 2556–2560.

Callen, E. O. 1965. Food habits of some Pre-Columbian Mexican Indians. *Economic Botany* 19: 335–343.

Camp, R. A. 1980. Yucatán's green gold. *Américas* 32: 3–8.

Cannon, W. A. 1906. Biological relations of certain cacti. *The American Naturalist* 40: 27–47.

Cannon, W. A. 1908. *The Topography of the Chlorophyll Apparatus in Desert Plants*. Publication 98, Carnegie Institution of Washington, Washington, D.C., pp. 1–42.

Cannon, W. A. 1911. *The Root Habits of Desert Plants*. Publication 131, Carnegie Institution of Washington, Washington, D.C., 96 pp.

Cannon, W. A. 1916. Distribution of the cacti with especial reference to the rôle played by the root response to soil temperature and soil moisture. *The American Naturalist* 50: 435–442.

Cannon, W. A. 1925. *Physiological Features of Roots, with Especial Reference to the Relation of Roots to Aeration of the Soil*. Publication 368, Carnegie Institution of Washington, Washington, D.C., 168 pp.

Carrion, R. M. 1981. Efecto de las aplicaciones de nitrógeno sobre la duración del ciclo productivo del henequén (*Agave fourcroydes*). *Agrotecnia de Cuba* 13: 45–49.

Chapman, H. D. (ed.). 1966. *Diagnostic Criteria for Plants and Soils*. University of California Division of Agricultural Sciences, Berkeley, 793 pp.

Chetti, M., and P. S. Nobel. 1987. High-temperature sensitivity and its acclimation for photosynthetic electron transport reactions of desert succulents. *Plant Physiology* 84: 1063–1067.

Chow, P. N., O. C. Burnside, and T. L. Lavy. 1966. Physiological studies with prickly pear. *Weeds* 14: 58–62.

Cockburn, W., I. P. Ting, and L. O. Sternberg. 1979. Relationships between stomatal behavior and internal carbon dioxide concentration in Crassulacean acid metabolism plants. *Plant Physiology* 63: 1029–1032.

Cody, M. L. 1984. Branching patterns in columnar cacti. *In* N. S. Margaris, M. Arianoustou-Farragitaki, and W. C. Oechel (eds.) *Being Alive on Land. Tasks for Vegetation Science, Volume 13*. Dr. W. Junk, The Hague, pp. 201–236.

Cody, M. L. 1986a. Structural niches in plant communities. *In* J. M. Diamond and T. J. Case (eds.) *Community Ecology*. Harper and Row, New York, pp. 381–405.

Cody, M. L. 1986b. Distribution and morphology of columnar cacti in tropical deciduous woodland, Jalisco, Mexico. *Vegetatio* 66: 137–146.

Cody, M. L. 1986c. Spacing patterns in Mojave Desert plant communities: Near-neighbor analyses. *Journal of Arid Environments* 11: 199–217.

Cogdell, R. J. 1983. Photosynthetic reaction centers. *Annual Review of Plant Physiology* 34: 21–45.

Comins, H. N., and G. D. Farquhar. 1982. Stomatal regulation and water economy in Crassulacean acid metabolism plants: an optimization model. *Journal of Theoretical Biology* 99: 263–284.

Conde, L. F. 1975. Anatomical comparisons of five species of *Opuntia* (Cactaceae). *Annals of the Missouri Botanical Garden* 62: 425–473.

Conde, L. F., and P. J. Kramer. 1975. The effect of vapor pressure deficit on diffusion resistance in *Opuntia compressa*. *Canadian Journal of Botany* 53: 2923–2926.

Contreras, S., and J. Toha C. 1984. Biogas production from a suspension of homogenized cladodes of the cactus *Opuntia cacti*. *Journal of Fermentation Technology* 62: 601–605.

Coville, F. V., and D. T. MacDougal. 1903. *Desert Botanical Laboratory of the Carnegie Institution*. Publication 6, Carnegie Institution of Washington, Washington, D.C., 58 pp.

Craig, H. 1957. Isotopic standards for carbon and oxygen and correction factors for mass spectrometric analysis of carbon dioxide. *Geochimica et Cosmochimica Acta* 12: 133–149.

Crane, H. R., and J. B. Griffin. 1958. University of Michigan radiocarbon dates III. *Science* 128: 1117–1123.

Cronquist, A. 1981. *An Integrated System of Classification of Flowering Plants*. Columbia University Press, New York, 1262 pp.

de la Cruz, C. J. A., S. Preciado S., A. Zarate L., P. Recio B., A. Mendoza A., R. del Toro W., and A. Huereca S. 1983. *El Nopal*. Comisión Tecnica para el

Programa de Empleo Rural-Agencia Coahuila, Saltillo, Coahuila.

Cruz-Ramos, C. A., R. Orellana, and M. I. Robert. 1985. Agave research progress in Yucatan. *Desert Plants* 7: 71–73, 80, 89–92.

Curtis, J. R. 1977. Prickly pear farming in the Santa Clara Valley, California. *Economic Botany* 31: 175–179.

Dahlgren, R. M. T., H. T. Clifford, and P. F. Yeo. 1985. *The Families of the Monocotyledons: Structure, Evolution, and Taxonomy*. Springer-Verlag, Berlin, 520 pp.

Darbishire, O. V. 1904. Observations on Mamillaria elongata. *Annals of Botany* 18: 375–416.

Darling, M. S. 1976. Interpretation of global differences in plant calorific values. The significance of desert and arid woodland vegetation. *Oecologia* 23: 127–139.

Dau, L., and L. G. Labouriau. 1974. Temperature control of seed germination in *Pereskia aculeata* Mill. *Annaes Academia brasileira de Ciências* 46: 311–322.

Davis, J. B., D. E. Kay, and V. Clark. 1983. *Plants Tolerant of Arid, or Semi-Arid, Conditions with Non-food Constituents of Potential Use*. Tropical Products Institute, London, 172 pp.

Dawson, E. Y. 1965. Further studies of Opuntia in the Galapagos Archipelago. *Cactus and Succulent Journal (U.S.)* 37: 135–148.

Dawson, E. Y. 1966. Cacti in the Galapagos Islands, with special reference to their relations with tortoises. *In* R. J. Bowman (ed.), *The Galápagos. Proceedings of the Symposium of the Galápagos International Scientific Project*. University of California Press, Los Angeles, pp. 209–214.

del Castillo Sanchez, R. F. 1982. Estudio Ecologico de *Ferocactus histrix* (DC.) Lindsay. Professional thesis. Universidad Nacional Autónoma de México, Mexico City, 228 pp.

Derby, R. W., and D. M. Gates. 1966. The temperature of tree trunks – calculated and observed. *American Journal of Botany* 53: 580–587.

Despain, D. G. 1974. The survival of saguaro (*Carnegiea gigantea*) seedlings on soils of differing albedo and cover. *Journal of the Arizona Academy of Science* 9: 102–107.

Despain, D. G., L. C. Bliss, and J. C. Boyer. 1970. Carbon dioxide exchange in *Saguaro* seedlings. *Ecology* 51: 912–914.

Díaz, M. 1983. Estudios fisioecológicos de cactáceas bajo condiciones naturales. Trabajo Especial de Grado. M. Sc. thesis, Instituto Venezolano de Investigaciones Científicas, Caracas, 147 pp.

Díaz, M., and E. Medina. 1984. Actividad CAM de cactaceas en condiciones naturales. *In* E. Medina (ed.) *Eco-Fisiologia de Plantas CAM*. Centro Internacional de Ecología Tropical, Caracas, Venezuela, pp. 98–113.

Didden-Zopfy, B., and P. S. Nobel. 1982. High temperature tolerance and heat acclimation of *Opuntia bigelovii*. *Oecologia* 52: 176–180.

Dinger, B. E., and D. T. Patten. 1972. Carbon dioxide exchange in selected species of *Echinocereus* (Cactaceae). *Photosynthetica* 6: 345–353.

Dinger, B. E., and D. T. Patten. 1974. Carbon dioxide exchange and transpiration in species of *Echinocereus* (Cactaceae), as related to their distribution within the Pinaleno Mountains, Arizona. *Oecologia* 14: 389–411.

Dittrich, P. 1976. Nicotinamide adenine dinucleotide-specific "malic" enzyme in *Kalanchoë daigremontiana* and other plants exhibiting crassulacean acid metabolism. *Plant Physiology* 57: 310–314.

Dodd, J. L., and W. K. Lauenroth. 1975. Responses of *Opuntia polyacantha* to water and nitrogen perturbations in the shortgrass prairie. *In* M. K. Wali (ed.) *Prairie: A Multiple View*. University of North Dakota Press, Grand Forks, pp. 229–240.

Domingues, O. 1963. Origem e introdução da palma forrageira no Nordeste. Institute Joaquim Nabuco de Pesquisas Sociais, Recife, Pernambuco, 73 pp.

Donkin, R. A. 1977. *Spanish Red. An Ethnogeographical Study of Cochineal and the Opuntia Cactus*. American Philosophical Society, Philadelphia, 84 pp.

Doussoulin E., E., and E. Acevedo H. 1984. Arquitectura y relaciones alométricas en *Opuntia ficus-indica* (L.) Mill. *Idesia (Chile)* 8: 33–45.

Downton, W. J. S., J. A. Berry, and J. R. Seemann. 1984. Tolerance of photosynthesis to high temperature in desert plants. *Plant Physiology* 74: 786–790.

Duisberg, P. C. 1952. Desert plant utilization. *The Texas Journal of Science* 4: 268–283.

Ebeling, W. 1986. *Handbook of Indian Foods and Fibers of Arid America*. University of California Press, Los Angeles, 971 pp.

Edwards, G. E., J. G. Foster, and K. Winter. 1982. Activity and intracellular compartmentation of enzymes of carbon metabolism in CAM plants. *In* I. P. Ting and M. Gibbs (eds.) *Crassulacean Acid Metabolism*. American Society of Plant Physiologists, Rockville (MD), pp. 92–111.

Ehleringer, J. 1981. Leaf absorptances of Mohave and Sonoran Desert plants. *Oecologia* 49: 366–370.

Ehleringer, J., and O. Björkman. 1977. Quantum yields for CO_2 uptake in C_3 and C_4 plants. *Plant Physiology* 59: 86–90.

Ehleringer, J., and D. House. 1984. Orientation and slope preference in barrel cactus (*Ferocactus acanthodes*) at its northern distribution limit. *Great Basin Naturalist* 44: 133–139.

Ehleringer, J., H. A. Mooney, S. L. Gulmon, and P. Rundel. 1980. Orientation and its consequences for *Copiapoa* (Cactaceae) in the Atacama Desert. *Oecologia* 46: 63–67.

Ehleringer, J., and R. W. Pearcy. 1983. Variation in quantum yield for CO_2 uptake among C_3 and C_4 plants. *Plant Physiology* 73: 555–559.

Ehrler, W. L. 1969. Daytime stomatal closure in *Agave americana* as related to enhanced water-use efficiency. *In* C. C. Hoff and M. L. Riedsel (eds.) *Physiological Systems in Semiarid Environments*. University of New Mexico Press, Albuquerque, pp. 239–247.

Ehrler, W. L. 1975. Environmental and plant factors influencing transpiration of desert plants. *In* N. F.

Hadley (ed.) *Environmental Physiology of Desert Organisms*, Dowden, Hutchinson, & Ross, Stroudsburg (PA), pp. 52–66.

Ehrler, W. L. 1983. The transpiration ratios of *Agave americana* L. and *Zea mays* L. as affected by soil water potential. *Journal of Arid Environments* 6: 107–113.

Eickmeier, W. G. 1978. Photosynthetic pathway distributions along an aridity gradient in Big Bend National Park, and implications for enhanced resource partitioning. *Photosynthetica* 12: 290–297.

Eickmeier, W. G. 1979. Eco-physiological differences between high and low elevation CAM species in Big Bend National Park, Texas. *American Midland Naturalist* 101: 118–126.

Eickmeier, W. G., and M. S. Adams. 1978. Gas exchange in *Agave lecheguilla* Torr. (Agavaceae) and its ecological implications. *The Southwestern Naturalist* 23: 473–486.

Eickmeier, W. G., and M. M. Bender. 1976. Carbon isotope ratios of Crassulacean acid metabolism species in relation to climate and phytosociology. *Oecologia* 25: 341–347.

El-Gamassy, A. M., A. A. Abd El-Rahman, M. Hassib, and M. S. Mandour. 1974. Vegetative patterns and water economy of *Agave sisalana* in saline soils. *Zeitschrift für Acker- und Pflanzenbau* 139: 165–171.

El-Meligy, M. M. 1980. Studies on the effect of urea and some micronutrient spray on the growth of *Echinocactus grusonii*. *Research Bulletin 0(1400)*, Ain Shams University Faculty of Agriculture, pp. 1–19.

Ellenberg, H. 1981. Ursachen des Vorkommens und Fehlens von Sukkulenten in den Trockengebieten der Erde. *Flora* 171: 114–169.

Epstein, E. 1972. *Mineral Nutrition of Plants: Principles and Perspectives*. John Wiley & Sons, New York, 412 pp.

Esau, K. 1977. *Anatomy of Seed Plants*, 2nd edition. John Wiley & Sons, New York, 550 pp.

Escobar A., H. A., V. M. Villalobos A., and A. Villegas M. 1986. Opuntia micropropagation by axillary proliferation. *Plant Cell, Tissue and Organ Culture* 7: 269–277.

Evans, E. H. 1967. They color things red. *Pacific Discovery* 20, Number 6: 24–25.

Evans, L. T. (ed.). 1975. *Crop Physiology: Some Case Histories*. Cambridge University Press, Cambridge, 374 pp.

Evenari, M., L. Shanan, and N. Tadmor. 1971. *The Negev: The Challenge of a Desert*. Harvard University Press, Cambridge, 345 pp.

Everitt, J. H., and M. A. Alaniz. 1981. Nutrient content of cactus and woody plant fruits eaten by birds and mammals in south Texas. *The Southwestern Naturalist* 26: 301–305.

Everitt, J. H., C. L. Gonzalez, M. A. Alaniz, and G. V. Latigo. 1981. Food habits of the collared peccary on south Texas rangelands. *Journal of Range Management* 34: 141–144.

Faraday, C. D., W. W. Thomson, and K. A. Platt-Aloia. 1982. Comparative ultrastructure of guard cells of C_3, C_4, and CAM plants. *In* I. P. Ting and M.

Gibbs (eds.) *Crassulacean Acid Metabolism*. American Society of Plant Physiologists, Rockville (MD), pp. 18–30.

Farquhar, G. D. 1983. On the nature of carbon isotope discrimination in C_4 species. *Australian Journal of Plant Physiology* 10: 205–226.

Felger, R. S. 1979. Ancient crops for the twenty-first century. *In* G. A. Ritchie (ed.) *New Agricultural Crops*, AAAS Selected Symposium 38, Westview Press, Boulder (CO), pp. 5–20.

Felger, R. S., and C. H. Lowe. 1967. Clinal variation in the surface-volume relationships of the columnar cactus *Lophocereus schottii* in northwestern Mexico. *Ecology* 48: 530–536.

Felger, R. S., and M. B. Moser. 1985. *People of the Desert and Sea; Ethnobotany of the Seri Indians*. University of Arizona Press, Tucson, 435 pp.

Fischer, R. A., and N. C. Turner. 1978. Plant productivity in arid and semi-arid zones. *Annual Review of Plant Physiology* 29: 277–317.

Fish, S. K., P. R. Fish, C. Miksicek, and J. Madsen. 1985. Prehistoric agave cultivation in southern Arizona. *Desert Plants* 7: 107–112, 100.

Flores Valdez, C. A., and J. R. Aguirre Rivera. 1979. El Nopal como Forraje. Universidad Autonoma Chapingo, Chapingo, Mexico, 91 pp.

Fogel, M. M. 1981. Precipitation in the desert. *In* D. D. Evans and J. L. Thames (eds.) *Water in Desert Ecosystems*. US/IBP Synthesis Series II. Dowden, Hutchinson & Ross, Stroudsburg (PA), pp. 219–234.

Fowler, N. 1986. The role of competition in plant communities in arid and semiarid regions. *Annual Review of Ecology and Systematics* 17: 89–110.

Foy, C. D., R. F. Chaney, and M. C. White. 1978. The physiology of metal toxicity in plants. *Annual Review of Plant Physiology* 29: 511–566.

Franco, A. C., and P. S. Nobel. 1988. Interactions between seedlings of *Agave deserti* and the nurse plant *Hilaria rigida*. Submitted.

Fraser, J. G., and R. D. Pieper. 1972. Growth characteristics of *Opuntia imbricata* [Haw.] DC. in New Mexico. *The Southwestern Naturalist* 17: 229–237.

Freeman, C. E. 1973. Some germination responses of lechuguilla (*Agave lecheguilla* Torr.). *The Southwestern Naturalist* 18: 125–134.

Freeman, C. E. 1975. Germination responses of a New Mexico population of Parry agave (*Agave parryi* Engelm. var. *parryi*) to constant temperature, water stress, and pH. *The Southwestern Naturalist* 20: 69–74.

Freeman, C. E., and W. H. Reid. 1985. Aspects of the reproductive biology of *Agave lechuguilla* Torr. *Desert Plants* 7: 75–80.

Freeman, C. E., R. S. Tiffany, and W. H. Reid. 1977. Germination responses of *Agave lecheguilla*, *A. parryi*, and *Fouquieria splendens*. *The Southwestern Naturalist* 22: 195–204.

Freeman, T. P. 1969. The developmental anatomy of *Opuntia basilaris*. I. Embryo, root, and transition zone. *American Journal of Botany* 56: 1067–1074.

Frego, K. A., and R. J. Staniforth. 1986. Vegetation

sequence on three boreal Manitoban rock outcrops and seral position of *Opuntia fragilis*. *Canadian Journal of Botany* 64: 77–84.

Ganong, W. F. 1895. Present problems in the anatomy, morphology, and biology of the Cactaceae. *Botanical Gazette* 20: 129–138.

Garcia de Cortázar, V., E. Acevedo, and P. S. Nobel. 1985. Modeling of PAR interception and productivity by *Opuntia ficus-indica*. *Agricultural and Forest Meteorology* 34: 145–162.

García de Cortázar, V., and P. S. Nobel. 1986. Modeling of PAR interception and productivity of a prickly pear cactus, *Opuntia ficus-indica* L., at various spacings. *Agronomy Journal* 78: 80–85.

Garcia de Fuentes, A., and A. de Sicilia. 1984. *El Mercado Mundial de las Fibras Duras*. Centro de Investigación Científica de Yucatán, A.C., Mérida, Yucatán, 105 pp.

Gates, D. M. 1980. *Biophysical Ecology*. Springer-Verlag, New York, 611 pp.

Gates, D. M., R. Alderfer, and E. Taylor. 1968. Leaf temperatures of desert plants. *Science* 159: 994–995.

Gates, D. M., B. R. Strain, and J. A. Weber. 1983. Ecophysiological effects of changing atmospheric CO_2 concentrations. *In* O. L. Lange, P. S. Nobel, C. B. Osmond, and H. Ziegler (eds.) *Physiological Plant Ecology IV, Encyclopedia of Plant Physiology, New Series, Volume 12 D*, Springer-Verlag, Berlin, pp. 503–526.

Gatti, V. 1977. Il fico d'India pianta utile nei paesi caldi. *Rivista di Agricoltura Subtropicale* 71: 141–146.

Geiger, R. 1965. *The Climate near the Ground*. Harvard University Press, Cambridge, 611 pp.

Geller, G. N., and P. S. Nobel. 1984. Cactus ribs: influence on PAR interception and CO_2 uptake. *Photosynthetica* 18: 482–494.

Geller, G. N., and P. S. Nobel. 1986. Branching patterns of columnar cacti: influences on PAR interception and CO_2 uptake. *American Journal of Botany* 73: 1193–1200.

Geller, G. N., and P. S. Nobel. 1987. Comparative cactus architecture and PAR interception. *American Journal of Botany* 74: 998–1005.

Gentry, H. S. 1972. *The Agave Family in Sonora*. Agricultural Handbook Number 399, United States Department of Agriculture, Washington, D.C., 195 pp.

Gentry, H. S. 1982. *Agaves of Continental North America*. The University of Arizona Press, Tucson, 670 pp.

Gentry, H. S. 1985. On the taxonomy of the genus *Agave*. *In* C. Cruz, L. del Castillo, M. Robert, and R. N. Ondarza (eds.) *Biología y Aprovechamiento Integral del Henequén y otros Agaves*. Centro de Investigación Científica de Yucatán, A.C., Mérida, Yucatán, pp. 29–38.

Gentry, H. S., and J. R. Sauck. 1978. The stomatal complex in *Agave*: groups Deserticolae, Campaniflorae, Umbelliflorae. *Proceedings of the California Academy of Sciences* 41: 371–387.

Gerwick, B. C.. and G. J. Williams, III. 1978.

Temperature and water regulation of gas exchange of *Opuntia polyacantha*. *Oecologia* 35: 149–159.

Gerwick, B. C., G. J. Williams, M. H. Spalding, and G. E. Edwards. 1978. Temperature responses of CO_2 fixation in isolated *Opuntia* cells. *Plant Science Letters* 13: 389–396.

Gerwick, B. C., G. J. Williams, III, and E. G. Uribe. 1977. Effects of temperature on the Hill reaction and photophosphorylation in isolated cactus chloroplasts. *Plant Physiology* 60: 430–432.

Gibbs, J. G., and D. T. Patten. 1970. Plant temperatures and heat flux in a Sonoran Desert ecosystem. *Oecologia* 5: 165–184.

Gibson, A. C., and P. S. Nobel. 1986. *The Cactus Primer*. Harvard University Press, Cambridge, 286 pp.

Gil, F. 1986. Origin of CAM as an alternative photosynthetic carbon fixation pathway. *Photosynthetica* 20: 494–507.

Gill, L. S. 1942. Death in the desert. *Natural History* 50: 22–26.

Goebel, K. 1895. Ueber die Einwirkung des Lichtes auf die Gestaltung der Kakteen und anderer Pflanzen. *Flora* 80: 96–116.

Gómez-Campo, C., and M. Casas-Builla. 1972. Some effects of ionizing radiation on plant phyllotaxis. *Radiation Botany* 12: 165–172.

Gonzalez, C. L., and J. H. Everitt. 1982. Nutrient contents of major food plants eaten by cattle in the south Texas plains. *Journal of Range Management* 35: 733–736.

González-Medrano, F., A. Castillo, G. R. Durán, C. Martínez del Río, and J. M. Quintanilla. 1981. Estimaciones de Biomasa a Partir de la Altura y la Cobertura de Plantas Xerófilas. General Technical Report WO, U.S.D.A. Forest Service, Washington, D.C., pp. 416–420.

Grant, V., and K. A. Grant. 1971. Natural hybridization between the cholla cactus species *Opuntia spinosior* and *Opuntia versicolor*. *Proceedings of the National Academy of Sciences, U.S.A.* 68: 1993–1995.

Grant, V., and K. A. Grant. 1979. The pollination spectrum in the southwestern American cactus flora. *Plant Systematics and Evolution* 133: 29–37.

Grant, V., and K. A. Grant. 1980. Clonal microspecies of hybrid origin in the *Opuntia lindheimeri* group. *Botanical Gazette* 141: 101–106.

Green, J. M., and G. J. Williams, III. 1982. The subdominant status of *Echinocereus viridiflorus* and *Mammillaria vivipara* in the shortgrass prairie: the role of temperature and water effects on gas exchange. *Oecologia* 52: 43–48.

Greene, R. A. 1936. The composition and uses of the fruit of the giant cactus (*Carnegiea gigantea*) and its products. *Journal of Chemical Education* 13: 309–312.

Griffiths, D. 1905. *The Prickly Pear and Other Cacti as Food for Stock*. Bulletin 74, United States Department of Agriculture Bureau of Plant Industry, Washington, D.C., 48 pp.

Griffiths, D. 1906. *Feeding Prickly Pear to Stock in*

Texas. Bulletin 91, United States Department of Agriculture Bureau of Animal Industry, Washington, D.C., 23 pp.

Griffiths, D. 1908. *The Prickly Pear as a Farm Crop*. Bulletin 124, United States Department of Agriculture Bureau of Plant Industry, Washington, D.C., 37 pp.

Griffiths, D. 1915a. *Yields of Native Prickly Pear in Southern Texas*. Bulletin 208, United States Department of Agriculture, Washington, D.C., 11 pp.

Griffiths, D. 1915b. Hardier spineless cacti. *The Journal of Heredity* 6: 182–191.

Griffiths, D., and R. E. Hare. 1906. Summary of recent investigations on the value of cacti as stock feed. Bulletin 102, United States Department of Agriculture Bureau of Plant Industry, Washington, D.C., pp. 7–18.

Gulmon, S. L., and A. J. Bloom. 1979. C_3 photosynthesis and high temperature acclimation of CAM in *Opuntia basilaris* Engelm. and Bigel. *Oecologia* 38: 217–222.

Gulmon, S. L., P. W. Rundel, J. R. Ehleringer, and H. A. Mooney. 1979. Spatial relationships and competition in a Chilean desert cactus. *Oecologia* 44: 40–43.

Gurr, E. 1965. *The Rational Uses of Dyes in Biology and General Staining Methods*. Williams & Wilkins, Baltimore, 422 pp.

Gustafson, F. G. 1932. Anaerobic respiration of cacti. *American Journal of Botany* 19: 823–834.

Hadač, E., and V. Hadačová. 1974. Osmotic values of some plant species from Cuba. *Folia Geobotanica & Phytotaxonomica* 9: 71–76.

Hadley, N. F. 1970. Micrometeorology and energy exchange in two desert arthropods. *Ecology* 51: 434–444.

Hadley, N. F. 1972. Desert species and adaptation. *American Scientist* 60: 338–347.

Hanscom, Z., III, and I. P. Ting. 1977. Physiological responses to irrigation in *Opuntia basilaris* Engelm. & Bigel. *Botanical Gazette* 138: 159–167.

Hanscom, Z., III, and I. P. Ting. 1978a. Irrigation magnifies CAM-photosynthesis in *Opuntia basilaris* (Cactaceae). *Oecologia* 33: 1–15.

Hanscom, Z., III, and I. P. Ting. 1978b. Responses of succulents to plant water stress. *Plant Physiology* 61: 327–330.

Harris, J. A., and J. V. Lawrence. 1917. Cryoscopic determinations on tissue fluids of plants of Jamaican coastal deserts. *The Botanical Gazette* 64: 285–305.

Harrison, D. G. 1985. The use of sisal by-products as feeds for ruminants. *In* C. Cruz, L. del Castillo, M. Robert, and R. N. Ondarza (eds.) *Biología y Aprovechamiento Integral del Henequén y Otros Agaves*. Centro de Investigación Científica de Yucatán, A.C., Mérida, Yucatán, pp. 159–168.

Hartsock, T. L., and P. S. Nobel. 1976. Watering converts a CAM plant to daytime CO_2 uptake. *Nature* 262: 574–576.

Hastings, J. R., and S. M. Alcorn. 1961. Physical determinations of growth and age in the giant cactus. *Journal of the Arizona Academy of Science* 2: 32–39.

Heins, R. D., A. M. Armitage, and W. H. Carlson. 1981. Influence of temperature, water stress and benzylamino purine on vegetative growth of *Schlumbergera truncata*. *HortScience* 16: 679–680.

Hernández Xolocotzi, E. 1970. Mexican experience. *In* H. E. Dregne (ed.) *Arid Lands in Transition*, Publication 90, American Association for the Advancement of Science, Washington, D.C., pp. 317–343.

Herz, J. E. 1985. Modificacion y recuperacion de estroides del henequen. *In* C. Cruz, L. del Castillo, M. Robert, and R. N. Ondarza (eds.) *Biología y Aprovechamiento Integral del Henequén y Otros Agaves*. Centro de Investigación Científica de Yucatán, A.C., Mérida, Yucatán, pp. 169–174.

Herzog, F. 1938. Formgestalt und Wärmehaushalt bei Sukkulenten. *Jahrbücher für wissenschafliche Botanik* 87: 211–243.

Hillel, D. 1971. *Soil and Water Physical Principles and Processes*. Academic Press, London, 288 pp.

Ho, Y.-S., K. C. Sanderson, and J. C. Williams. 1985. Effect of chemicals and photoperiod on the growth and flowering of Thanksgiving cactus. *Journal of the American Society for Horticultural Science* 110: 658–662.

Hoagland, D. R., and D. I. Arnon. 1950. The water-culture method for growing plants without soil. *California Agricultural Experiment Station Circular* 347: 1–32.

Hodáňová, D. 1979. Sugar beet canopy photosynthesis as limited by leaf age and irradiance. Estimation by models. *Photosynthetica* 13: 376–385.

Hoffmann, J. J. 1983. Arid land plants as feedstocks for fuels and chemicals. *CRC Critical Reviews in Plant Sciences* 1: 95–116.

Holdsworth, M. 1971. Carbon dioxide uptake by succulents. *Canadian Journal of Botany* 49: 1520–1522.

Holm, L. G., D. L. Plucknett, J. V. Pancho, and J. P. Herberger. 1977. *The World's Worst Weeds: Distribution and Biology*. University of Hawaii Press, Honolulu, 609 pp.

Holthe, P. A., and S. R. Szarek. 1985. Physiological potential for survival of propagules of Crassulacean acid metabolism species. *Plant Physiology* 79: 219–224.

Hopkinson, D. 1969. Leguminous cover crops for maintaining soil fertility in sisal in Tanzania I. Effects on growth and yield. *Experimental Agriculture* 5: 283–294.

Houston, W. R. 1963. Plains pricklypear, weather, and grazing in the northern Great Plains. *Ecology* 44: 569–574.

Howell, D. J. 1979. Flock foraging in nectar-feeding bats. *The American Naturalist* 114: 23–49.

Howell, D. J., and B. S. Roth. 1981. Sexual reproduction in agaves: the benefits of bats; the cost of semelparous advertising. *Ecology* 62: 1–7.

Huber, B. 1932. Einige Grundfragen des Wärmehaushalts der Pflanzen. I. Die Ursache der hohen Sukkulenten-Temperaturen. *Berichte der Deutschen botanischen Gesselschaft* 50: 68–76.

Humphrey, R. R. 1936. Growth habits of barrel cacti. *Madroño* 3: 281–290.

Humphrey, R. R. 1958. The desert grassland. A history of vegetational change and an analysis of causes. *The Botanical Review* 24: 193–252.

Hunt, E. R., Jr., and P. S. Nobel. 1987a. A two-dimensional model for water uptake by desert succulents: implications of root distribution. *Annals of Botany* 59: 559–569.

Hunt, E. R., Jr., and P. S. Nobel. 1987b. Allometric root/shoot relationships and predicted water uptake for desert succulents. *Annals of Botany* 59: 571–577.

Hunt, E. R., Jr., and P. S. Nobel. 1987c. Non-steady-state water flow for three desert perennials with different capacitances. *Australian Journal of Plant Physiology* 14: 363–375.

Hunt, E. R., Jr., N. J. D. Zakir, and P. S. Nobel. 1987. Water costs and water revenues for established and rain-induced roots of *Agave deserti*. *Functional Ecology* 1: 125–129.

Hunt, R. 1982. *Plant Growth Curves. The Functional Approach to Plant Growth Analysis*. University Park Press, Baltimore, 248 pp.

Hutto, R. L., J. R. McAuliffe, and L. Hogan. 1986. Distributional associates of the saguaro (*Carnegiea gigantea*). *The Southwestern Naturalist* 31: 469–476.

Idso, S. B., R. D. Jackson, W. L. Ehrler, and S. T. Mitchell. 1969. A method for determination of infrared emittance of leaves. *Ecology* 5: 899–902.

Idso, S. B., B. A. Kimball, M. G. Anderson, and S. R. Szarek. 1986. Growth response of a succulent plant, *Agave vilmoriniana*, to elevated CO_2. *Plant Physiology* 80: 796–797.

Janzen, D. H. 1986. Chihuahuan Desert nopaleras: defaunated big mammal vegetation. *Annual Review of Ecology and Systematics* 17: 595–636.

Jarvis, P. G. 1975. Water transport in plants. *In* D. A. de Vries and N. H. Afgan (eds.) *Heat and Mass Transfer in the Biosphere. Part I. Transfer Processes in the Plant Environment*. Scripta, Washington, D.C., pp. 369–394.

Jarvis, P. G., and J. W. Leverenz. 1983. Productivity of temperate, deciduous and evergreen forests. *In* O. L. Lange, P. S. Nobel, C. B. Osmond, and H. Ziegler (eds.) *Encyclopedia of Plant Physiology, New Series, Volume 12D*. Springer-Verlag, Berlin, pp. 233–280.

Jarvis, P. G., and T. A. Mansfield. 1981. *Stomatal Physiology*. Cambridge University Press, Cambridge, 295 pp.

Jewer, P. C., L. D. Incoll, and G. L. Howarth. 1981. Stomatal responses in isolated epidermis of the crassulacean acid metabolism plant *Kalanchoe daigremontiana* Hamet et Perr. *Planta* 153: 238–245.

Johnson, D. S. 1924. The influence of insolation on the distribution and on the developmental sequence of the flowers of the giant cactus of Arizona. *Ecology* 5: 70–82.

Johnson, J. L., and E. R. Emino. 1979. In vitro propagation of *Mammillaria elongata*. *HortScience* 14: 605–606.

Jordan, P. W., and P. S. Nobel. 1979. Infrequent establishment of seedlings of *Agave deserti* (Agavaceae) in the northwestern Sonoran Desert. *American Journal of Botany* 66: 1079–1084.

Jordan, P. W., and P. S. Nobel. 1981. Seedling establishment of *Ferocactus acanthodes* in relation to drought. *Ecology* 62: 901–906.

Jordan, P. W., and P. S. Nobel. 1982. Height distributions of two species of cacti in relation to rainfall, seedling establishment, and growth. *Botanical Gazette* 143: 511–517.

Jordan, P. W., and P. S. Nobel. 1984. Thermal and water relations of roots of desert succulents. *Annals of Botany* 54: 705–717.

Kaplan, A., J. Gale, and A. Poljakoff-Mayber. 1976. Resolution of net dark fixation of carbon dioxide into its respiration and gross fixation components in *Bryophyllum daigremontianum*. *Journal of Experimental Botany* 27: 220–230.

Kappen, L. 1981. Ecological significance of resistance to high temperature. *In* O. L. Lange, P. S. Nobel, C. B. Osmond, and H. Ziegler (eds.) *Physiological Plant Ecology I Responses to the Physical Environment. Encyclopedia of Plant Physiology, New Series, Volume 12A*. Springer-Verlag, Berlin, pp. 439–474.

Kausch, W. 1965. Beziehungen zwischen Wurzelwachstum, Transpiration und CO_2-Gaswechsel bei einigen Kakteen. *Planta* 66: 229–238.

Kee, S. C., and P. S. Nobel. 1985. Fatty acid composition of chlorenchyma membrane fractions from three desert succulents grown at moderate and high temperatures. *Biochimica et Biophysica Acta* 820: 100–106.

Kee, S. C., and P. S. Nobel. 1986. Concomitant changes in high temperature tolerance and heat-shock proteins in desert succulents. *Plant Physiology* 80: 596–598.

Kenya Sisal Board. 1984. The Sisal Industry in Kenya. Kenya Sisal Board, Nairobi, 19 pp.

Key, J. L., W. B. Gurley, R. T. Nagao, E. Czarnecka, and M. A. Mansfield. 1985. Multigene families of heat shock proteins. *In* L. van Vloten-Doting, G. S. P. Groot, and T. C. Hall (eds.) *Molecular Form and Function of the Plant Genome, NATO Advance Science Institutes, Series A, Life Sciences, Volume 83*. Plenum Press, New York, pp. 81–100.

Kinraide, T. B. 1978. The ecological distribution of cholla cactus (*Opuntia imbricata* (Haw.) DC.) in El Paso County, Colorado. *The Southwestern Naturalist* 23: 117–134.

Kirby, R. H. 1963. *Vegetable Fibres. Botany, Cultivation, and Utilization*. Interscience, New York, 464 pp.

Klingman, G. L. 1979. Winter hardiness of selected American desert plants. *HortScience* 14: 131.

Kluge, M., A. Fischer, and I. C. Buchanan-Bollig. 1982. Metabolic control of CAM. *In* I. P. Ting and M. Gibbs (eds.) *Crassulacean Acid Metabolism*. American Society of Plant Physiologists, Rockville (MD), pp. 31–50.

Kluge, M., O. L. Lange, M. V. Eichmann, and R. Schmid. 1973. Diurnaler Säurerhythmus bei *Tillandsia usneoides*: Untersuchungen über den Weg des kohlenstoffs sowie die Abhängigkeit des CO_2-

Gaswechsels von Lichtintensität, Temperatur und Wassergehalt der Pflanze. *Planta* 112: 357–372.

Kluge, M., and I. P. Ting. 1978. *Crassulacean Acid Metabolism: Analysis of an Ecological Adaptation.* Ecological Studies Series, Volume 30. Springer-Verlag, Berlin, 209 pp.

Koch, K. E., and R. A. Kennedy. 1980. Effects of seasonal changes in the Midwest on Crassulacean acid metabolism (CAM) in *Opuntia humifusa* Raf. *Oecologia* 45: 390–395.

Köhlein, F. 1976. Winterharte Agaven. *Gartenpraxis* 10: 498–499.

Kolar, Z., J. Bartek, and B. Vyskot. 1976. Vegetative propagation of the cactus *Mamillaria woodsii* Craig through tissue cultures. *Experientia* 32: 668–669.

Konis, E. 1950. On the temperature of *Opuntia* joints. *Palestine Journal of Botany, Jerusalem Series*, 5: 46–55.

Körner, Ch., and P. Cochrane. 1983. Influence of plant physiognomy on leaf temperature on clear midsummer days in the Snowy Mountains, south-eastern Australia. *Acta Œcologica/Œcologia Plantarum* 4: 117–124.

Kristen, U. 1969. Untersuchungen über den Zusammenhang zwischen dem CO_2-Gaswechsel und der Luftwegigkeit an den CAM-Sukkulenten *Bryophyllum daigremontianum* Berg. und *Agave americana* L. *Flora, Abteilung* A. *Physiologie und Biochemie* 160: 127–138.

Kuraishi, S., and N. Nito. 1980. The maximum leaf surface temperatures of the higher plants observed in the Inland Sea area. *The Botanical Magazine, Tokyo*, 93: 209–220.

Lange, O. L., P. S. Nobel, C. B. Osmond, and H. Ziegler (eds.). 1981 to 1983. *Physiological Plant Ecology I* (*Responses to the Physical Environment*; 1981, 625 pp.), *II* (*Water Relations and Carbon Assimilation*; 1982, 747 pp.), *III* (*Responses to the Chemical and Biological Environment*; 1983, 799 pp.) and *IV* (*Ecosystem Processes: Mineral Cycling, Productivity and Man's Influence*; 1983, 644 pp.), *Encyclopedia of Plant Physiology, New Series, Volumes 12A–D.* Springer-Verlag, Berlin.

Larcher, W. 1980. *Physiological Plant Ecology, 2nd Edition.* Springer-Verlag, Berlin, 252 pp.

Lehninger, A. L. 1982. *Principles of Biochemistry.* Worth, New York, 1011 pp.

Le Houerou, H. N. 1970. North Africa: Past, present, future. *In* H. E. Dregne (ed.) *Arid Lands in Transition.* Publication 90, American Association for the Advancement of Science, Washington, D.C., pp. 227–278.

Lemon, E. R. (ed.). 1983. *CO_2 and Plants: The Response of Plants to Rising Levels of Atmospheric Carbon Dioxide.* American Association for the Advancement of Sciences Selected Symposium 84, AAAS, Westview Press, Boulder (CO), 280 pp.

Lerman, J. C. 1975. How to interpret variations in the carbon isotope ratio of plants: biologic and environmental effects. *In* R. Marcelle (ed.) *Environmental and Biological Control of Photosynthesis.* W. Junk, The Hague, pp. 323–335.

Levitt, J. 1978. An overview of freezing injury and survival, and its interrelationships to other stresses. *In* P. H. Li and A. Sakai (eds.) *Plant Cold Hardiness and Freezing Stress.* Academic Press, New York, pp. 3–15.

Levitt J. 1980. *Responses of Plants to Environmental Stresses, 2nd Ed., Volume 1, Chilling, Freezing, and High Temperature Stresses.* Academic Press, New York, 497 pp.

Lewis, D. A., and P. S. Nobel. 1977. Thermal energy exchange model and water loss of a barrel cactus, *Ferocactus acanthodes. Plant Physiology* 60: 609–616.

Lieth, H., and R. H. Whittaker (eds.). 1975. *Primary Productivity of the Biosphere. Ecological Studies, Volume 14.* Springer-Verlag, New York, 339 pp.

Lindsay, G. 1952. The use of cactus as stock food. *Desert Plant Life* 24: 4–6.

List, R. T. 1963. *Smithsonian Meteorological Tables, 6th Edition.* Smithsonian Institution Press, Washington, D.C., 527 pp.

Littlejohn, R. O., Jr. 1983. Environmental regulation of photosynthesis and characterization of phase II Crassulacean acid metabolism in three *Opuntia* species. Ph.D. thesis, Washington State University, Pullman, 149 pp.

Littlejohn, R. O., Jr., J. M. Green, and G. J. Williams, III. 1982. A morning peak in acidity during phase II Crassulacean acid metabolism in *Opuntia erinacea, Echinocereus viridiflorus,* and *Mammillaria vivipara. Plant Science Letters* 27: 43–49.

Littlejohn, R. O., and M. S. B. Ku. 1984. Characterization of early morning Crassulacean acid metabolism in *Opuntia erinaceae* var Columbiana (Griffiths) L. Benson. *Plant Physiology* 74: 1050–1054.

Littlejohn, R. O., and M. S. B. Ku. 1985. Light and temperature regulation of early morning Crassulacean acid metabolism in *Opuntia erinaceae* var Columbiana (Griffiths) L. Benson. *Plant Physiology* 77: 489–491.

Littlejohn, R. O., Jr., and G. J. Williams, III. 1983. Diurnal and seasonal variations in activity of Crassulacean acid metabolism and plant water status in a northern latitude population of *Opuntia erinaceae. Oecologia* 59: 83–87.

Livingston, B. E. 1906. *The Relation of Desert Plants to Soil Moisture, and to Evaporation.* Publication 50, Carnegie Institution of Washington, Washington, D.C., 78 pp.

Livingston, B. E. 1907. Relative transpiration in cacti. *The Plant World* 10: 110–114.

Lock, G. W. 1962. *Sisal: Twenty-five Years Sisal Research.* Longmans, London, 355 pp.

Lock, G. W. 1985. On the scientific and practical aspects of sisal (*Agave sisalana*) cultivation. *In* C. Cruz, L. del Castillo, M. Robert, and R. N. Ondarza (eds.), *Biología y Aprovechamiento Integral del Henequén y Otros Agaves.* Centro de Investigación Científica de Yucatán, A.C., Mérida, Yucatán, pp. 99–119.

Lonard, R. I., and F. W. Judd. 1985. Effects of a severe freeze on native woody plants in the lower Rio

Grande Valley, Texas. *The Southwestern Naturalist* 30: 397–403.

Long, E. R. 1915. Acid accumulation and destruction in large succulents. *The Plant World* 18: 261–272.

Loomis, R. S. 1983. Productivity of agricultural systems. *In* O. L. Lange, P. S. Nobel, C. B. Osmond, and H. Ziegler (eds.) *Physiological Plant Ecology IV. Ecosystem Processes: Mineral Cycling, Productivity and Man's Influence, Encyclopedia of Plant Physiology, New Series, Volume 12D.* Springer-Verlag, Berlin, pp. 151–172.

Loomis, R. S., and W. A. Williams. 1969. Productivity and the morphology of crop stands: patterns with leaves. *In* J. O. Eastin, F. A. Haskins, C. Y. Sullivan, and C. H. M. van Bavel (eds.) *Physiological Aspects of Crop Yield.* American Society of Agronomy and Crop Society of America, Madison, pp. 27–47.

Loomis, R. S., W. A. Williams, and A. E. Hall. 1971. Agricultural productivity. *Annual Review of Plant Physiology* 22: 431–468.

Lopez G., J. J., J. Gastó C., and R. Nava C. 1981. Analisis Cuantitativo de la Arquitectura de *Opuntia streptacantha* Lemaire en Poblaciones Naturales. *Monografia Técnico-Científica*, Vol. 7, No. 3. Universidad Autonoma Agraria Antonio Narro, Saltillo, Coahuila, pp. 127–169.

Lopez G., J. J., J. M. Gasto C., R. Nava C., and J. G. Medina T. 1977. Ecosistema *Opuntia streptacantha* Lemaire. *Monografia Técnico-Científica*, Vol. 3, No. 5. Universidad Autonoma Agraria Antonio Narro, Saltillo, Coahuila, pp. 388–545.

López G., J. J., R. Nava C., and J. Gastó C. 1978. Ecosistemas Naturales y Diseñados en *Opuntia engelmannii* y *Opuntia cantabrigiensis. Monografia Técnico-Científica*, Vol. 4, No. 1. Universidad Autonoma Agraria Antonio Narro, Saltillo, Coahuila, pp. 1–117.

Loucks, M., and J. D. Ownby. 1978. Effect of pH and metabolic inhibitors on stomatal opening in *Crassula argentea. Botanical Gazette* 139: 381–384.

Lowe, C. H., and D. S. Hinds. 1971. Effect of paloverde (*Cercidium*) trees on the radiation flux at ground level in the Sonoran Desert in the winter. *Ecology* 52: 916–922.

Lozano, R., and W. Reid. 1982. Claret cup cactus at White Sands National Monument. *Cactus & Succulent Journal (U.S.)* 54: 196–201.

Lumholtz, C. 1912. *New Trails in Mexico.* Scribner & Sons, New York, 411 pp.

Lüttge, U., and E. Ball. 1980. $2H^+:1$ malate^{2-} stoichiometry during Crassulacean acid metabolism is unaffected by lipophilic cations. *Plant, Cell and Environment* 3: 195–200.

Lüttge, U., and P. S. Nobel. 1984. Day-night variations in malate concentration, osmotic pressure, and hydrostatic pressure in *Cereus validus. Plant Physiology* 75: 804–807.

Lüttge, U., and J. A. C. Smith. 1984. Mechanism of passive malic-acid efflux from vacuoles of the CAM plant *Kalanchoe daigremontiana. Journal of Membrane Biology* 81: 149–158.

Lüttge, U., J. A. C. Smith, and G. Marigo. 1982.

Membrane transport, osmoregulation, and the control of CAM. *In* I. P. Ting and M. Gibbs (eds.) *Crassulacean Acid Metabolism.* American Society of Plant Physiologists, Rockville (MD), pp. 69–91.

MacCallum, W. B. 1908. The flowering stalk of the century plant. *The Plant World* 11: 141–147.

MacDougal, D. T. 1903. The influence of light and darkness upon growth and development. *Memoirs of the New York Botanical Garden* 2: 1–319.

MacDougal, D. T. 1908. *Botanical Features of North American Deserts.* Publication 99, Carnegie Institution of Washington, Washington, D.C., 111 pp.

MacDougal, D. T. 1910. *Variations of the Water-Balance.* Carnegie Institution of Washington, Washington, D.C., pp. 45–77.

MacDougal, D. T. 1912. The water-balance of desert plants. *Annals of Botany* 26: 71–93.

MacDougal, D. T. 1921. A new high temperature record for growth. *Science, New Series* 53: 370–372.

MacDougal, D. T. 1924. *Growth in Trees and Massive Organs of Plants – Dendrographic Measurements.* Publication 350, Carnegie Institution of Washington, Washington, D.C., pp. 1-88.

MacDougal, D. T., and W. A. Cannon. 1910. *The Conditions of Parasitism in Plants.* Publication 129, Carnegie Institution of Washington, Washington, D.C., 60 pp.

MacDougal, D. T., E. R. Long, and J. G. Brown. 1915. End results of desiccation and respiration in succulent plants. *Physiological Researches* 1: 289–325.

MacDougal, D. T., and E. S. Spalding. 1910. *The Water-Balance of Succulent Plants.* Publication 141, Carnegie Institution of Washington, Washington, D.C., pp. 1–44.

MacDougal, D. T., and E. B. Working. 1921. Another high-temperature record for growth and endurance. *Science, New Series* 54: 152–153.

MacDougal, D. T., and E. B. Working. 1922. A new high-temperature record for growth. *Carnegie Institution of Washington Year Book* 20: 47–48.

MacDougal, D. T., and E. B. Working. 1933. *The Pneumatic System of Plants, Especially Trees.* Publication 441, Carnegie Institution of Washington, Washington, D.C., 87 pp.

MacEwan, R. C. 1973. Freeze hardiness effect in agaves. *American Horticulturist* 52: 17.

McAuliffe, J. R. 1984a. Sahuaro-nurse tree associations in the Sonoran Desert: competitive effects of sahuaros. *Oecologia* 64: 319–321.

McAuliffe, J. R. 1984b. Prey refugia and the distributions of two Sonoran Desert cacti. *Oecologia* 65: 82–85.

McAuliffe, J. R., and F. J. Janzen. 1986. Effects of intraspecific crowding on water uptake, water storage, apical growth, and reproductive potential in the sahuaro cactus, *Carnegiea gigantea. Botanical Gazette* 147: 334–341.

McDaniel, R. G. 1985. Field evaluation of agave in Arizona. *Desert Plants* 7: 57–60, 101.

McDonough, W. T. 1964. Germination responses of *Carnegiea gigantea* and *Lemaireocereus thurberi. Ecology* 45: 155–159.

McGarvie, D., and H. Parolis. 1981. The mucilage of

Opuntia aurantiaca. Carbohydrate Research 94: 67–71.

McGee, J. M. 1916. The effect of position upon the temperature and dry weight of joints of *Opuntia. Carnegie Institution of Washington Year Book* 15: 73–74.

McGinnies, W. G. 1981. *Discovering the Desert: Legacy of the Carnegie Desert Botanical Laboratory.* University of Arizona Press, Tucson, 276 pp.

McGregor, S. E., S. M. Alcorn, E. B. Kurtz, Jr., and G. D. Butler, Jr. 1959. Bee visitors to saguaro flowers. *Journal of Economic Entomology* 52: 1002–1004.

McIntosh, R. P. 1983. Pioneer support for ecology. *BioScience* 33: 107–112.

McKelvey, S. D., and K. Sax. 1933. Taxonomic and cytological relationships of *Yucca* and *Agave. Journal of the Arnold Arboretum, Harvard University*, 14: 76–80.

McLaughlin, S. P., and J. E. Bowers. 1982. Effects of wildfire on a Sonoran Desert community. *Ecology* 63: 246–248.

McLeod, M. G. 1975. A new fleshy-fruited prickly-pear in California. *Madroño* 23: 96–98.

Martin, C. E., C. A. Eades, and R. A. Pitner. 1986. Effects of irradiance on Crassulacean acid metabolism in the epiphyte *Tillandsia usneoides* L. (Bromeliaceae). *Plant Physiology* 80: 23–26.

Martin, C. E., A. E. Lubbers, and J. A. Teeri. 1982. Variability in Crassulacean acid metabolism: a survey of North Carolina succulent species. *Botanical Gazette* 143: 491–497.

Martinez Davila, M. 1980. Alternativas para el manejo y transformacion de *Opuntia engelmannii* Salm-Dyck y *O. cantabrigiensis* Lynch en el noreste de Zacatecas. Professional thesis, Universidad Autonoma Agraria "Antonio Narro," Saltillo, Coahuila, 58 pp.

Martinez Morales, R. 1985. Demografia en una Poblacion Silvestre de Maguey Mezcalero (*Agave salmiana* ssp. *crassispina* (Trel.) Gentry), bajo Condiciones de Utilizacion Intensiva. Professional thesis, Universidad Veracruzana. Cordoba, Veracruz, 61 pp.

Martinez-Morales, R., and S. E. Meyer. 1985. A demographic study of maguey verde (*Agave salmiana* ssp. *crassispina*) under conditions of intensive utilization. *Desert Plants* 7: 61–64, 101–102.

Master, R. W. P. 1959. Organic acid and carbohydrate metabolism in *Nopalea cochinellifera. Experientia* 15: 30–31.

Mathur, D. D., N. J. Natarella, and H. M. Vines. 1978. Element analysis of Crassulacean acid metabolism plant tissue. *Communications in Soil Science and Plant Analysis* 9: 127–129.

Mathur, D. D., and H. M. Vines. 1982. Study of multiple characteristics of Crassulacean acid metabolism in a plant. *Indian Journal of Plant Physiology* 25: 49–54.

Mauseth, J. D. 1979. A new method for the propagation of cacti: sterile culture of axillary buds. *Cactus & Succulent Journal (U.S.)* 51: 186–187.

Mauseth, J. D. 1983a. Introduction to cactus anatomy. Part 2. Apical meristems. *Cactus & Succulent Journal (U.S.)* 55: 18–21, 42.

Mauseth, J. D. 1983b. Introduction to cactus anatomy, Part 3. Cell structure. *Cactus & Succulent Journal (U.S.)* 55: 84–89.

Mauseth, J. D., and K. J. Niklas. 1979. Constancy of relative volumes of zones in shoot apical meristems in Cactaceae: implications concerning meristem size, shape, and metabolism. *American Journal of Botany* 66: 933–939.

Medina, J. G., E. Acuña, and J. A. De la Cruz. 1988. Opuntia revegetation: an agroecological restoration alternative for deteriorated rangelands in Coahuila, Mexico. *In* E. E. Whitehead, C. F. Hutchinson, B. N. Timmermann, and R. G. Varady (eds.) *Arid Lands: Today and Tomorrow.* Westview Press, Boulder (CO), pp. 127–136.

Mekhtiev, T. A. 1972. [Frost resistance of some subtropical decorative plants] (in Azerbian). *Doklady Akademic Nauk Azerbaidzhanskoi SSR (Azarbaijan SSR Elmlar Akademiyasy Doklady)* 28: 62–64.

de Menezes, T. J. B., and A. Azzini. 1985. Utilization of agroindustrial products for the manufacture of ethanol. *In* C. Cruz, L. del Castillo, M. Robert, and R. N. Ondarza (eds.) *Biología y Aprovechamiento Integral del Henequén y Otros Agaves.* Centro de Investigación Científica de Yucatán, A.C., Mérida, Yucatán, pp. 139–148.

Metral, J. J. 1965. Les cactées fourragères dans le nord-est du Brésil plus particulièrement dans l'état du Ceara. *L'Agronomie Tropicale* 20: 248–261.

Meyer, M. W., and R. D. Brown. 1985. Seasonal trends in the chemical composition of ten range plants in south Texas. *Journal of Range Management* 38: 154–157.

Mindt, L., K. Saag, G. R. Sanderson, P. Moyna, and G. Ramos. 1975. Cactaceae mucilage composition. *Journal of the Science of Food and Agriculture* 26: 993–1000.

Molz, F. J., and Ferrier, J. M. 1982. Mathematical treatment of water movement in plant cells and tissues: a review. *Plant, Cell and Environment* 5: 191–206.

Monjauze, A., and H. N. Le Houérou. 1965. Le rôle des Opuntia dans l'économie agricole Nord Africaine. *Bulletin École National Supérieure de Tunisi* 8–9: 85–164.

Monteith, J. L. 1977. Climate and the efficiency of crop production in Britain. *Philosophical Transactions of the Royal Society of London, Series B*, 281: 277–294.

Mooney, H., J. H. Troughton, and J. A. Berry. 1974. Arid climates and photosynthetic systems. *Carnegie Institution Year Book 73*: 793–805.

Mooney, H. A., P. J. Weisser, and S. L. Gulmon. 1977. Environmental adaptations of the Atacaman Desert cactus *Copiapoa haseltoniana. Flora* 166: 117–124.

Mozingo, H. N., and P. L. Comanor. 1975. Implications of the thermal response of *Ferocactus acanthodes. Supplemental Volume of the Cactus & Succulent Journal (U.S.)* 47: 22–28.

Mukerji, S. K., G. G. Sanwal, and P. S. Krishnan. 1964. Four-hourly variations in enzymic activities in the cactus (*Nopalea dejecta*) plant. *Indian Journal of Biochemistry* 1: 36–40.

Mukerji, S. K., and I. P. Ting. 1968. Intracellular

localization of CO_2 metabolism enzymes in cactus phylloclades. *Phytochemistry* 7: 903–911.

Nagano, T., S. Kuraishi, and N. Nito. 1980. Temperature of cacti grown in the cactus-house in Aichi Prefecture. *Environmental Control in Biology* 18: 11–19.

Nagel, L. 1975. *SPICE: A Computer Program to Simulate Semiconductor Circuits*. ERL-M520. Electronics Research Laboratory, University of California, Berkeley, 429 pp.

Nardina, N. S., and G. M. Mukhammedov. 1973. Kul'tura vidov Opuntia Mill. v tsentral'nykh Karakumakh. *Problemy Osvoeniya Pustyn'* 5: 60–61.

Neales, T. F. 1970. Effect of ambient carbon dioxide concentration on the rate of transpiration of *Agave americana* in the dark. *Nature* 228: 880–882.

Neales, T. F. 1973. The effect of night temperature on CO_2 assimilation, transpiration, and water use efficiency in *Agave americana* L. *Australian Journal of Biological Sciences* 26: 705–714.

Neales, T. F. 1975. The gas exchange pattern of CAM plants. *In* R. Marcelle (ed.) *Environmental and Biological Control of Photosynthesis*. Dr. W. Junk, The Hague, pp. 299–310.

Neales, T. F., A. A. Patterson, and V. J. Hartney. 1968. Physiological adaptation to drought in the carbon assimilation and water loss of xerophytes. *Nature* 219: 469–472.

Niering, W. A., and C. H. Lowe. 1984. Vegetation of the Santa Catalina Mountains: community types and dynamics. *Vegetatio* 58: 3–28.

Niering, W. A., R. H. Whittaker, and C. H. Lowe. 1963. The saguaro: a population in relation to environment. *Science* 142: 15–23.

Nisbet, R. A., and D. T. Patten. 1974. Seasonal temperature acclimation of a prickly-pear cactus in south-central Arizona. *Oecologia* 15: 345–352.

Nobel, P. S. 1974. Boundary layers of air adjacent to cylinders. Estimation of effective thickness and measurements on plant material. *Plant Physiology* 54: 177–181.

Nobel, P. S. 1975. Effective thickness and resistance of the air boundary layer adjacent to spherical plant parts. *Journal of Experimental Botany* 26: 120–130.

Nobel, P. S. 1976. Water relations and photosynthesis of a desert CAM plant, *Agave deserti*. *Plant Physiology* 58: 576–582.

Nobel, P. S. 1977a. Water relations of flowering of *Agave deserti*. *Botanical Gazette* 138: 1–6.

Nobel, P. S. 1977b. Water relations and photosynthesis of a barrel cactus, *Ferocactus acanthodes*, in the Colorado Desert. *Oecologia* 27: 117–133.

Nobel, P. S. 1978. Surface temperatures of cacti – influences of environmental and morphological factors. *Ecology* 59: 986–996.

Nobel, P. S. 1980a. Interception of photosynthetically active radiation by cacti of different morphology. *Oecologia* 45: 160–166.

Nobel, P. S. 1980b. Leaf anatomy and water use efficiency. *In* N. C. Turner and P. J. Kramer (eds.) *Adaptation of Plants to Water and High Temperature Stress*. John Wiley & Sons, New York, pp. 43–55.

Nobel, P. S. 1980c. Morphology, surface temperatures, and northern limits of columnar cacti in the Sonoran Desert. *Ecology* 61: 1–7.

Nobel, P. S. 1980d. Morphology, nurse plants, and minimum apical temperatures for young *Carnegiea gigantea*. *Botanical Gazette* 141: 188–191.

Nobel, P. S. 1980e. Influences of minimum stem temperatures on ranges of cacti in southwestern United States and central Chile. *Oecologia* 47: 10–15.

Nobel, P. S. 1981a. Influence of freezing temperatures on a cactus, *Coryphantha vivipara*. *Oecologia* 48: 194–198.

Nobel, P. S. 1981b. Influences of photosynthetically active radiation on cladode orientation, stem tilting, and height of cacti. *Ecology* 62: 982–990.

Nobel, P. S. 1982a. Low-temperature tolerance and cold hardening of cacti. *Ecology* 63: 1650–1656.

Nobel, P. S. 1982b. Orientations of terminal cladodes of platyopuntias. *Botanical Gazette* 143: 219–224.

Nobel, P. S. 1982c. Orientation, PAR interception, and nocturnal acidity increases for terminal cladodes of a widely cultivated cactus, *Opuntia ficus-indica*. *American Journal of Botany* 69: 1462–1469.

Nobel, P. S. 1982d. Interaction between morphology, PAR interception, and nocturnal acid accumulation in cacti. *In* I. P. Ting and M. Gibbs (eds.) *Crassulacean Acid Metabolism*. American Society of Plant Physiologists, Rockville (MD), pp. 260–277.

Nobel, P. S. 1983a. *Biophysical Plant Physiology and Ecology*. W. H. Freeman, San Francisco/New York, 608 pp.

Nobel, P. S. 1983b. Spine influences on PAR interception, stem temperature, and nocturnal acid accumulation by cacti. *Plant, Cell and Environment* 6: 153–159.

Nobel, P. S. 1983c. Low and high temperature influences on cacti. *In* R. Marcelle, H. Clijsters, and M. van Poucke (eds.) *Effect of Stress on Photosynthesis*. Martinus Nijhoff/Dr. W. Junk, The Hague, pp. 165–174.

Nobel, P. S. 1983d. Nutrient levels in cacti – relation to nocturnal acid accumulation and growth. *American Journal of Botany* 70: 1244–1253.

Nobel, P. S. 1984a. Productivity of *Agave deserti*: measurement by dry weight and monthly prediction using physiological responses to environmental parameters. *Oecologia* 64: 1–7.

Nobel, P. S. 1984b. PAR and temperature influences on CO_2 uptake by desert CAM plants. Proceedings, VI International Congress on Photosynthesis. *Advances in Photosynthesis Research* IV. 3: 193–200.

Nobel, P. S. 1984c. Extreme temperatures and the thermal tolerances for seedlings of desert succulents. *Oecologia* 62: 310–317.

Nobel, P. S. 1985a. PAR, water and temperature limitations on the productivity of cultivated *Agave fourcroydes* (henequen). *Journal of Applied Ecology* 22: 157–173.

Nobel, P. S. 1985b. Environmental responses of agaves: a case study with *Agave deserti*. *In* C. Cruz, L. del Castillo, M. Robert, and R. N. Ondarza (eds.) *Biología y Aprovechamiento Integral del Henequén y*

Otros Agaves. Centro de Investigación Científica de Yucatán, A.C., Mérida, Yucatán, pp. 55–66.

Nobel, P. S. 1985c. Water relations and carbon dioxide uptake of *Agave deserti* – special adaptations to desert climates. *Desert Plants* 7: 51–56, 70.

Nobel, P. S. 1986a. Relation between monthly growth of *Ferocactus acanthodes* and an environmental productivity index. *American Journal of Botany* 73: 541–547.

Nobel, P. S. 1986b. Form and orientation in relation to PAR interception by cacti and agaves. *In* T. J. Givnish (ed.), *On the Economy of Plant Form and Function*. Cambridge University Press, Cambridge, pp. 83–103.

Nobel, P. S. 1987a. Water relations and plant size aspects of flowering for *Agave deserti*. *Botanical Gazette* 148: 79–84.

Nobel, P. S. 1987b. Transpiration analysis using resistances and capacitances. *In* D. W. Newman and K. G. Wilson (eds.) *Models in Plant Physiology and Biochemistry, Volume III*. CRC Press, Boca Raton (FL), pp. 37–39.

Nobel, P. S. 1988. Productivity of desert succulents. *In* E. E. Whitehead, C. F. Hutchinson, B. N. Timmermann, and R. G. Varady (eds.) *Arid Lands: Today and Tomorrow*. Westview Press, Boulder (CO), pp. 137–148.

Nobel, P. S., and W. L. Berry. 1985. Element responses of agaves. *American Journal of Botany* 72: 686–694.

Nobel, P. S., and V. García de Cortázar. 1987. PAR interception and predicted productivity for agave rosettes. *Photosynthetica, in press*.

Nobel, P. S., and G. N. Geller. 1987. Temperature modelling of wet and dry desert soils. *Journal of Ecology* 75: 247–258.

Nobel, P. S., G. N. Geller, S. C. Kee, and A. D. Zimmerman. 1986. Temperatures and thermal tolerances for cacti exposed to high temperatures near the soil surface. *Plant, Cell and Environment* 9: 279–287.

Nobel, P. S., and T. L. Hartsock. 1978. Resistance analysis of nocturnal carbon dioxide uptake by a Crassulacean acid metabolism succulent, *Agave deserti*. *Plant Physiology* 61: 510–514.

Nobel, P. S., and T. L. Hartsock. 1979. Environmental influences on open stomates of a Crassulacean acid metabolism plant, *Agave deserti*. *Plant Physiology* 63: 63–66.

Nobel, P. S., and T. L. Hartsock. 1981. Shifts in the optimal temperature for nocturnal CO_2 uptake caused by changes in growth temperature for cacti and agaves. *Physiologia Plantarum* 53: 523–527.

Nobel, P. S., and T. L. Hartsock. 1983. Relationships between photosynthetically active radiation, nocturnal acid accumulation, and CO_2 uptake for a Crassulacean acid metabolism plant, *Opuntia ficus-indica*. *Plant Physiology* 71: 71–75.

Nobel, P. S., and T. L. Hartsock. 1984. Physiological responses of *Opuntia ficus-indica* to growth temperature. *Physiologia Plantarum* 60: 98–105.

Nobel, P. S., and T. L. Hartsock. 1986a. Leaf and stem CO_2 uptake in the three subfamilies of the Cactaceae. *Plant Physiology* 80: 913–917.

Nobel, P. S., and T. L. Hartsock. 1986b. Influence of nitrogen and other nutrients on the growth of *Agave deserti*. *Journal of Plant Nutrition* 9: 1273–1288.

Nobel, P. S., and T. L. Hartsock. 1986c. Temperature, water, and PAR influences on predicted and measured productivity of *Agave deserti* at various elevations. *Oecologia* 68: 181–185.

Nobel, P. S., and T. L. Hartsock. 1986d. Environmental influences on the productivity of three desert succulents in the southwestern United States. *Plant, Cell and Environment* 9: 741–749.

Nobel, P. S., and T. L. Hartsock. 1986e. Short-term and long-term responses of Crassulacean acid metabolism plants to elevated CO_2. *Plant Physiology* 82: 604–606.

Nobel, P. S., and T. L. Hartsock. 1987. Drought-induced shifts in daily CO_2 uptake patterns for leafy cacti. *Physiologia Plantarum* 70: 114–118.

Nobel, P. S., and P. W. Jordan. 1983. Transpiration stream of desert species: resistances and capacitances for a C_3, a C_4, and a CAM plant. *Journal of Experimental Botany* 34: 1379–1391.

Nobel, P. S., D. J. Longstreth, and T. L. Hartsock. 1978. Effect of water stress on the temperature optimum of net CO_2 exchange for two desert species. *Physiologia Plantarum* 44: 97–101.

Nobel, P. S., U. Lüttge, S. Heuer, and E. Ball. 1984. Influence of applied NaCl on Crassulacean acid metabolism and ionic levels in a cactus, *Cereus validus*. *Plant Physiology* 75: 799–803.

Nobel, P. S., and R. G. McDaniel. 1988. Low temperature tolerances, nocturnal acid accumulation, and biomass increases for seven species of *Agave*. *Journal of Arid Environments*, in press.

Nobel, P. S., and S. E. Meyer. 1985. Field productivity of a CAM plant, *Agave salmiana*, estimated using daily acidity changes under various environmental conditions. *Physiologia Plantarum* 65: 397–404.

Nobel, P. S., and E. Quero. 1986. Environmental productivity indices for a Chihuahuan Desert CAM plant, *Agave lechuguilla*. *Ecology* 67: 1–11.

Nobel, P. S., C. E. Russell, P. Felker, J. G. Medina, and E. Acuña. 1987. Nutrient relations and productivity of prickly pear cacti. *Agronomy Journal* 79: 550–555.

Nobel, P. S., and J. Sanderson. 1984. Rectifier-like activities of roots of two desert succulents. *Journal of Experimental Botany* 35: 727–737.

Nobel, P. S., and S. D. Smith. 1983. High and low temperature tolerances and their relationships to distribution of agaves. *Plant, Cell and Environment* 6: 711–719.

Nobel, P. S., and A. G. Valenzuela. 1987. Environmental responses and productivity of the CAM plant, *Agave tequilana*. *Agricultural and Forest Meteorology* 39: 319–334.

Nobel, P. S., L. J. Zaragoza, and W. K. Smith. 1975. Relation between mesophyll surface area, photosynthetic rate, and illumination level during development for leaves of *Plectranthus parviflorus* Henckel. *Plant Physiology* 55: 1067–1070.

Norman, F., and C. E. Martin. 1986. Effects of spine removal on *Coryphantha vivipara* in central Kansas. *The American Midland Naturalist* 116: 118–124.

Nose, A., S. Matake, K. Miyazato, and S. Murayama. 1985. Studies on matter production in pineapple plants III. Effects of nitrogen nutrition on the gas exchange of plant tops. *Japanese Journal of Crop Science* 54: 195–204.

Nose, A., M. Shiroma, K. Miyazato, and S. Murayama. 1977. Studies on matter production in pineapple plants I. Effects of light intensity in light period on the CO_2 exchange and CO_2 balance of pineapple plants. *Japanese Journal of Crop Science* 46: 580–587.

Noy-Meir, I. 1973. Desert ecosystems: environment and producers. *Annual Review of Ecology and Systematics* 4: 25–51.

Nuernbergk, E. L. 1961. Endogener Rhythmus und CO_2-Stoffwechsel bei Pflanzen mit diurnalem Säurerhythmus. *Planta* 56: 28–70.

O'Leary, M. H. 1981. Carbon isotope fractionation in plants. *Phytochemistry* 20: 553–567.

O'Leary, M. H., and C. B. Osmond. 1980. Diffusional contribution to carbon isotope fractionation during dark CO_2 fixation in CAM plants. *Plant Physiology* 66: 931–934.

Odum, E. P. 1971. *Fundamentals of Ecology, 3rd Ed.* W. B. Saunders, Philadelphia, 574 pp.

Oeschager, H., and J. Lerman. 1970. Die Radiokarbonmethode nach 20 Jahren Annendung. *Chemische Rundschau* 23: 585–588.

Oke, T. R. 1978. *Boundary Layer Climates*. Methuen, London, 372 pp.

Olszyk, D. M., A. Bytnerowicz, and C. A. Fox. 1987. Sulfur dioxide effects on plants exhibiting Crassulacean acid metabolism. *Environmental Pollution* 43: 47–62.

Onwueme, I. C. 1979. Rapid, plant-conserving estimate of heat tolerance in plants. *Journal of Agricultural Science, Cambridge*, 92: 527–536.

Osmond, C. B. 1978. Crassulacean acid metabolism, a curiosity in context. *Annual Review of Plant Physiology* 29: 379–414.

Osmond, C. B. 1981. Photorespiration and photoinhibition – some implications for the energetics of photosynthesis. *Biochimica et Biophysica Acta* 639: 77–98.

Osmond, C. B. 1982. Carbon cycling and stability of the photosynthetic apparatus in CAM. *In* I. P. Ting and M. Gibbs (eds.) *Crassulacean Acid Metabolism*. American Society of Plant Physiologists, Rockville (MD), pp. 112–127.

Osmond, C. B., W. W. Adams, III, and S. D. Smith. 1988. Crassulacean acid metabolism. *In* R. A. Pearcy, J. R. Ehleringer, H. A. Mooney, and P. W. Rundel (eds.) *Physiological Plant Ecology: Methods and Instrumentation*, Chapman and Hall, New York, *in press*.

Osmond, C. B., W. G. Allaway, B. G. Sutton, J. H.

Troughton, Q. Queiroz, U. Lüttge, and K. Winter. 1973. Carbon isotope discrimination in photosynthesis of CAM plants. *Nature* 246: 41–42.

Osmond, C. B., O. Björkman, and D. J. Anderson. 1980. *Physiological Processes in Plant Ecology. Toward a Synthesis with Atriplex. Ecological Studies Series 36*. Springer-Verlag, Berlin, 468 pp.

Osmond, C. B., M. M. Ludlow, R. Davis, I. R. Cowan, S. B. Powles, and K. Winter. 1979a. Stomatal responses to humidity in *Opuntia inermis* in relation to control of CO_2 and H_2O exchange patterns. *Oecologia* 41: 65–76.

Osmond, C. B., and J. Monro. 1981. Prickly pear. *In* D. J. Carr and S. M. Carr (eds.) *Man and Plants in Australia*. Academic Press, Sydney, pp. 194–222.

Osmond, C. B., D. L. Nott, and P. M. Firth. 1979b. Carbon assimilation patterns and growth of the introduced CAM plant *Opuntia inermis* in eastern Australia. *Oecologia* 40: 331–350.

Padilla R., M. en C., and P. Fuentes R. 1985. Laminados polietileno – henequén. *In* C. Cruz, L. del Castillo, M. Robert, and R. N. Ondarza (eds.) *Biología y Aprovechamiento Integral del Henequén y Otros Agaves*. Centro de Investigación Científica de Yucatán, A.C., Mérida, Yucatán, pp. 223–230.

Pandey, O. P., and G. G. Sanwal. 1984. Diurnal variations of some glycolate pathway enzymes in *Nopalea dejecta*: a CAM plant. *Indian Journal of Plant Physiology* 27: 334–339.

Parker, K. C. 1987. Site-related demographic patterns of organ pipe cactus populations in southern Arizona. *Bulletin of the Torrey Botanical Club* 114: 149–155.

Patten, D. T., and B. E. Dinger. 1969. Carbon dioxide exchange patterns of cacti from different environments. *Ecology* 50: 686–688.

Penningsfeld, F. 1972. Macro and micro nutrient requirements of pot plants in peat. *Acta Horticulturae* 26: 81–101.

Peralta Mata, V. M. 1983. Caracterizacion Fenologica y Morphologica de Formas de Nopal (*Opuntia* spp) de Fruto (Tuna) en el Altiplano Potosino-Zacatecano. Professional thesis. Universidad Autónoma de Aguascalientes, Aguascalientes, Aguascalientes, 88 pp.

Phillips, R. D., and D. H. Jennings. 1976. Succulence, cations and organic acids in leaves of *Kalanchoe daigremontiana* grown in long and short days in soil and water culture. *New Phytologist* 77: 599–611.

Pieper, R. D., K. H. Rea, and J. G. Fraser. 1974. Ecological characteristics of walkingstick cholla. *New Mexico Agricultural Station Bulletin, Volume 623*, Las Cruces (NM), pp. 1–20.

Pinkava, D. J., and M. A. Baker. 1985. Chromosome and hybridization studies of agaves. *Desert Plants* 7: 93–100.

Pinkava, D. J., M. G. McLeod, L. A. McGill, and R. C. Brown. 1973. Chromosome numbers in some cacti of western North America.–II. *Brittonia* 25: 2–9.

Pinkava, D. J., and B. D. Parfitt. 1982. Chromosome numbers in some cacti of western North America IV. *Bulletin of the Torrey Botanical Club* 109: 121–128.

Pinkerton, A. 1971. Some micronutrient deficiencies in sisal (*Agave sisalana*). *Experimental Agriculture* 7: 113–122.

Potter, R. L., J. L. Petersen, and D. N. Ueckert. 1984. Germination responses of *Opuntia* spp. to temperature, scarification, and other seed treatments. *Weed Science* 32: 106–110.

Potter, R. L., J. L. Petersen, and D. N. Ueckert. 1986. Seasonal trends of total nonstructural carbohydrates in lindheimer pricklypear (*Opuntia lindheimeri*). *Weed Science* 34: 361–365.

Preston, C. E. 1900. Observations of the root system of certain Cactaceae. *Botanical Gazette* 30: 348–351.

Pushkaren, M., T. S. Suryanarayanan, P. Jayaraman, and K. R. Purohit. 1980. Influence of photoperiod and nutrient on the vegetative growth of *Echinopsis* sp. (Cactaceae). *Indian Journal of Botany* 3: 160–162.

Queiroz, O., and J. Brulfert. 1982. Photoperiod-controlled induction and enhancement of seasonal adaptation to drought. *In* I. P. Ting and M. Gibbs (eds.) *Crassulacean Acid Metabolism*. American Society of Plant Physiologists, Rockville (MD), pp. 208–230.

Quero, E., and P. S. Nobel. 1987. Predictions of field productivity for *Agave lechuguilla*. *Journal of Applied Ecology, in press.*

Rabinowitch, E. I. 1951. *Photosynthesis and Related Processes, Volume II, Part 1*. Interscience, New York, 606 pp.

Racine, C. H., and J. F. Downhower. 1974. Vegetation and reproductive strategies of *Opuntia* (Cactaceae) in the Galapagos Islands. *Biotropica* 6: 175–186.

Raison, J. K., J. A. Berry, P. A. Armond, and C. S. Pike. 1980. Membrane properties in relation to the adaptation of plants to temperature stress. *In* N. C. Turner and P. J. Kramer (eds.) *Adaptation of Plants to Water and High Temperature Stress*. John Wiley & Sons, New York, pp. 271–273.

Rajashekar, C., L. V. Gusta, and M. J. Burke. 1979. Membrane structural transitions: probable relation to frost damage in hardy herbaceous species. *In* J. M. Lyons, D. Graham, and J. K. Raison (eds.) *Low Temperature Stress in Crop Plants*. Academic Press, New York, pp. 255–274.

Ramachandra Reddy, A., and V. S. Rama Das. 1978. The decarboxylating systems in fourteen taxa exhibiting CAM pathway. *Zeitschrift für Pflanzenphysiologie* 86: 141–146.

Ramírez, E. A. 1985. *El Ixtle. Un Sistema Sociotécnico.* Centro de Investigación en Química Aplicada, Saltillo, Coahuila, 241 pp.

Raphael, D. O., and P. S. Nobel. 1986. Growth and survivorship of ramets and seedlings of *Agave deserti*: influences of parent-ramet connections. *Botanical Gazette* 147: 78–83.

Rayder, L., and I. P. Ting. 1981. Carbon metabolism in two species of *Pereskia* (Cactaceae). *Plant Physiology* 68: 139–142.

Reñasco, G. 1976. *Cultivo de Tunales*. Boletín Divulgativo Number 44, Servicio Agrícola y Ganadero. Santiago, Chile, 35 pp.

Richards, H. M. 1915. *Acidity and Gas Interchange in Cacti*. Publication 209, Carnegie Institution of Washington, Washington, D.C., 107 pp.

Richards, L. A. (ed.). 1954. *Diagnosis and Improvement of Saline and Alkali Soils*. United States Department of Agriculture Handbook 60, U.S. Government Printing Office, Washington, D.C., 160 pp.

Robberecht, R., and P. S. Nobel. 1983. A Fibonacci sequence in rib number for a barrel cactus. *Annals of Botany* 51: 153–155.

Robert, M., and A. Garcia. 1985. El cultivo de tejidos vegetales y su posible aplicacion en el mejoramiento genetico de las agavaceas. *In* C. Cruz, L. del Castillo, M. Robert, and R. N. Ondarza (eds.) *Biología y Aprovechamiento Integral del Henequén y Otros Agaves*. Centro de Investigación Científica de Yucatán, A.C., Mérida, Yucatán, pp. 83–89.

Robert, M. L., J. L. Herrera, F. Contreras, and K. N. Scorer. 1987. In vitro propagation of *Agave fourcroydes* Lem. (Henequen). *Plant Cell, Tissue and Organ Culture* 8: 37–48.

Rogers, G. F. 1985. Mortality of burned *Cereus giganteus*. *Ecology* 66: 630–632.

Rössner, H. and M. Popp. 1986. Ionic patterns in some Crassulaceae from Austrian habitats. *Flora* 178: 1–10.

Royal Gardens, Kew. 1898. *Bulletin of Miscellaneous Information Additional Series, II. Selected Papers from the Kew Bulletin. I. Vegetable Fibres*. Kew, London, 280 pp.

Rundel, P. W. 1974. *Trichocereus* in the Mediterranean zone of central Chile. *Cactus & Succulent Journal (U.S.)* 46: 86–88.

Rundel, P. W., J. Ehleringer, H. A. Mooney, and S. L. Gulmon. 1980. Patterns of drought response in leaf-succulent shrubs of the coastal Atacama Desert in northern Chile. *Oecologia* 46: 196–200.

Rundel, P. W., and M. Mahu. 1976. Community structure and diversity in a coastal fog desert in northern Chile. *Flora* 165: 493–505.

Rünger, W. 1968. Über die Blütenbildung von Mammillaria longicoma. *Die Gartenbauwissenschaft* 33: 463–468.

Rünger, W. 1969. Blütenbildung von Rebutia marsoneri. *Gartenbauwissenschaft* 34: 511–515.

Rünger, W. 1979. Vegetatives Wachstum von Schlumbergera (Zygocactus). *Gartenbauwissenschaft* 44: 241–246.

Rünger, W., and H. Führer. 1981. Tageslänge, Temperatur und Blühreaktion bei Schlumbergera (Zygocactus). *Gartenbauwissenschaft* 46: 209–213.

Russell, C. E. 1985. Cactus, ecology and range management during drought. *In* R. D. Brown (ed.) *Livestock and Wildlife Management during Drought*. Texas A & I University, Kingsville, pp. 59–69.

Russell, R. S., and J. Sanderson. 1967. Nutrient uptake by different parts of intact roots of plants. *Journal of Experimental Botany* 18: 491–508.

Sale, P. J. M., and T. F. Neales. 1980. Carbon dioxide assimilation by pineapple plants, *Ananas comosus*

(L.) Merr. I. Effects of daily irradiance. *Australian Journal of Plant Physiology* 7: 363–373.

Salisbury, F. B., and C. W. Ross. 1985. *Plant Physiology, 3rd Edition.* Wadsworth, Belmont (CA), 540 pp.

Samish, Y. B., and S. J. Ellern. 1975. Titratable acids in *Opuntia ficus-indica* L. *Journal of Range Management* 28: 365–369.

Sanada, Y., and K. Nishida. 1982. The presence of pyruvate, orthophosphate dikinase in CAM plants. *Zeitschrift für Pflanzenphysiologie* 105: 189–192.

Sanchez Marroquin, A. 1979. *Los Agaves de Mexico en la Industria Alimentaria.* Centro de Estudios Economicos y Sociales del Tercer Mundo, Mexico City, 526 pp.

Sánchez-Mejorada R., H. 1982. *Some Prehispanic Uses of Cacti among the Indians of Mexico.* Secretaria de Desarrollo Agropecuario, Touca, México, 42 pp.

Sanderson, K. C., Y.-S. Ho, W. C. Martin, Jr., and B. N. Reed. 1986. Effect of photoperiod and growth regulators on growth of three Cactaceae. *HortScience* 21: 1381–1382.

Sauer, C. O. 1965. Cultural factors in plant domestication in the New World. *Euphytica* 14: 301–306.

Schaffer, W. M., and M. V. Schaffer. 1977. The reproductive biology of Agavaceae: I. Pollen and nectar production in four Arizona agaves. *The Southwestern Naturalist* 22: 157–168.

Schill, R., and W. Barthlott. 1973. Kakteendornen als wasserabsorbierende Organe. *Naturwissenschaften* 60: 202–203.

Schill, R., W. Barthlott, and N. Ehler. 1973. Cactus spines under the electron scanning microscope. *Cactus & Succulent Journal (U.S.)* 45: 175–185.

Schmidt, J. O., and S. L. Buchmann. 1986. Floral biology of the Saguaro (*Cereus giganteus*) I. Pollen harvest by *Apis mellifera. Oecologia* 69: 491–498.

Schnetter, R. 1971. Untersuchungen zum Wärme- und Wasserhaushalt ausgewählter Pflanzenarten des Trockengebietes von Santa Marta (Kolumbien). *Beiträge zur Biologie der Pflanzen* 47: 155–213.

Schroeder, C. A. 1975. Diurnal size fluctuations in Opuntia stem. *Botanical Gazette* 136: 94–98.

Schulze, E.-D. 1986. Carbon dioxide and water vapor exchange in response to drought in the atmosphere and in the soil. *Annual Review of Plant Physiology* 37: 247–274.

Seemann, J. R., J. A. Berry, and W. J. S. Downton. 1984. Photosynthetic response and adaptation to high temperature in desert plants. A comparison of gas exchange and fluorescence methods for studies of thermal tolerance. *Plant Physiology* 75: 364–368.

Seeni, S., and A. Gnanam. 1980. Photosynthesis in cell suspension cultures of the CAM plant *Chamaecereus sylvestrii* (Cactaceae). *Physiologia Plantarum* 49: 465–472.

Serey, I., and J. Simonetti. 1981. Ségrégation en populations des cactacees. *Acta Œcologica/Œcologia Plantarum* 2: 3–6.

Sheldon, S. 1980. Ethnobotany of Agave lecheguilla and

Yucca carnerosana in Mexico's Zona Ixtlera. *Economic Botany* 34: 376–390.

Shoop, M. C., E. J. Alford, and H. F. Mayland. 1977. Plains pricklypear is a good forage for cattle. *Journal of Range Management* 30: 12–17.

Shreve, E. B. 1915. An investigation of the causes of autonomic movements in succulent plants. *The Plant World* 18: 331–343.

Shreve, E. B. 1916. An analysis of the causes of variations in the transpiring power of cacti. *Physiological Researches* 2: 73–127.

Shreve, F. 1910. The rate of establishment of the giant cactus. *The Plant World* 13: 235–240.

Shreve, F. 1911. The influence of low temperatures on the distribution of the giant cactus. *The Plant World* 14: 136–146.

Shreve, F. 1931a. *The Cactus and Its Home.* William & Wilkins, Baltimore. 195 pp.

Shreve, F. 1931b. Physical conditions in sun and shade. *Ecology* 12: 96–104.

Shreve, F. 1935. The longevity of cacti. *Cactus and Succulent Journal of the Cactus and Succulent Society of America* 7: 66–68.

Shreve, F., and I. L. Wiggins. 1964. *Vegetation and Flora of the Sonoran Desert, Volumes I and II.* Stanford University Press, Stanford, 1740 pp.

Silverman, F. P., D. R. Young, and P. S. Nobel. 1988. Effects of applied NaCl on *Opuntia humifusa. Physiologia Plantarum*, in press.

Sinclair, J. G. 1922. Temperatures of the soil and air in the desert. *Monthly Weather Review* 50: 142–144.

Slavik, B. 1974. *Methods for Measuring Plant Water Relations.* Chapman and Hall, London, 449 pp.

Smith, B. N., and S. Madhaven. 1982. Carbon isotope ratios in obligate and facultative CAM plants. *In* I. P. Ting and M. Gibbs (eds.) *Crassulacean Acid Metabolism.* American Society of Plant Physiologists, Rockville (MD), pp. 231–243.

Smith, C. E., Jr., and M. L. Cameron. 1977. Ethnobotany in the Puuc, Yucatan. *Economic Botany* 31: 93–110.

Smith, H. H. 1929. *Sisal: Production and Preparation.* John Bale, Sons & Danielsson, London, 384 pp.

Smith, J. A. C. 1984. Water relations in CAM plants. *In* E. Medina (ed.) *Eco-Fisiologia de Plantas CAM.* Centro Internacional de Ecología Tropical, Caracas, Venezuela, pp. 30–51.

Smith, J. A. C., H. Griffiths, M. Bassett, and N. M. Griffiths. 1985. Day-night changes in the leaf water relations of epiphytic bromeliads in the rain forests of Trinidad. *Oecologia* 67: 475–485.

Smith, J. A. C. and P. S. Nobel. 1986. Water movement and storage in a desert succulent: anatomy and rehydration kinetics for leaves of *Agave deserti. Journal of Experimental Botany* 37: 1044–1053.

Smith, J. A. C., P. J. Schulte, and P. S. Nobel. 1987. Water flow and water storage in *Agave deserti:* osmotic implications of crassulacean acid metabolism. *Plant, Cell and Environment,* 10: 639–648.

Smith, S. D., B. Didden-Zopfy, and P. S. Nobel. 1984.

High-temperature responses of North American cacti. *Ecology* 65: 643–651.

Smith, W. K. 1978. Temperatures of desert plants: another perspective on the adaptability of leaf size. *Science* 201: 614–616.

Sober, H. A. 1968. *Handbook of Biochemistry. Selected Data for Molecular Biology*. The Chemical Rubber Co., Cleveland, 1042 pp.

Soule, O. H., and C. H. Lowe. 1970. Osmotic characteristics of tissue fluids in the sahuaro cactus (*Cereus giganteus*). *Annals of the Missouri Botanical Garden* 57: 265–351.

Spalding, E. S. 1905. Mechanical adjustment of the sahuaro (*Cereus giganteus*) to varying quantities of stored water. *Bulletin of the Torrey Botanical Club* 32: 57–68.

Spoehr, H. A. 1919. *The Carbohydrate Economy of Cacti*. Publication 287, Carnegie Institution of Washington, Washington, D.C., 79 pp.

Sprague, M. A., W. J. Hanna, and W. E. Chappell. 1978. Alternatives to a monoculture of henequen in Yucatán: The agriculture, climate, soil, and weed control. *Interciencia* 3: 285–290.

Stadelmann, E. J., and H. Kinzel. 1972. Vital staining of plant cells. *In* D. M. Prescott (ed.) *Methods in Cell Physiology, Volume 5*. Academic Press, New York, pp. 357–359.

Starling, R. 1985. In vitro propagation of *Leuchtenbergia principis*. *Cactus & Succulent Journal (U.S.)* 57: 114–115.

Steenbergh, W. F. 1972. Lightning-caused destruction in a desert plant community. *The Southwestern Naturalist* 16: 419–429.

Steenbergh, W. F., and C. H. Lowe. 1969. Critical factors during the first years of life of the saguaro (*Cereus giganteus*) at Saguaro National Monument, Arizona. *Ecology* 50: 825–834.

Steenbergh, W. F., and C. H. Lowe. 1976. Ecology of the saguaro: I. The role of freezing weather in a warm-desert population. *In Research in the Parks*, National Park Service Symposium Series, Number 1. U.S. Government Printing Office, Washington, D.C., pp. 49–92.

Steenbergh, W. F., and C. H. Lowe. 1977. *Ecology of the Saguaro. II. Reproduction, Germination, Establishment, Growth, and Survival of the Young Plant*. National Park Service Scientific Monograph Series, Number 8. U.S. Government Printing Office, Washington, D.C., 242 pp.

Steenbergh, W. F., and C. H. Lowe. 1983. *Ecology of the Saguaro: III. Growth and Demography*. National Park Service Scientific Monograph Series, Number 17. U.S. Government Printing Office, Washington, D.C., 228 pp.

Stefanis, J. P., and R. W. Langhans. 1980. Factors influencing the culture and propagation of xerophytic succulent species. *HortScience* 15: 504–505.

Stelfox, J. G., and H. G. Vriend. 1977. Prairie fires and pronghorn use of cactus. *The Canadian Field-Naturalist* 91: 282–285.

Steponkus, P. L. 1981. Responses to extreme temperatures. Cellular and subcellular bases. *In* O. L. Lange, P. S. Nobel, C. B. Osmond, and H. Ziegler (eds.) *Physiological Plant Ecology I Responses to the Physical Environment. Encyclopedia of Plant Physiology, New Series, Volume 12A*. Springer-Verlag, Berlin, pp. 371–402.

Stewart, A. 1911. Expedition of the California Academy of Sciences to the Galapagos Islands, 1905–1906. II. A botanical survey of the Galapagos Islands. *Proceedings of the California Academy of Sciences, Fourth Series*, 1: 7–288.

Sutton, B. G., I. P. Ting, and R. Sutton. 1981. Carbohydrate metabolism of cactus in a desert environment. *Plant Physiology* 68: 784–787.

Sutton, B. G., I. P. Ting, and J. H. Troughton. 1976. Seasonal effects on carbon isotope composition of cactus in a desert environment. *Nature* 261: 42–43.

Szarek, S. R. 1979. The occurrence of Crassulacean acid metabolism: a supplementary list during 1976 to 1979. *Photosynthetica* 13: 467–473.

Szarek, S. R., P. A. Holthe, and I. P. Ting. 1987. Minor physiological response to elevated CO_2 by the CAM plant *Agave vilmoriniana*. *Plant Physiology* 83: 938–940.

Szarek, S. R., H. B. Johnson, and I. P. Ting. 1973. Drought adaptation in *Opuntia basilaris*. Significance of recycling carbon through Crassulacean acid metabolism. *Plant Physiology* 52: 539–541.

Szarek, S. R., and I. P. Ting. 1974a. Respiration and gas exchange in stem tissue of *Opuntia basilaris*. *Plant Physiology* 54: 829–834.

Szarek, S. R., and I. P. Ting. 1974b. Seasonal patterns of acid metabolism and gas exchange in *Opuntia basilaris*. *Plant Physiology* 54: 76–81.

Szarek, S. R., and I. P. Ting. 1975a. Physiological responses to rainfall in *Opuntia basilaris* (Cactaceae). *American Journal of Botany* 62: 602–609.

Szarek, S. R., and I. P. Ting. 1975b. Photosynthetic efficiency of CAM plants in relation to C_3 and C_4 plants. *In* R. Marcelle (ed.) *Environmental and Biological Control of Photosynthesis*. Dr. W. Junk, The Hague, pp. 289–297.

Szarek, S. R., and I. P. Ting. 1977. The occurrence of Crassulacean acid metabolism among plants. *Photosynthetica* 11: 330–342.

Szarek, S. R., and J. H. Troughton. 1976. Carbon isotope ratios in Crassulacean acid metabolism plants – Seasonal patterns from plants in natural stands. *Plant Physiology* 58: 367–370.

Taylor, G. M. 1972. Rooting imported plants. *The National Cactus and Succulent Journal* 27: 82–83.

Taylor, S. A., and G. L. Ashcroft. 1972. *Physical Edaphology. The Physics of Irrigated and Nonirrigated Soils*. W. H. Freeman, San Francisco, 533 pp.

Taylor, W. W. 1966. Archaic cultures adjacent to the northwestern frontiers of Mesoamerica. *In* G. F. Ekholm and G. R. Willey (eds.) *Archaeological Frontiers and External Connections, Volume 4, Handbook of Middle American Indians*. University of Texas Press, Austin, pp. 59–94.

Teeri, J. A. 1982. Carbon isotopes and the evolution of C$_4$ photosynthesis and Crassulacean acid metabolism. *In* M. H. Nitecki (ed.) *Biochemical Aspects of Evolutionary Biology*. University of Chicago Press, Chicago, pp. 93–130.

Teeri, J. A., L. G. Stowe, and D. A. Murawski. 1978. The climatology of two succulent plant families: Cactaceae and Crassulaceae. *Canadian Journal of Botany* 56: 1750–1758.

Teles, F. F. F., J. W. Stull, W. H. Brown, and F. M. Whiting. 1984. Amino and organic acids of the prickly pear cactus (*Opuntia ficus-indica* L.). *Journal of the Science of Food and Agriculture* 35: 421–425.

Tello Balderas, J. J. 1983. Utilizacion del Maguey (*Agave* spp) en el Altiplano Potosino-Zacatecano. Professional thesis, Universidad Autonoma de San Luis Potosi, San Luis Potosi, San Luis Potosi, 125 pp.

Tello-Balderas, J. J., and E. García-Moya. 1985. The mezcal industry in the Altiplano Potosino-Zacatecano of North-Central Mexico. *Desert Plants* 7: 81–87.

Terblanche, I. L., A. M. Mulder, and J. W. Rossouw. 1971. Die invloed van voginhoud op die droë materiaalinname en verteerbaarheid by doringlose turksuyblaaie. *Agroanimalia* 3: 73–78.

Thomas, M., and S. L. Ranson. 1954. Physiological studies on acid metabolism in green plants. III. Further evidence of CO$_2$ fixation during dark acidification of plants showing crassulacean acid metabolism. *The New Phytologist* 53: 1–27.

Thornber J. J. 1910. The grazing ranges of Arizona. Bulletin 65, University of Arizona Agricultural Experiment Station, Tucson, pp. 245–360.

Thornber, J. J. 1911. Native cacti as emergency forage plants. Bulletin 67, University of Arizona Agricultural Experiment Station, Tucson, pp. 457–508.

Thorne, R. F. 1981. Phytochemistry and angiosperm phylogeny: a summary statement, including a synopsis of the class Angiospermae (Annonopsida). *In* D. A. Young and D. S. Seigler (eds.) *Phytochemistry and Angiosperm Phylogeny*. Praeger, New York, pp. 233–295.

Thorne, R. F. 1983. Proposed new realignments in the angiosperms. *Nordic Journal of Botany* 3: 85–117.

Timmons, F. L. 1942. The dissemination of prickly pear seed by jack rabbits. *Journal of the American Society of Agronomy* 34: 513–520.

Ting, I. P. 1976a. Succulents. *In* I. P. Ting and B. Jennings (eds.) *Deep Canyon, a Desert Wilderness for Science*. Philip L. Boyd Deep Canyon Desert Research Center, University of California, Riverside, pp. 125–130.

Ting, I. P. 1976b. Crassulacean acid metabolism in natural ecosystems in relation to annual CO$_2$ uptake patterns and water utilization. *In* R. H. Burris and C. C. Black (eds.) *CO$_2$ Metabolism and Plant Productivity*. University Park Press, Baltimore, pp. 251–268.

Ting, I. P. 1985. Crassulacean acid metabolism. *Annual Review of Plant Physiology* 36: 595–622.

Ting, I. P., and W. M. Dugger, Jr. 1968. Non-autotrophic carbon dioxide metabolism in cacti. *Botanical Gazette* 129: 9–15.

Ting, I. P., and Z. Hanscom, III. 1977. Induction of acid metabolism in *Portulacaria afra*. *Plant Physiology* 59: 511–514.

Ting, I. P., and B. Jennings (eds.). 1976. *Deep Canyon, a Desert Wilderness for Science*. The Philip L. Boyd Deep Canyon Desert Research Center, University of California, Riverside, 177 pp.

Ting, I. P., and S. R. Szarek. 1975. Drought adaptation in crassulacean acid metabolism plants. *In* N. F. Hadley (ed.) *Environmental Physiology of Desert Organisms*. Dowden, Hutchinson & Ross, Stroudsburg (PA), pp. 152–167.

Trachtenberg, A., and A. M. Mayer. 1981. Calcium oxalate crystals in *Opuntia ficus-indica* (L.) Mill.: Development and relation to mucilage cells – a stereological analysis. *Protoplasma* 109: 271–283.

Trachtenberg, S., and A. M. Mayer. 1982a. Mucilage cells, calcium oxalate crystals and soluble calcium in *Opuntia ficus-indica*. *Annals of Botany* 50: 549–557.

Trachtenberg, S., and A. M. Mayer. 1982b. Biophysical properties of *Opuntia ficus-indica* mucilage. *Phytochemistry* 21: 2835–2843.

Tranquillini, W. 1982. Frost-drought and its ecological significance. *In* O. L. Lange, P. S. Nobel, C. B. Osmond, and H. Ziegler (eds.) *Physiological Plant Ecology II Water Relations and Carbon Assimilation. Encyclopedia of Plant Physiology, New Series, Volume 12B*. Springer-Verlag, Berlin, pp. 329–400.

Troll, W. 1937. *Vergleichende Morphologie der höheren Pflanzen, Band 1, Vegetationsorgane, Teil 1*. Gebrüder Bortraeger, Berlin, 955 pp.

Troughton, J. H., K. A. Card, and C. H. Hendy. 1974. Photosynthetic pathways and carbon isotope discrimination by plants. *Carnegie Institution Year Book* 73: 768–780.

Troughton, J. H., P. V. Wells, and H. A. Mooney. 1974. Photosynthetic mechanisms in ancient C$_4$ and CAM species. *Carnegie Institution Year Book* 73: 812–816.

Trujillo Argueta, S. 1982. Estudio sobre algunos Aspectos Ecologicos de *Echinocactus platyacanthus* Lk. & O. en el Estado de San Luis Potosi. Professional thesis. Universidad Nacional Autónoma de México, Mexico City, 126 pp.

Turnage W. V., and A. L. Hinckley. 1938. Freezing weather in relation to plant distribution in the Sonoran Desert. *Ecological Monographs* 8: 529–550.

Turner, N. C. 1981. Techniques and experimental approaches for the measurement of plant water status. *Plant and Soil* 58: 339–366.

Turner, R. M., S. M. Alcorn, and G. Olin. 1969. Mortality of transplanted saguaro seedlings. *Ecology* 50: 835–844.

Turner, R. M., S. M. Alcorn, G. Olin, and J. A. Booth. 1966. The influence of shade, soil, and water on saguaro seedling establishment. *Botanical Gazette* 127: 95–102.

Tyree, M. T., and H. T. Hammel. 1972. The measurement of the turgor pressure and the water relations of plants by the pressure-bomb technique. *Journal of Experimental Botany* 23: 267–282.

United States Department of the Interior. 1985. *Quality of Water Colorado River Basin*. Progress Report No. 12, United States Department of the Interior, Washington, D.C., 220 pp.

Uphof, J. C. Th. 1916. *Cold-Resistance in Spineless Cacti*. Bulletin 79, University of Arizona Agricultural Experiment Station, Tucson, 144 pp.

Valenzuela A. G. 1985. The tequila industry in Jalisco, Mexico. *Desert Plants* 7: 65–70.

Vandermeer J. 1980. Saguaros and nurse trees: a new hypothesis to account for population fluctuations. *The Southwestern Naturalist* 25: 357–360.

Varlet-Grancher, C., M. Chartier, G. Gosse, and R. Bonhomme. 1981. Rayonnement utile pour la photosynthèse des végétaux en conditions naturelles: caractérisation et variations. *Acta Œcologica/ Œcologia Plantarum* 2: 189–202.

Vinson, A. E. 1911. Nutritive value of cholla fruit. Bulletin 67, University of Arizona Agricultural Experiment Station, Tucson, pp. 509–519.

Wallace, R. H., and H. H. Clum. 1938. Leaf temperatures. *American Journal of Botany* 25: 83–97.

Walter, H. 1931. *Die Hydratur der Pflanze*. Gustav Fischer, Jena. 174 pp.

Walter, H. 1985. *Vegetation of the Earth and Ecological Systems of the Geo-biosphere*. Springer-Verlag, Berlin, 318 pp.

Walter, H., and E. Stadelmann. 1974. A new approach to the water relations of desert plants. *In* C. W. Brown, Jr. (ed.) *Desert Biology, Volume 2*. Academic Press, New York, pp. 213–310.

Watson, A. N. 1933. Preliminary study of the relation between thermal emissivity and plant temperatures. *Ohio Journal of Science* 33: 435–450.

Wattendorff, J., and P. J. Holloway. 1980. Studies on the ultrastructure and histochemistry of plant cuticles: the cuticular membrane of *Agave americana* L. *in situ*. *Annals of Botany* 46: 13–28.

Wattendorff, J., and P. J. Holloway. 1982. Studies on the ultrastructure and histochemistry of plant cuticles: isolated cuticular membrane preparations of *Agave americana* L. and the effects of various extraction procedures. *Annals of Botany* 49: 769–804.

Weedin, J. F., and A. M. Powell. 1978. Chromosome numbers in Chihuahuan Desert Cactaceae. Trans-Pecos Texas. *American Journal of Botany* 65: 531–537.

Weindling, L. 1947. *Long Vegetable Fibers. Manila · Sisal · Jute · Flax and Related Fibers of Commerce*. Columbia University Press, New York, 311 pp.

Whiting, B. H., and E. E. Campbell. 1984. Effect of moisture supply on CAM in *Opuntia aurantiaca* (jointed cactus). *South African Journal of Plant and Soil* 1: 87–91.

Whiting, B. H., H. A. van de Venter, and J. G. C. Small. 1979. Crassulacean acid metabolism in jointed cactus (*Opuntia aurantiaca*) Lindley. *Agroplantae* 11: 41–43.

Wiebe, H. H., and H. A. Al-Saadi. 1976. Matric bound water of water tissue from succulents. *Physiologia Plantarum* 36: 47–51.

Wienk, J. F. 1970. The long fibre agaves and their improvement through breeding in East Africa. *In* C. L. A. Learey (ed.) *Crop Improvement in East Africa*. Commonwealth Agricultural Bureax, Farnham Royal (England), pp. 209–230.

Wiggins, I. L., and D. W. Focht. 1967. Seeds and seedlings of *Opuntia echios* J. T. Howell var. *gigantea* Dawson. *Cactus and Succulent Journal of the Cactus and Succulent Society of America* 39: 26–30.

Wiggins, I. L. and D. M. Porter. 1971. *Flora of the Galápagos Islands*. Stanford University Press, Stanford, 998 pp.

Wilder, J. C. 1967. The years of a desert laboratory. *The Journal of Arizona History* 8: 179–199.

Willmer, C. M., and J. E. Pallas, Jr. 1973. A survey of stomatal movements and associated potassium fluxes in the plant kingdom. *Canadian Journal of Botany* 51: 37–42.

Winter, K. 1985. Crassulacean acid metabolism. *In* J. Barber and N. R. Baker (eds.) *Photosynthetic Mechanisms and the Environment, Volume 6*. Elsevier, Amsterdam, pp. 329–387.

Winter, K., G. Schröppel-Maier, and M. M. Caldwell. 1986. Respiratory CO_2 as carbon source for nocturnal acid synthesis at high temperatures in three species exhibiting Crassulacean acid metabolism. *Plant Physiology* 81: 390–394.

Winter, K., and D. J. von Willert. 1972. NaCl-induzierter Crassulaceansäurestoffwechsel bei *Mesembryanthemum crystallinum*. *Zeitschrift für Pflanzenphysiologie* 67: 166–170.

Wood, I. M., and J. F. Angus. 1974. *A Review of Prospective Crops for the Ord Irrigation Area. II. Fiber Crops*. CSIRO Australian Division of Land Use Research Technical Papers Number 36, CSIRO, Melbourne, 27 pp.

Woodhouse, R. M., J. G. Williams, and P. S. Nobel. 1980. Leaf orientation, radiation interception, and nocturnal acidity increases by the CAM plant *Agave deserti* (Agavaceae). *American Journal of Botany* 67: 1179–1185.

Woodhouse, R. M., J. G. Williams, and P. S. Nobel. 1983. Simulation of plant temperature and water loss by the desert succulent, *Agave deserti*. *Oecologia* 57: 291–297.

Yeaton, R. I. 1978. A cyclical relationship between *Larrea tridentata* and *Opuntia leptocaulis* in the northern Chihuahuan Desert. *Journal of Ecology* 66: 651–656.

Yeaton, R. I. 1982. Ecomorphology and habitat utilization of *Echinocereus engelmannii* and *E. triglochidiatus* (Cactaceae) in southeastern California. *Great Basin Naturalist* 42: 353–359.

Yeaton, R. I., and M. L. Cody. 1976. Competition and spacing in plant communities: the northern Mohave Desert. *Journal of Ecology* 64: 689–696.

Yeaton, R. I., and M. L. Cody. 1979. The distribution of cacti along environmental gradients in the Sonoran and Mohave Deserts. *Journal of Ecology* 67: 529–541

Yeaton, R. I., R. Karban, and H. B. Wagner. 1980. Morphological growth patterns of saguaro (*Carnegiea gigantea*: Cactaceae) on flats and slopes in Organ Pipe Cactus National Monument, Arizona. *The Southwestern Naturalist* 25: 339–349.

Yeaton, R. I., E. Layendecker, K. S. Sly, and R. Eckert. 1983. Microhabitat differences between *Opuntia parryi* and *O. littoralis* (Cactaceae) in the mixed chaparral-inland coastal sage association. *The Southwestern Naturalist* 28: 215–220.

Yeaton, R. I., and A. R. Manzanares. 1986. Organization of vegetation mosaics in the *Acacia schaffneri-Opuntia streptacantha* association, southern Chihuahuan Desert, Mexico. *Journal of Ecology* 74: 211–217.

Yeaton, R. I., J. Travis, and E. Gilinsky. 1977. Competition and spacing in plant communities: the Arizona upland association. *Journal of Ecology* 65: 587–595.

Yensen, N. P., M. R. Fontes, E. P. Glenn, and R. S. Felger. 1981. New salt tolerant crops for the Sonoran Desert. *Desert Plants* 3: 111–118.

Young, D. R., and P. S. Nobel. 1986. Predictions of soil-water potentials in the north-western Sonoran Desert. *Journal of Ecology* 74: 143–154.

Zeiger, E. 1983. The biology of stomatal guard cells. *Annual Review of Plant Physiology* 34: 441–475.

Zimmer, K. 1968a. Untersuchungen über den Einfluss der temperatur auf die Keimung von Kakteen-Saatgut VI. Über die Keimung einiger mexikanischer Arten. *Die Gartenbauwissenschaft* 33: 167–175.

Zimmer, K. 1968b. Untersuchungen über den Einfluss der Temperatur auf die Keimung von Kakteen-Saatgut VII. Über die Keimung einiger argentinischer Arten. *Die Gartenbauwissenschaft* 33: 335–344.

Zimmer, K. 1969a. Über die Keimung von Kakteensamen (I) Quellung und Keimung bei konstanten Temperatur. *Kakteen und andere Sukkulenten* 20: 105–107.

Zimmer, K. 1969b. Über die Keimung von Kaktensamen (III) Die Bedeutung des Lichtes. *Kakteen und andere Sukkulenten* 20: 144–147.

Zimmer, K. 1972. Untersuchungen über den Einfluss Temperatur auf die Keimung von Kakteen-Saatgut VIII. Zum Kältebedürfnis bei der Keimung von *Maihuenia poeppigii* (Otto) Web. *Gartenbauwissenschaft* 37: 109–121.

Zimmer, K. 1980a. Einfluss der Temperatur auf die Keimung von Kakteensaatgut X. Keimung einiger Ferocactus-Arten. *Gartenbauwissenschaft* 45: 121–123.

Zimmer, K. 1980b. Untersuchungen über den Einfluss der Temperatur auf die Keimung von Kakteensaatgut XI. Keimung einiger Epiphyllinae und Rhipsalidinae. *Gartenbauwissenschaft* 45: 205–207.

Zimmermann, H. G., and V. C. Moran. 1982. Ecology and management of cactus weeds in South Africa. *South African Journal of Science* 78: 314–320.

Index

A page number in **boldface** indicates where a species is illustrated or a term is defined (discussion also may occur on that page).